Boundary and Eigenvalue Problems in Mathematical Physics

BY HANS SAGAN

PROFESSOR OF MATHEMATICS
NORTH CAROLINA STATE UNIVERSITY
RALEIGH, N.C.

Dover Publications, Inc., NEW YORK

Published in Canada by General Publishing Company, Ltd., 30 Lesmill Road, Don Mills, Toronto, Ontario.
Published in the United Kingdom by Constable and Company, Ltd., 10 Orange Street, London WC2H 7EG.

This Dover edition, first published in 1989, is an unabridged, slightly corrected republication of the fourth printing of the work originally published by John Wiley & Sons, Inc., New York, 1961.

Manufactured in the United States of America
Dover Publications, Inc., 31 East 2nd Street, Mineola, N.Y. 11501

Library of Congress Cataloging-in-Publication Data

Sagan, Hans.
Boundary and eigenvalue problems in mathematical physics / by Hans Sagan.
p. cm.
Reprint. Originally published: New York : Wiley, 1961.
Bibliography: p.
Includes index.
ISBN 0-486-66132-6
1. Mathematical physics. 2. Boundary value problems. 3. Eigenvalues. I. Title.
QA401.S24 1989
530.1'5535—dc20 89-35689
CIP

Photograph Kobé, Vienna
Dr. Johann Radon (1887–1956)

TO THE MEMORY OF

DR. JOHANN RADON

LATE PROFESSOR OF MATHEMATICS
UNIVERSITY OF VIENNA

PREFACE

This book contains the material which I have taught to seniors and beginning graduate students in mathematics, applied mathematics, physics, and engineering for the past five years. It is not the purpose of this book to present a vast number of seemingly unrelated mathematical techniques and tricks that are used in the mathematical treatment of problems which arise in physics and engineering. On the contrary, the material contained in this book is developed from a few basic concepts, namely, Hamilton's principle together with the theory of the first variation and Bernoulli's separation method for the solution of linear homogeneous partial differential equations. These concepts, which must be thoroughly understood, lend unity to the variety of topics that present themselves quite naturally in the pursuit of a complete answer to a few basic problems such as the problems of the vibrating string, the vibrating membrane, and heat conduction.

The concept of eigenvalues of self-adjoint boundary value problems, which evolves as one of major importance, is introduced from an elementary standpoint by a method which was first developed by H. Pruefer (Chapter V). Later on (Chapter VII), eigenvalues are characterized by their extremum properties, and the equivalence of the Sturm-Liouville boundary value problem and Rayleigh's principle is established. This duality in approach is used not only to provide a thorough understanding of this important matter, but also to utilize the apparatus of the theory of the first variation as introduced in the beginning (Chapter I) and develop a useful numerical method, the method of Rayleigh and Ritz, for the computation of eigenvalues—without burdening the reader with a great deal of extraneous material.

The material contained in Chapter VII (Characterization of Eigenvalues by a Variational Principle) can really be taken up any time after completion of Chapter V (Self-adjoint Boundary Value Problems). There are, however, two reasons for putting it where it is now: to collect more illustrative material (Legendre functions and Bessel functions) and to break the monotony in the discussion of orthogonal systems as given in Chapters VI and VIII.

The problems which are to be found at the end of most sections are in part purely routine problems designed to give the student an opportunity to acquaint himself with new techniques and acquire a working knowledge of the theory presented; in part, problems of a complementary nature and thus essential for a good understanding of the text; and lastly, problems of a supplementary nature with the purpose of rounding off the treatment. Problems with which my students have had difficulties are designated by an asterisk. Answers and hints to even numbered problems are to be found at the back of the book except for cases where the answer is obvious from the context and/or the hint which is supplied following the statement of the problem appears to suffice.

The student who wishes to take this course with a reasonable chance to succeed should be familiar with the topics that are condensed in the appendix. A course in advanced calculus will usually suffice for this purpose. The appendix consists of three parts; these parts are subdivided into sections which are enumerated by capital letters, and these sections in turn are subdivided into subsections which are enumerated by arabic numerals. Reference to the appendix in the main text appears in the form, e.g., (**A**I.B3), denoting Appendix, part I, section B, subsection 3.

The course, as presented here, is designed as a two-semester, three-credit course. To a class of very able students, however, the same material can be presented in a one-semester, five-credit course. Chapters II (with omission of §2, subsections 2 and 3), III, IV, and V can be taught in a one-semester, three-credit course; Chapters I, V, and VII in a one-semester, two-credit course. In the latter case, the instructor will have to supplement the material to provide for a smooth transition between chapters.

It was my aim in preparing this textbook to give the student a feeling for the usefulness and necessity of the mathematical methods that are developed here and at the same time not to obscure the leading ideas with too many technicalities. I hope that I have also created the impression that the material dealt with is part of a living and growing discipline.

HANS SAGAN

Moscow, Idaho
February, 1961

ACKNOWLEDGMENTS

It is my pleasure to express my sincere appreciation to Professor Peter Lax (New York University) for his many valuable suggestions, encouragement, and constructive criticism at a time when the manuscript was at an early stage of its development; to my colleague Professor Anthony E. Labarre, Jr., for his attentive reading of the manuscript in its next-to-final form; and to Professor Wolfgang Wasow (University of Wisconsin) for scrutinizing the final manuscript and rendering valuable suggestions for its improvement. Finally, I wish to express my gratitude to my friend Professor Karl Prachar (University of Vienna) for the effort and care he exercised in reading a major portion of the galley proofs, and last but not least I wish to thank my publishers for their patience and cooperation in preparing this text.

H. S.

CONTENTS

I

HAMILTON'S PRINCIPLE AND THE THEORY OF THE FIRST VARIATION

§1. VARIATIONAL PROBLEMS IN ONE INDEPENDENT VARIABLE

1. Newton's Equations of Motion

Let us consider a field of force, which can be represented by the gradient of a point function $U(x, y, z)$. (See **AI.B3**—Appendix, Part I, section B, subsection 3.) We call such a field *conservative* because the law of conservation of energy is satisfied therein. (see problem I.1). We will assume that $U(x, y, z)$ and its first- and second-order derivatives are continuous. The components of the force in the directions of the coordinate axes are then

$$f_{(x)} = \frac{\partial U}{\partial x},$$

$$f_{(y)} = \frac{\partial U}{\partial y},$$

$$f_{(z)} = \frac{\partial U}{\partial z},$$

where the subscripts x, y, z designate the x, y, and z components.

If we consider a mass point of mass m moving within this field, it experiences at any point (x, y, z) a force

$$\mathbf{f} = f_{(x)}\mathbf{i} + f_{(y)}\mathbf{j} + f_{(z)}\mathbf{k}.$$

According to the fundamental principle of vectorial mechanics, the mass point m moves in a path

$$\mathbf{s}(t) = x\mathbf{i} + y\mathbf{j} + z\mathbf{k},$$

which is defined by the condition that the *external force* exerted by the field at any point is equal to the *inertial force* $m(d^2\mathbf{s}/dt^2)$ exerted by the mass point. This gives *Newton's* well-known *equations of motion*

$$m\frac{d^2x}{dt^2} = \frac{\partial U}{\partial x},$$

$$m\frac{d^2y}{dt^2} = \frac{\partial U}{\partial y},$$

$$m\frac{d^2z}{dt^2} = \frac{\partial U}{\partial z}.$$

The point function $U(x, y, z)$, whose derivatives with respect to the space coordinates are the force components acting at each point of the field, is nothing but the negative *work function* or *potential energy*, since

$$U = \int_\Gamma \frac{\partial U}{\partial x}\,dx + \frac{\partial U}{\partial y}\,dy + \frac{\partial U}{\partial z}\,dz$$

is the work to be performed in order to move a unit mass along a certain path Γ:

$$x = x(t),$$

$$y = y(t),$$

$$z = z(t),$$

through this field.

By introducting the usual notation $U = -V$ we may write Newton's equations of motion in the form

$$m\frac{d^2\mathbf{s}}{dt^2} = -\text{grad}\, V(x, y, z).$$

We have found these equations by taking into account what forces are exerted upon the mass point and what forces are exerted by the mass point.

The great German philosopher *Leibnitz* suggested using—instead of inertial and external force—the following two scalar quantities for the description of a mechanical system:

The *vis viva* or *living force* (which is essentially the *kinetic energy* except for a factor 2) instead of the inertial force and the work function (or potential energy) instead of the external force. The question is now the following one: What relation has to hold between these two quantities; i.e., what principle must those two quantities obey in order to characterize the actual motion of the particle of mass m in a field given by a potential function $V(x, y, z)$?

The *kinetic energy* of a point with mass m is defined as

$$T = \frac{m}{2}\mathbf{v}^2$$

where $\mathbf{v} = \dot{x}\mathbf{i} + \dot{y}\mathbf{j} + \dot{z}\mathbf{k}$ is the velocity of the mass point at any time t. (The dots denote differentiation with respect to the time t.)

Let us now assume that the mass point starts its motion at the point P_1 at a time t_1 and arrives at a point P_2 at a time t_2. We consider now all possible trial paths joining the space-time points P_1, t_1 and P_2, t_2 and assume that we have evaluated for all those paths the quantity

$$A = \int_{t_1}^{t_2} (T - V)\,dt,$$

which is called the *action*.

In accordance with the literature we denote the integrand by L,

$$T - V = L.$$

(*Euler* and *Lagrange* were the first to formulate the ideas put forth in this subsection, and one calls L, in honor of the latter, the *Lagrange function*.)

If we consider all possible continuous paths with a continuous derivative joining the space-time points P_1, t_1 and P_2, t_2, we can be sure that one of them is the path actually taken by the mass point under consideration. The action will in general have different values for the different paths we consider, and if there is one which yields a "minimum value" for the action, then it is the one actually taken by the mass point. (We put minimum value in quotes because we will have to modify this statement a little later.)

This constitutes the so-called "principle of least action," formulated at first by *Euler* and *Lagrange* for conservative fields and later generalized by *Hamilton* for nonconservative fields.[1] The way *Euler* and *Lagrange* came to formulate this principle can probably be understood on the basis of the general philosophical and religious background of their time, when it was generally believed that God made the world in the most economical way and therefore everything had to obey minimum principles of some kind. The establishment of this principle may appear less artificial in this light. We are now going to show that this principle is equivalent to the leading principle of vectorial mechanics, which guided our considerations at the beginning of this subsection, insofar as it also leads to *Newton's* equations

[1] For conservative fields we have, because of $T + V = C$ (constant),

$$T - V = 2T - C.$$

Hence, in Euler's investigation for the case of conservative fields the minimum principle amounted to a minimization of the time integral of twice the kinetic energy.

of motion. Let us reformulate this principle in a closed form before we draw conclusions from it:

Principle of Least Action (Euler, Lagrange, Hamilton). *A mass point with mass m and potential energy $V(x, y, z)$ takes, in the time interval $t_1 \leq t \leq t_2$, that path joining the points P_1 and P_2 which gives the integral*

$$A = \int_{t_1}^{t_2} (T - V)\, dt$$

its smallest value, compared with the values it assumes for any other continuous path with a continuous derivative joining the same end points.

If we substitute for T and V their respective expressions, we face the following mathematical problem:

The integral

$$A = \int_{t_1}^{t_2} \left[\frac{m}{2} (\dot{x}^2 + \dot{y}^2 + \dot{z}^2) - V(x, y, z) \right] dt \tag{I.1}$$

is to be made a minimum by proper choice of a curve

$$x = x(t),$$
$$y = y(t),$$
$$z = z(t)$$

joining the points $P_1(x_1, y_1, z_1)$ and $P_2(x_2, y_2, z_2)$, i.e., satisfying the boundary conditions

$$\begin{aligned} x(t_1) &= x_1, & x(t_2) &= x_2, \\ y(t_1) &= y_1, & y(t_2) &= y_2, \\ z(t_1) &= z_1; & z(t_2) &= z_2. \end{aligned} \tag{I.2}$$

Even though we develop in the next subsection the theory which leads to the solution of more general problems of this type, we will attempt to solve this particular problem at this time, for it will teach us what steps we have to take in the general discussion.

To obtain a necessary condition, we will assume in accordance with *Euler* and *Lagrange* that

$$C \begin{cases} x = \bar{x}(t), \\ y = \bar{y}(t), \\ z = \bar{z}(t) \end{cases} \tag{I.3}$$

is the curve which minimizes (I.1) and satisfies the boundary conditions (I.2). One calls such a curve C an *extremal*.

Let us now consider a path which differs from (I.3) but joins the same end points, such as

$$C_\epsilon \begin{cases} x = \bar{x}(t) + \epsilon\xi(t), \\ y = \bar{y}(t), \\ z = \bar{z}(t) \end{cases} \tag{I.4}$$

with $\xi(t_1) = \xi(t_2) = 0$. This latter condition expresses the fact that (I.4) passes through the same end points as (I.3).

C_ϵ is called a *variation*[2] of C for obvious reasons. Hence the name *calculus of variations.*

We assume that the variation (I.4) of (I.3) is such that $\xi(t)$ is continuous and has a continuous derivative.

Substituting (I.4) into (I.1) makes the action a function of ϵ:

$$A(\epsilon) = \int_{t_1}^{t_2} \left[\frac{m}{2} (\dot{\bar{x}}^2 + \epsilon^2\dot{\xi}^2 + 2\epsilon\dot{\bar{x}}\dot{\xi} + \dot{\bar{y}}^2 + \dot{\bar{z}}^2) - V(\bar{x} + \epsilon\xi, \bar{y}, \bar{z}) \right] dt.$$

Since we have assumed that (I.3) is the solution of our problem, this function of ϵ has to yield the smallest value for $\epsilon = 0$; i.e., the first necessary condition for the existence of an extreme value of a function of one variable has to be satisfied:

$$\left. \frac{dA(\epsilon)}{d\epsilon} \right|_{\epsilon=0} = 0.$$

One calls the expression $\left. \frac{dA(\epsilon)}{d\epsilon} \right|_{\epsilon=0}$ the *first variation* of A and denotes it frequently by δA.

If we differentiate A with respect to ϵ and set $\epsilon = 0$, we obtain

$$\left. \frac{dA(\epsilon)}{d\epsilon} \right|_{\epsilon=0} = \int_{t_1}^{t_2} \left(m\dot{\bar{x}}\dot{\xi} - \xi \frac{\partial V}{\partial x} \right) dt.$$

We apply integration by parts, according to *Lagrange*, to the first term of the integrand:

$$\left. \frac{dA(\epsilon)}{d\epsilon} \right|_{\epsilon=0} = (m\dot{\bar{x}}\xi) \Big|_{t_1}^{t_2} - \int_{t_1}^{t_2} \xi \left(m\ddot{\bar{x}} + \frac{\partial V}{\partial x} \right) dt.$$

From the boundary condition imposed on $\xi(t)$ it follows that $(m\dot{\bar{x}}\xi) \Big|_{t_1}^{t_2} = 0.$

Thus we obtain the condition

$$\int_{t_1}^{t_2} \xi \left(m\ddot{\bar{x}} + \frac{\partial V}{\partial x} \right) dt = 0$$

for all possible functions $\xi(t)$ which are continuous, have a continuous derivative, and vanish at the end points.

[2] A more general variation will be discussed in subsection 2 of this section.

This is possible only if

$$m\ddot{x} + \frac{\partial V}{\partial x} = 0,$$

as we will see in the next subsection in a more general form (Fundamental Lemma of the Calculus of Variations).

The reader will immediately recognize in this equation the first one of Newton's equations of motion. We easily obtain the remaining two equations by going through the same process, but varying now the y-coordinate, then the z-coordinate of the path (I.3).

At this time we can already see how we have to modify the expression "smallest value" which we used in formulating the principle of least action. We know from the theory of extreme values of functions of one variable that the condition $\left.\frac{dA(\epsilon)}{d\epsilon}\right|_{\epsilon=0} = 0$ guarantees only a horizontal tangent at the point $\epsilon = 0$ and nothing else. This does not necessarily mean that our function has at this point a minimum value. On the other hand we have seen that the condition $\left.\frac{dA(\epsilon)}{d\epsilon}\right|_{\epsilon=0} = 0$ suffices to obtain *Newton's* equations of motion. Thus, we actually do not have to find a path along which A assumes a minimum, but rather a path for which the first variation $\left.\frac{dA(\epsilon)}{d\epsilon}\right|_{\epsilon=0}$ vanishes, or, as we will express it, the action takes on a *stationary value*.

We formulated the principle of least action earlier for the mechanics of points. According to Hamilton it is also valid in the mechanics of continua and we will state it now in the more general form taking the preceding modification of "minimum value" into account:

Hamilton's Principle. *A mechanical system with the kinetic energy T and the potential energy V behaves within a time interval $t_1 \leq t \leq t_2$ for a given initial and end position such that*

$$A = \int_{t_1}^{t_2} L\,dt, \qquad L = T - V$$

assumes a stationary value.

The principle in this form will enable us to derive in the next section and in the next chapter some partial differential equations of physics in a very elegant form.

Problems I.1–6

***1.** (a) Show that $E = \frac{1}{2}m\dot{s}^2 + V = \text{constant}$ if $m\ddot{s} + \text{grad } V = 0$.
(b) State the condition for the components of a conservative field.

2. Is Newton's gravitational field, which is given by

$$F = \frac{\mathbf{r}}{r^3}, \qquad \text{where } \mathbf{r} = x\mathbf{i} + y\mathbf{j} + z\mathbf{k} \quad \text{and } r = |\mathbf{r}|,$$

conservative?

3. If problem 2 permits a positive answer, find the potential function which characterizes Newton's gravitational field.

4. Find the work one has to perform if one wants to push a unit mass along the path

$$\left.\begin{aligned} x &= \cos\phi \\ y &= \sin\phi \\ z &= 0 \end{aligned}\right\}, \qquad 0 \leq \phi \leq \frac{\pi}{2},$$

through the field $\mathbf{f} = x\mathbf{i} + (x^2 + y^2)\mathbf{j} + \mathbf{k}$.

5. Derive the second of Newton's equations of motion by varying the y-component in (I.3) analogously to the process carried out in the text.

***6.** Find the third of Newton's equations of motion by considering the following variation:

$$\begin{aligned} x &= \bar{x}(t), \\ y &= \bar{y}(t), \\ z &= z(t, \epsilon). \end{aligned}$$

What conditions are to be imposed on z?

2. The Euler-Lagrange Equation

We consider now a problem which is in a certain sense more general than the one considered in the preceding subsection, and in another sense somewhat simpler. Namely, we seek a continuous function with a continuous derivative

$$y = y(x)$$

which satisfies the boundary conditions

$$\begin{aligned} y(x_1) &= y_1, \\ y(x_2) &= y_2 \end{aligned} \tag{I.5}$$

such that

$$I = \int_{x_1}^{x_2} f(x, y, y')\, dx \tag{I.6}$$

yields a minimum value.

To be on the safe side, we assume that f, a function of the three variables x, y, y', is continuous and has continuous first- and second-order derivatives with respect to each variable.

In our attempt to solve this problem we let our guide be the experience we gained in the preceding subsection and assume that

$$E: \quad y = \bar{y}(x)$$

is the solution of our problem (or an extremal in our previously introduced terminology).

Now we consider a one-parameter variation of this curve (see Fig. 1)

$$E_\epsilon\colon \quad y = \bar{y}(x, \epsilon), \qquad \frac{\partial \bar{y}}{\partial x}, \frac{\partial \bar{y}}{\partial \epsilon}, \frac{\partial^2 \bar{y}}{\partial x\, \partial \epsilon} \text{ continuous,} \tag{I.7}$$

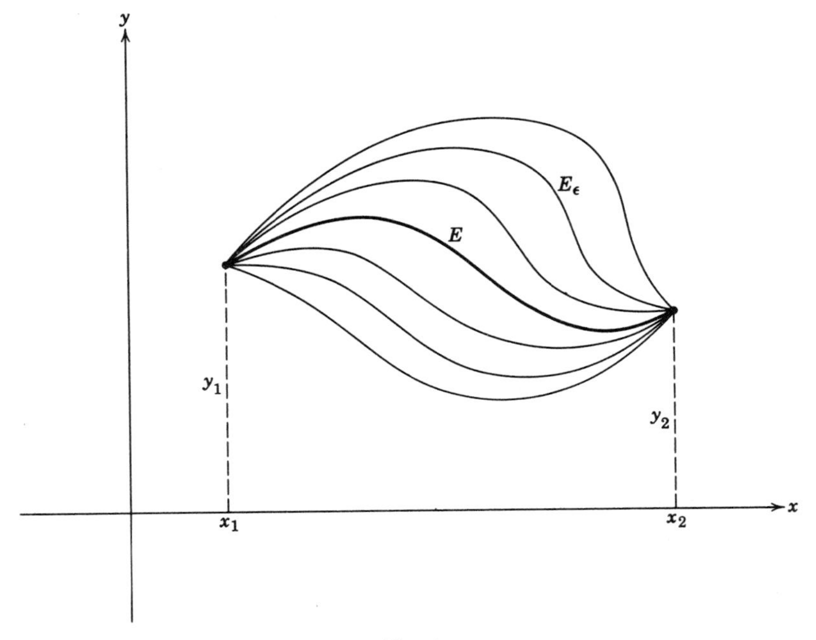

Fig. 1

which becomes E for $\epsilon = 0$:

$$\bar{y}(x, 0) = \bar{y}(x), \tag{I.8}$$

and joins the same end points $P_1(x_1, y_1)$ and $P_2(x_2, y_2)$,

$$\begin{aligned} \bar{y}(x_1, \epsilon) &= y_1, \\ \bar{y}(x_2, \epsilon) &= y_2. \end{aligned} \tag{I.9}$$

For reasons of convenience we introduce the following abbreviation:

$$\frac{\partial \bar{y}(x, \epsilon)}{\partial \epsilon} = \eta(x, \epsilon).$$

From (I.9) it follows by differentiation with respect to ϵ that

$$\eta(x_1, \epsilon) = 0, \quad \eta(x_2, \epsilon) = 0. \tag{I.10}$$

If we substitute (I.7) into (I.6), we obtain I as a function of ϵ

$$I(\epsilon) = \int_{x_1}^{x_2} f(x, \bar{y}(x, \epsilon), \bar{y}'(x, \epsilon))\, dx.$$

For (I.6) to assume a minimum value, the first variation $\left.\dfrac{dI(\epsilon)}{d\epsilon}\right|_{\epsilon=0}$ has to vanish. Differentiation of the integral above with respect to ϵ yields

$$\frac{dI(\epsilon)}{d\epsilon} = \int_{x_1}^{x_2} \left(\frac{\partial f}{\partial y}\eta + \frac{\partial f}{\partial y'}\eta'\right) dx. \tag{I.11}$$

If we apply integration by parts to the second term in the integrand in analogy to the procedure we followed in subsection 1 of this section, we obtain

$$\frac{dI(\epsilon)}{d\epsilon} = \left(\eta \frac{\partial f}{\partial y'}\right)\Bigg|_{x_1}^{x_2} - \int_{x_1}^{x_2} \eta\left(-\frac{\partial f}{\partial y} + \frac{d}{dx}\frac{\partial f}{dy'}\right) dx.$$

In view of (I.10) the integrated term vanishes, which leaves us with the following condition for a minimum of (I.6):

$$\left.\frac{dI(\epsilon)}{d\epsilon}\right|_{\epsilon=0} = -\int_{x_1}^{x_2} \eta\left(\frac{\partial f}{\partial y} - \frac{d}{dx}\frac{\partial f}{\partial y'}\right) dx = 0 \tag{I.12}$$

for all η which are continuous, have a continuous derivative with respect to x, and satisfy (I.10). Such functions η we will call *permissible variations.*

The continuity of η' has to be assumed since we could not otherwise carry out the above integration by parts and interchange differentiation with respect to x and ϵ in the case of $\bar{y}(x, \epsilon)$.

This integration by parts has another consequence, since $\dfrac{d}{dx}\dfrac{\partial f}{\partial y'}$ also has to be continuous and since

$$\frac{d}{dx}\frac{\partial f}{\partial y'} = \frac{\partial^2 f}{\partial y'\,\partial x} + \frac{\partial^2 f}{\partial y'\,\partial y} y' + \frac{\partial^2 f}{\partial y'\,\partial y'} y''.$$

The left side is continuous, and y' and the second-order derivatives of f with respect to all variables are already assumed to be continuous; hence y'' has to be continuous. This excludes automatically all curves with a discontinuous curvature.

We will further assume that $\partial^2 f/\partial y'\partial y'$ does not vanish identically in $x_1 \leq x \leq x_2$. Otherwise the differential expression above degenerates to one of the first order. A problem where such is the case is called a *singular variational problem*, while a problem with $\partial^2 f/\partial y'\partial y' \not\equiv 0$ is called a *regular variational problem*.

As already announced in the preceding subsection, we will now prove a lemma which will enable us to deduce from the vanishing of (I.12) for all permissible η the vanishing of the integrand itself.

Fundamental Lemma of the Calculus of Variations. *If $M(x)$ is a continuous function in $x_1 \leq x \leq x_2$ and $\eta(x)$ is any function which is continuous, has a continuous derivative, and vanishes at x_1 and x_2 and if*

$$\int_{x_1}^{x_2} \eta(x)M(x)\,dx = 0 \tag{I.13}$$

for all possible functions $\eta(x)$ which satisfy the stated conditions, it follows necessarily that

$$M(x) = 0$$

identically in $x_1 \leq x \leq x_2$.

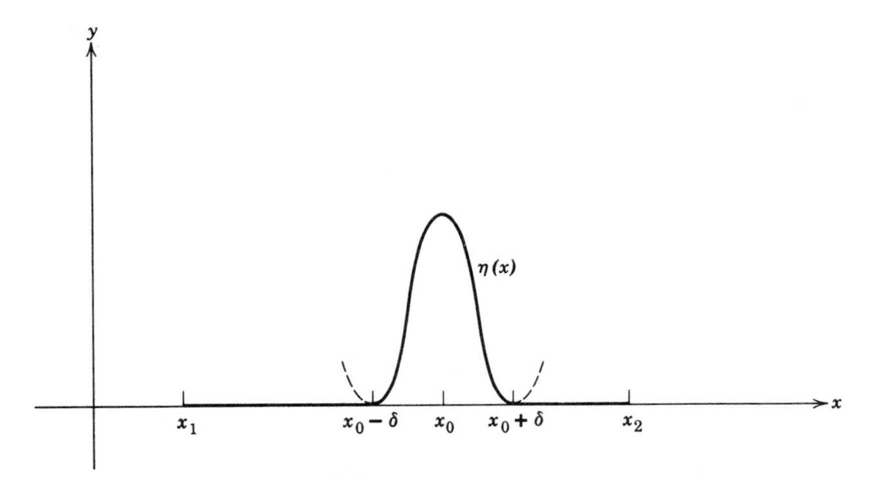

Fig. 2

Proof (*by contradiction*). Let us assume that $M(x)$ is not everywhere zero. Therefore, there has to exist a point x_0 where $M(x)$ is either positive or negative. We assume without loss of generality for the following proof that

$$M(x_0) > 0.$$

Since $M(x)$ is continuous in $x_1 \leq x \leq x_2$, we can certainly find a $\delta > 0$ such that

$$M(x) > 0 \quad \text{for all } x \text{ in } |x - x_0| < \delta.$$

Now, all we have to do is find a permissible function $\eta(x)$ which is zero everywhere except for $|x - x_0| < \delta$, where we assume it to be positive. A very simple function of this type can be pieced together with $\eta = 0$ for $|x - x_0| > \delta$ and that part of a fourth-order parabola in $|x - x_0| \leq \delta$ which has double zeros at $x = x_0 \pm \delta$ and is positive in $|x - x_0| < \delta$ (see Fig. 2).

Such a function, represented by

$$\eta(x) = \begin{cases} 0 & \text{for } |x - x_0| > \delta, \\ (x - x_0 + \delta)^2(x - x_0 - \delta)^2 & \text{for } |x - x_0| \leq \delta, \end{cases}$$

certainly satisfies all the requirements which we formulated in the lemma and has the property that

$$\eta(x) > 0 \quad \text{for all } x \text{ in } |x - x_0| < \delta.$$

Thus we find that

$$\int_{x_1}^{x_2} \eta(x)M(x)\,dx = \int_{x_0-\delta}^{x_0+\delta} (x - x_0 + \delta)^2(x - x_0 - \delta)^2 M(x)\,dx > 0$$

and not zero, as it is supposed to be, q.e.d.

In view of our assumptions about f and y,

$$M(x) = \frac{\partial f}{\partial y} - \frac{d}{dx}\frac{\partial f}{\partial y'}$$

is continuous in $x_1 \leq x \leq x_2$. We can therefore apply the fundamental lemma to (I.12) and obtain the necessary condition for a minimum value of (I.6) as

$$\frac{\partial f}{\partial y} - \frac{d}{dx}\frac{\partial f}{\partial y'} = 0. \tag{I.14}$$

Let us summarize our result as:

Theorem I.1. *Let $f(x, y, y')$ be a function which is continuous and has continuous derivatives of the first and second order with respect to each variable, and let $y = y(x)$ be a continuous function with a continuous first- and second-order derivative, and let*

$$y(x_1) = y_1,$$

$$y(x_2) = y_2.$$

Then, in order to yield a minimum for

$$I = \int_{x_1}^{x_2} f(x, y, y')\,dx$$

it is necessary that $y(x)$ satisfy the second-order differential equation

$$\frac{\partial f}{\partial y} - \frac{d}{dx}\frac{\partial f}{\partial y'} = 0.$$

Remark: It is possible to derive equation (I.14) without any assumption about y'' by means of the lemma of Dubois-Reymond.[3]

Equation (I.14) was first established by the Swiss mathematician *Leonhard Euler* in 1744 by a rather tedious—however, ingenious—process in which he approximated the integral (I.6) by a sum and the extremal $\bar{y}(x)$ by its ordinates, varying those ordinates one at a time. *Lagrange* arrived at the same equation in 1755 (at the tender age of 19 years) by a method which is essentially reproduced in this discussion. In honor of these two great mathematicians of the 18th century, we call equation (I.14) the *Euler-Lagrange equation.*

Before we can apply this theory to the problem of Newtonian motion, as we have formulated it in the preceding subsection, we have to generalize it with respect to the number of unknown functions.

Even though three unknown functions would suffice for this particular purpose, we will consider here the general case with n unknown functions, since this by no means makes the problem more complicated.

We seek n functions

$$y_k = y_k(x), \qquad k = 1, 2, \cdots, n,$$

which satisfy the boundary conditions

$$y_k(x_1) = y_k^{(1)}, \quad y_k(x_2) = y_k^{(2)}, \qquad k = 1, 2, \cdots, n, \tag{I.15}$$

such that the integral

$$I = \int_{x_1}^{x_2} f(x, y_1, y_2, \cdots, y_n, y_1', y_2', \cdots, y_n')\, dx \tag{I.16}$$

yields a minimum.

We assume that f is continuous and has continuous first- and second-order derivatives with respect to each variable and follow the same procedure as before, inasmuch as we assume that

$$y_k = \bar{y}_k(x), \qquad k = 1, 2, \cdots, n,$$

are the solutions of our problem, and vary one of them at a time:

$$\begin{aligned} y_1 &= \bar{y}_1(x), \\ &\vdots \\ y_k &= y_k(x, \epsilon), \\ &\vdots \\ y_n &= \bar{y}_n(x). \end{aligned}$$

[3] See R. Courant and D. Hilbert: *Methods of Mathematical Physics,* Vol. I, Interscience Publishers, New York, 1953, p. 200.

The substitution of this variation into (I.16) makes this integral a function of ϵ, and the vanishing condition for the first variation leads to the n equations

$$\frac{\partial f}{\partial y_k} - \frac{d}{dx}\frac{\partial f}{\partial y_k'} = 0, \qquad k = 1, 2, \cdots, n, \tag{I.17}$$

which are a simple generalization of the Euler-Lagrange equation (I.14). Again, the y_k'' have to be assumed to be continuous.

In the problem of Newtonian motion we had

$$f = \frac{m}{2}(\dot{x}^2 + \dot{y}^2 + \dot{z}^2) - V(x, y, z),$$

which, in the notation of this subsection, reads

$$f = \frac{m}{2}(y_1'^2 + y_2'^2 + y_3'^2) - V(y_1, y_2, y_3).$$

Therefore, we obtain

$$\frac{\partial f}{\partial y_k} = -\frac{\partial V}{\partial y_k},$$

$$\frac{\partial f}{\partial y_k'} = my_k',$$

$$\frac{d}{dx}\frac{\partial f}{\partial y_k'} = my_k'',$$

which yields the Newtonian equations of motion

$$my_k'' = -\frac{\partial V}{\partial y_k}, \qquad k = 1, 2, 3.$$

Problems I.7–19

7. State Hamilton's principle for the motion of a mass point in a Newtonian gravitational field and derive the Euler-Lagrange equations.

8. Give a geometric interpretation of the variational problem

$$\int_0^1 \sqrt{1 + y'^2}\, dx \to \text{minimum}$$

with the boundary conditions $y(0) = 0$, $y(1) = 1$.

***9.** Solve problem 8 for the extremal, find the stationary value of the integral, and compare it with the values of the integral which are obtained for curves that join the same end points but are, however, different from the extremal.

10. Find the Euler-Lagrange equation for

(a) $f = x^2y^2 - y'^2$, (b) $f = \sqrt{xy} + y'^2$,

(c) $f = \sin(xy')$, (d) $f = \dfrac{x^2y'}{\sqrt{1 + y'^2}}$.

11. Let $f = f(x, y')$, i.e., not be explicitly dependent on y.
Show, that the Euler-Lagrange equation reduces in this case to

$$\frac{\partial f}{\partial y'} = C$$

where C is an arbitrary constant.

***12.** Let $f = f(y, y')$, i.e., not be explicitly dependent on x.
Show that the Euler-Lagrange equation reduces in this case to

$$f - y'\frac{\partial f}{\partial y'} = C$$

where C is an arbitrary constant. (*Hint:* Consider y as an independent variable and replace d/dx by $\dfrac{d}{dy} \cdot y'$ in (I.14), and note that $df = (\partial f/\partial y)\,dy + (\partial f/\partial y')\,dy'$.)

13. Find the differential equation of a path down which a particle will fall from one given point to another in the shortest possible time. (*Hint:* It follows from the equation of motion of free-falling bodies that $v = \sqrt{2gy}$, where v is the velocity and y the height. Since $ds/dt = v$, we have

$$t = \int_{x_1}^{x_2} \frac{ds}{\sqrt{2gy}}$$

where x_1 and x_2 are the x-coordinates of the beginning and the end point of motion.)

Remark: The solution of this problem (a *cycloid*), which was first proposed "to the mathematicians of the world to give their consideration" by *John Bernoulli* in 1696, is called *Brachistochron* ($\beta\rho\alpha\chi\iota\sigma\tau o\varsigma$ = shortest, $\chi\rho\acute{o}\nu o\varsigma$ = time).

14. Find the differential equation of a curve through two given points which generates upon rotation about the x-axis a surface of smallest surface area.

Remark: Such surfaces are called *minimal surfaces* and are characterized by the vanishing of their *mean curvature*, which is expressed by their Euler-Lagrange equation.

15. Formulate the proof for the fundamental lemma of the calculus of variations for the case where $M(x)$ is to be assumed as nowhere positive.

16. Find the Euler-Lagrange equation of the variational problem

$$\int_{x_1}^{x_2} f(x, y, y', y'')\,dx \rightarrow \text{minimum}$$

with the boundary conditions

$$y(x_1) = y_1, \qquad y(x_2) = y_2,$$
$$y'(x_1) = y_1', \qquad y'(x_2) = y_2'.$$

17. Find the Euler-Lagrange equation of the variational problem

$$\int_{x_1}^{x_2} f(x, y, y', y'', \cdots, y^{(n)})\, dx \to \text{minimum}$$

and state suitable boundary conditions.

18. The potential energy (deformation energy) of an elastic (laterally movable) rod is given by

$$V = \int_0^l \kappa^2\, dx$$

where κ is the curvature of the rod $y = y(x)$, and the interval $0 \le x \le l$ is the projection of the rod onto the x-axis.

Find the Euler-Lagrange equation for the variational problem

$$V \to \text{minimum}$$

and state suitable boundary conditions. (See problem 16.)

19. In problem 18 consider only those deformations of the rod for which dy/dx is small of first order, and neglect all terms which are small of second and higher order in the expression for V.

Find the Euler-Lagrange equation of the problem $V \to$ minimum under this simplifying assumption.

§2. VARIATIONAL PROBLEMS IN TWO AND MORE INDEPENDENT VARIABLES

1. Vibrations of a Stretched String

We consider a string stretched between two fixed points. We assume that the string is uniformly covered with mass of constant density ρ and is perfectly flexible. We choose our coordinate system so that the string at rest coincides with the x-axis, the beginning point with the origin, and the end point with the point (1, 0).

Let $u = u(x, t)$ represent the displacement of the string at the distance x from the origin at the time t. We consider only deformations of the string in which $\partial u/\partial x$ is small, and which are such that all terms of higher order are neglectible compared with ones of lower order.

If ρ represents the constant mass density of the string, the kinetic energy dT of an element of length ds will be

$$dT = \frac{dm}{2} v^2 = \frac{\rho\, ds}{2}\left(\frac{\partial u}{\partial t}\right)^2.$$

If we expand ds/dx according to the binomial law and neglect terms of the second and higher order

$$\frac{ds}{dx} = \sqrt{1 + (\partial u/\partial x)^2} = 1 + \frac{1}{2}\left(\frac{\partial u}{\partial x}\right)^2 + \cdots,$$

we simply obtain

$$dT = \frac{\rho\, dx}{2}\left(\frac{\partial u}{\partial t}\right)^2,$$

and therefore for the whole string

$$T = \frac{\rho}{2}\int_0^1 \left(\frac{\partial u}{\partial t}\right)^2 dx.$$

The total potential energy is given by the product of the total external force and the increase in length. As the only external force, we will consider the tension τ which is exerted upon the end points. Hence, we obtain

$$V = \tau\int_0^1 (\sqrt{1 + (\partial u/\partial x)^2} - 1)\, dx = \tau\int_0^1 \left(\frac{1}{2}\left(\frac{\partial u}{\partial x}\right)^2 + \cdots\right) dx.$$

In order to find the motion (vibration) of the string in a certain time interval $t_1 \leq t \leq t_2$ we have to find—according to Hamilton's principle—the stationary value of the integral

$$I = \int_{t_1}^{t_2} L\, dt = \frac{1}{2}\int_{t_1}^{t_2}\int_0^1 \left[\rho\left(\frac{\partial u}{\partial t}\right)^2 - \tau\left(\frac{\partial u}{\partial x}\right)^2\right] dx\, dt,$$

subject to the side conditions that the end points remain at rest,

$$u(0, t) = 0,$$
$$u(1, t) = 0,$$

and the initial and end positions of the string are given by

$$u(x, t_1) = u_1(x),$$
$$u(x, t_2) = u_2(x).$$

We see from that if we invoke Hamilton's principle in considering a one-dimensional mechanical continuum we are led to a variational problem with the structure

$$\int_{t_1}^{t_2}\int_{x_1}^{x_2} f\left(x, t, u(x, t), \frac{\partial u}{\partial x}, \frac{\partial u}{\partial t}\right) dx\, dt \rightarrow \text{minimum}$$

where u has given values for all x and t along a closed curve (in the present case a rectangle) in the x, t–plane.

2. The Euler-Lagrange Equation for the Two-Dimensional Problem

Choosing a more convenient notation, we can formulate the problem obtained in the foregoing subsection in the following form:

Find the function $z = z(x, y)$ which assumes over a closed curve C in the x, y–plane given values

$$z(x, y) = z_0(x, y) \quad \text{on } C \tag{I.18}$$

and is such that the integral

$$I = \iint\limits_R f\left(x, y, z(x, y), \frac{\partial z}{\partial x}, \frac{\partial z}{\partial y}\right) dx\, dy \tag{I.19}$$

yields a minimum value, where R is the region bounded by C.

We will assume that R is a simply connected region and C a rectifiable curve. In order to solve this problem, we proceed in a way analogous to that of §1 of this chapter:

We assume that

$$E: \quad z = \bar{z}(x, y), \quad z, \frac{\partial z}{\partial x}, \frac{\partial z}{\partial y} \text{ continuous,}$$

is the solution which satisfies the boundary condition (I.18)

$$\bar{z}(x, y) = z_0(x, y) \quad \text{on } C$$

and consider a one-parameter variation of the form

$$E_\epsilon: \quad z = z(x, y, \epsilon), \tag{I.20}$$

which contains E as a member for $\epsilon = 0$,

$$z(x, y, 0) = \bar{z}(x, y), \tag{I.21}$$

and satisfies the given boundary condition

$$z(x, y, \epsilon) = z_0(x, y) \quad \text{on } C. \tag{I.22}$$

If, for convenience, we put

$$\frac{\partial z}{\partial \epsilon} = \zeta,$$

we obtain, by differentiating (I.22) with respect to ϵ,

$$\zeta(x, y, \epsilon) = 0 \quad \text{on } C. \tag{I.23}$$

We will again assume that ζ and $\partial\zeta/\partial x$, $\partial\zeta/\partial y$ are continuous, for reasons which will become quite obvious in the following process.

Substituting the variation (I.20) for $z(x, y)$ in (I.19) gives

$$I(\epsilon) = \iint\limits_R f\left(x, y, z(x, y, \epsilon), \frac{\partial z(x, y, \epsilon)}{\partial x}, \frac{\partial z(x, y, \epsilon)}{\partial y}\right) dx\, dy. \tag{I.24}$$

If we differentiate (I.24) with respect to ϵ and then put $\epsilon = 0$, the vanishing condition for the first variation $\left.\frac{dI(\epsilon)}{d\epsilon}\right|_{\epsilon=0}$ appears in the form

$$\left.\frac{dI(\epsilon)}{d\epsilon}\right|_{\epsilon=0} = \iint_R \left(\frac{\partial f}{\partial z}\zeta + \frac{\partial f}{\partial z_x}\cdot\frac{\partial \zeta}{\partial x} + \frac{\partial f}{\partial z_y}\cdot\frac{\partial \zeta}{\partial y}\right) dx\, dy = 0. \qquad (I.25)$$

The expression on the right side of (I.25) is the two-dimensional analog of (I.11) in the preceding subsection. The reader will recall that we reduced (I.11) to a form in which it was accessible to the fundamental lemma of the calculus of variations by carrying out an integration-by-parts process of the term which contained the derivative of η with respect to x.

It is to be expected that a procedure which will change the terms containing $\partial\zeta/\partial x$ and $\partial\zeta/\partial y$ into terms containing ζ itself will likewise lead to a form which is accessible to a generalization of the fundamental lemma. Indeed, application of Green's theorem will bring about the desired result. Green's theorem reads

$$\iint_R \operatorname{div} \mathbf{v}\, dx\, dy = \oint_C \mathbf{v} \times d\mathbf{s}, \qquad d\mathbf{s} = dx\mathbf{i} + dy\mathbf{j},$$

where C is the boundary of the region R. (See **A**I.C2; note in particular that $\mathbf{v} \times d\mathbf{s}$ is not a vector.)

For our specific purpose, let

$$\mathbf{v} = \zeta\frac{\partial f}{\partial z_x}\mathbf{i} + \zeta\frac{\partial f}{\partial z_y}\mathbf{j}.$$

Then

$$\iint_R \left(\frac{\partial}{\partial x}\zeta\frac{\partial f}{\partial z_x} + \frac{\partial}{\partial y}\zeta\frac{\partial f}{\partial z_y}\right) dx\, dy = \oint_C \zeta\left(\frac{\partial f}{\partial z_x}dy - \frac{\partial f}{\partial z_y}dx\right).$$

Since

$$\frac{\partial}{\partial x}\zeta\frac{\partial f}{\partial z_x} + \frac{\partial}{\partial y}\zeta\frac{\partial f}{\partial z_y} = \zeta\left(\frac{\partial}{\partial x}\frac{\partial f}{\partial z_x} + \frac{\partial}{\partial y}\frac{\partial f}{\partial z_y}\right) + \frac{\partial f}{\partial z_x}\cdot\frac{\partial \zeta}{\partial x} + \frac{\partial f}{\partial z_y}\cdot\frac{\partial \zeta}{\partial y},$$

we obtain

$$\iint_R \left(\frac{\partial f}{\partial z_x}\frac{\partial \zeta}{\partial x} + \frac{\partial f}{\partial z_y}\frac{\partial \zeta}{\partial y}\right) dx\, dy$$

$$= -\iint_R \zeta\left(\frac{\partial}{\partial x}\frac{\partial f}{\partial z_x} + \frac{\partial}{\partial y}\frac{\partial f}{\partial z_y}\right) dx\, dy + \oint_C \zeta\left(\frac{\partial f}{\partial z_x}dy - \frac{\partial f}{\partial z_y}dx\right).$$

The line integral along C vanishes in view of (I.23)—this is in analogy to the vanishing of the term $\eta \left.\dfrac{\partial f}{\partial y'}\right|_{x_1}^{x_2}$ in the one-dimensional problem—and we obtain

$$\iint_R \zeta\left(\frac{\partial f}{\partial z} - \frac{\partial}{\partial x}\frac{\partial f}{\partial z_x} - \frac{\partial}{\partial y}\frac{\partial f}{\partial z_y}\right) dx\, dy = 0 \tag{I.26}$$

as the first necessary condition for a minimum. This relation has to hold for all permissible functions ζ. (In analogy to the definition of a permissible function η in the one-dimensional problem, we define a permissible function $\zeta(x, y)$ as a function which is continuous, has continuous first-order derivatives with respect to x and y, and vanishes along C.)

From this we conclude again that

$$\frac{\partial f}{\partial z} - \frac{\partial}{\partial x}\frac{\partial f}{\partial z_x} - \frac{\partial}{\partial y}\frac{\partial f}{\partial z_y} = 0 \tag{I.27}$$

has to be satisfied.

The step taken from (I.26) to (I.27) again requires a lemma analogous to the fundamental lemma of the calculus of variations, which we proved in the preceding section. We are not going to give a proof here for such a generalized lemma but we will indicate how the reader can establish such a proof himself. Since we have to assume the integrand of (I.26) to be continuous (to permit the application of Green's theorem), there is certainly a point (x_0, y_0) in R and a neighborhood of that point throughout which the integrand is positive (or negative)—if it is not identically zero. All we have to do now is to choose a surface ζ which goes through the curve C and is everywhere zero except in the neighborhood of (x_0, y_0), where we can assume it to be positive (or negative). This particular choice of a permissible function ζ will make (I.26) positive (or negative) and not zero, as it is supposed to be.

We state our result as:

Theorem I.2. *If $f = f(x, y, z(x, y), \partial z/\partial x, \partial z/\partial y)$ is continuous and has continuous first- and second-order derivatives with respect to each variable, and if $z = z(x, y)$ is continuous and has continuous first- and second-order derivatives with respect to x and y, and satisfies the boundary condition*

$$z(x, y) = z_0(x, y) \quad \text{on } C,$$

where C is a closed rectifiable curve and yields a minimum for

$$I = \iint_R f\left(x, y, z(x, y), \frac{\partial z}{\partial x}, \frac{\partial z}{\partial y}\right) dx\, dy,$$

where R is a simply connected region, encompassed by C, then it is necessary that $z(x, y)$ satisfy the following differential equation:

$$\frac{\partial f}{\partial z} - \frac{\partial}{\partial x}\frac{\partial f}{\partial z_x} - \frac{\partial}{\partial y}\frac{\partial f}{\partial z_y} = 0.$$

Remark: It is to be observed that the symbols $\partial/\partial x$, $\partial/\partial y$ refer to differentiation with respect to x and y, wherever x and y occur, while the symbols $\partial/\partial z$, $\partial/\partial z_x$, $\partial/\partial z_y$ refer to differentiation with respect to the indicated variables, which only occur explicitly. Thus, the derivatives of the first kind appear to be total derivatives in a certain sense while the derivatives of the second kind are genuine partial derivatives. Some authors choose to indicate the difference in the nature of these derivatives by retaining the symbols $\partial/\partial x$, $\partial/\partial y$ for the derivatives of the first kind, while designating the derivatives of the second kind by subscripts, as f_z, f_{z_x}, f_{z_y}. In this text we use both notations interchangeably.

Equation (I.27) enables us to write down the equation for the vibrations of a stretched string as we formulated the problem in subsection 1. We have

$$\frac{\partial f}{\partial u} = 0, \quad \frac{\partial f}{\partial u_x} = -2\tau\frac{\partial u}{\partial x}, \quad \frac{\partial f}{\partial u_t} = 2\rho\frac{\partial u}{\partial t}.$$

Therefore, the Euler-Lagrange equation has the form

$$\tau\frac{\partial^2 u}{\partial x^2} = \rho\frac{\partial^2 u}{\partial t^2}.$$

In the next chapter we are going to derive this equation anew by using the fundamental principle of vectorial mechanics. At this later stage we will also give a discussion concerning the solution.

The reader can already guess that the consideration of a two-dimensional or n-dimensional mechanical continuum will lead to a variational problem of 3 or $n + 1$ independent variables, and we will actually develop such a problem in the next chapter (vibrations of a membrane, see Chapter II, §2.2).

Without going into details, we state here the Euler-Lagrange equation for such an n-dimensional problem and refer the reader for a derivation either to his own resources or to the appropriate literature.[4]

[4] See R. Weinstock: *Calculus of Variations*, McGraw-Hill Book Co., New York, 1952, pp. 135 and 206; or R. Courant and D. Hilbert: *Methods of Mathematical Physics*, Interscience Publishers, New York, 1953, p. 191.

A necessary condition for an n-dimensional hypersurface $z = z(x_1, x_2, \cdots, x_n)$ which is supposed to pass through the $n - 1$ dimensional hypersurface S_{n-1} to minimize the integral

$$\int_{S_n}\cdots\int f\left(x_1, x_2, \cdots, x_n, z(x_1, x_2, \cdots, x_n), \frac{\partial z}{\partial x_1}, \frac{\partial z}{\partial x_2}, \cdots, \frac{\partial z}{\partial x_n}\right) dx_1\, dx_2, \cdots, dx_n,$$

where S_n is an n-dimensional region enclosed by S_{n-1} in the $n + 1$ dimensional space $(x_1, \cdots, x_n, z)$, is

$$\frac{\partial f}{\partial z} - \sum_{k=1}^{n} \frac{\partial}{\partial x_k} \frac{\partial f}{\partial z_{x_k}} = 0. \tag{I.28}$$

Problems I.20–22

20. Find the Euler-Lagrange equation for the following variational problem:

$$\iint_R \sqrt{1 + (\partial u/\partial x)^2 + (\partial u/\partial y)^2}\, dx\, dy \to \text{minimum}$$

with the boundary condition

$$u(x, y) = 1 \quad \text{on } x^2 + y^2 = 1.$$

***21.** Prove a generalization of the fundamental lemma of the calculus of variations for two dimensions, namely:

If $M(x, y)$ is continuous in R and if

$$\iint_R \zeta(x, y) M(x, y)\, dx\, dy = 0$$

for all permissible functions $\zeta(x, y)$, then $M(x, y) = 0$ identically in R. (*Hint:* If $M(x, y)$ is not identically zero in R, then there is a point (x_0, y_0) where it is either positive or negative. Translate the coordinate system in such a way that (x_0, y_0) becomes the new coordinate origin and rotate a curve like the one we used in constructing a permissible function for the proof of the fundamental lemma in one dimension about the z-axis.)

22. Study the result of problem 17 (p. 15) and equation (I.28) on p. 21 and make a conjecture as to the structure of the Euler-Lagrange equation for the following variational problem:

$$\int_{S_n}\cdots\int f\left(x_1, x_2, \cdots, x_n, z, \frac{\partial z}{\partial x_1}, \frac{\partial z}{\partial x_2}, \cdots, \frac{\partial z}{\partial x_n}, \frac{\partial^2 z}{\partial x_1\, \partial x_1}, \frac{\partial^2 z}{\partial x_1\, \partial x_2}, \cdots, \frac{\partial^2 z}{\partial x_n\, \partial x_n}\right) dx_1, dx_2, \cdots, dx_n \to \text{minimum}$$

with the appropriate boundary conditions.

§3. THE ISOPERIMETRIC PROBLEM

1. The Problem of Dido

We are now going to develop the elements of a theory which we will need at a later time in computing the so-called "eigenvalues" of boundary-value problems of a certain type. We do not want to postpone this discussion until this later time since it concerns again a variational problem of a rather elementary nature.

We will discuss a problem which is purported to have been solved for the first time by *Queen Dido of Carthage*—of course, in a purely intuitive way, as is characteristic of female reasoning, especially if tricking a man and making considerable gain is the goal. To be more specific, it is purported that *Queen Dido*, being a refugee from *Tyria*, asked a North African chieftain named *Jarbas* for as much land as she could encompass with the hide of a cow. This poor man, falsely thinking that this could not possibly amount to much, gave his consent to the deal, whereupon *Queen Dido* proceeded to slice the hide into numerous small strips and lay these out along a semicircle, using the North African coast as the supplementary boundary. Whether she knew it or not, she obtained in this way the maximal area she could possibly obtain by such a procedure, and established within that area the State of Carthage, in 850 B.C.

Let us now translate this shrewd procedure into mathematical language, modifying it slightly. There is given a perimeter for a curve (the total length of all strips of the hide) and the problem is to find the shape of a closed curve of this given length which bounds the largest possible area among all areas that can be encompassed by a curve of this length. This problem is called the *isoperimetric problem* since it deals with curves of equal ('ισος) perimeter.

If we choose the x-axis as a part of the boundary, say between the points (0, 0) and (1, 0), then Dido's problem is to find a curve $y = y(x)$ which shall join those two points so that the enclosed area

$$I = \int_0^1 y\,dx$$

becomes a maximum, while the length of the curve

$$C = \int_0^1 \sqrt{1 + y'^2}\,dx = \frac{\pi}{2}$$

has a given value (for sake of convenience, we choose the value $\pi/2$).

A more general form of a problem of this type will be:
Find a function $y = y(x)$ such that

$$y(x_1) = y_1, \quad y(x_2) = y_2 \tag{I.29}$$

and

$$I = \int_{x_1}^{x_2} f(x, y, y')\, dx \to \text{maximum (minimum)}, \tag{I.30}$$

with

$$C = \int_{x_1}^{x_2} g(x, y, y')\, dx = L \text{ (constant)} \tag{I.31}$$

given. This is the problem to be dealt with in the next subsection.

2. The Euler-Lagrange Equation for the Isoperimetric Problem in One Independent Variable

We make the usual assumptions about f and its derivatives and like assumptions about the function g. We assume in accordance with our standard procedure that $y = \bar{y}(x)$ is the solution of (I.29, 30, 31) and consider a variation of this solution. However, a one-parameter variation, as before, will not suffice, as a very simple argument will show. Suppose we have only one parameter ϵ at our disposal. This parameter ϵ will change not only the value of I but also the value of C. Since C has to be kept at the constant value L, we have to use a second parameter to counterbalance the effect of the first one. Hence, we will consider a two-parameter variation as

$$y = \bar{y}(x) + \epsilon_1 \eta_1(x) + \epsilon_2 \eta_2(x) \tag{I.32}$$

between the same end points (x_1, y_1) and (x_2, y_2); i.e., $\eta_1(x_i) = \eta_2(x_i) = 0$, $i = 1, 2$.

If we substitute the variation (I.32) into (I.30) and (I.31), we obtain for $\epsilon_1 = \epsilon_2 = 0$ the following system of equations:

$$\begin{aligned} I(\epsilon_1, \epsilon_2)\big|_{\epsilon_1=\epsilon_2=0} &= E, \\ C(\epsilon_1, \epsilon_2)\big|_{\epsilon_1=\epsilon_2=0} &= L, \end{aligned} \tag{I.33}$$

where E shall stand for the extreme value of I.

It is necessary, in order for E to be an extreme value, that the *Jacobian*

$$\Delta = \frac{\partial(I, C)}{\partial(\epsilon_1, \epsilon_2)}\bigg|_{\epsilon_1=\epsilon_2=0} = 0, \tag{I.34}$$

for the following reason: If $\Delta \neq 0$, according to the theorem on implicit functions (see **AIII.A**) we could solve the system (I.33) in a neighborhood of $\epsilon_1 = \epsilon_2 = 0$,

$$\epsilon_1 = \epsilon_1(E, L),$$
$$\epsilon_2 = \epsilon_2(E, L).$$

Now, if E is the minimum (maximum) of I, we will obtain, according to our assumption, $\epsilon_1 = \epsilon_2 = 0$. However, we will also obtain solutions (ϵ_1, ϵ_2) for smaller (larger) values than E, which means that E cannot possibly be the minimum (maximum).

Let us now evaluate the Jacobian Δ and see what consequences the condition (I.34) has.

Since

$$\Delta = \left.\frac{\partial(I, C)}{\partial(\epsilon_1, \epsilon_2)}\right|_{\epsilon_1=\epsilon_2=0} = \begin{vmatrix} \left.\dfrac{\partial I}{\partial \epsilon_1}\right|_{\epsilon_1=\epsilon_2=0} & \left.\dfrac{\partial I}{\partial \epsilon_2}\right|_{\epsilon_1=\epsilon_2=0} \\ \left.\dfrac{\partial C}{\partial \epsilon_1}\right|_{\epsilon_1=\epsilon_2=0} & \left.\dfrac{\partial C}{\partial \epsilon_2}\right|_{\epsilon_1=\epsilon_2=0} \end{vmatrix}$$

we have to find $\left.\dfrac{\partial I}{\partial \epsilon_i}\right|_{0,0}$ and $\left.\dfrac{\partial C}{\partial \epsilon_i}\right|_{0,0}$ for $i = 1, 2$. These variations can be found in the same way as in §1.2 and we obtain

$$\left.\frac{\partial I}{\partial \epsilon_i}\right|_{0,0} = \int_{x_1}^{x_2} \eta_i\left(\frac{\partial f}{\partial y} - \frac{d}{dx}\frac{\partial f}{\partial y'}\right) dx,$$
$$\left.\frac{\partial C}{\partial \epsilon_i}\right|_{0,0} = \int_{x_1}^{x_2} \eta_i\left(\frac{\partial g}{\partial y} - \frac{d}{dx}\frac{\partial g}{\partial y'}\right) dx.$$

(All the integrated parts vanish in view of $\eta_1(x_i) = \eta_2(x_i) = 0$.) Hence

$$\Delta = \begin{vmatrix} \displaystyle\int_{x_1}^{x_2} \eta_1\left(\frac{\partial f}{\partial y} - \frac{d}{dx}\frac{\partial f}{\partial y'}\right) dx, & \displaystyle\int_{x_1}^{x_2} \eta_2\left(\frac{\partial f}{\partial y} - \frac{d}{dx}\frac{\partial f}{\partial y'}\right) dx \\ \displaystyle\int_{x_1}^{x_2} \eta_1\left(\frac{\partial g}{\partial y} - \frac{d}{dx}\frac{\partial g}{\partial y'}\right) dx, & \displaystyle\int_{x_1}^{x_2} \eta_2\left(\frac{\partial g}{\partial y} - \frac{d}{dx}\frac{\partial g}{\partial y'}\right) dx \end{vmatrix}$$

We have to assume for the following discussion that $\bar{y}(x)$ is not a solution of

$$\frac{\partial g}{\partial y} - \frac{d}{dx}\frac{\partial g}{\partial y'} = 0.$$

(This assumption appears to be quite reasonable since it is not likely that the value assigned to C in (I.31) is at the same time its extreme value, and even if it is, the function letting (I.31) attain its extreme value is not very likely to also make (I.30) an extreme.)

We can, therefore, certainly choose from all permissible functions an $\eta_2(x)$ such that

$$\int_{x_1}^{x_2} \eta_2\left(\frac{\partial g}{\partial y} - \frac{d}{dx}\frac{\partial g}{\partial y'}\right) dx \neq 0. \tag{I.35}$$

Once we have chosen $\eta_2(x)$ we call

$$\int_{x_1}^{x_2} \eta_2\left(\frac{\partial g}{\partial y} - \frac{d}{dx}\frac{\partial g}{\partial y'}\right) dx = \lambda_1,$$

$$\int_{x_1}^{x_2} \eta_2\left(\frac{\partial f}{\partial y} - \frac{d}{dx}\frac{\partial f}{\partial y'}\right) dx = \lambda_2.$$

Thus, (I.34) can be written in the form

$$\lambda_1\int_{x_1}^{x_2} \eta_1\left(\frac{\partial f}{\partial y} - \frac{d}{dx}\frac{\partial f}{\partial y'}\right) dx - \lambda_2\int_{x_1}^{x_2} \eta_1\left(\frac{\partial g}{\partial y} - \frac{d}{dx}\frac{\partial g}{\partial y'}\right) dx = 0.$$

Division by λ_1, which certainly is different from zero according to (I.35), and introduction of a new parameter λ according to

$$\lambda = -\frac{\lambda_2}{\lambda_1}$$

yields

$$\int_{x_1}^{x_2} \eta_1\left[\frac{\partial f}{\partial y} - \frac{d}{dx}\frac{\partial f}{\partial y'} + \lambda\left(\frac{\partial g}{\partial y} - \frac{d}{dx}\frac{\partial g}{\partial y'}\right)\right] dx = 0$$

for all permissible functions η_1. We can now apply the fundamental lemma of the calculus of variations and obtain as a necessary condition for the solution of the isoperimetric problem

$$\frac{\partial}{\partial y}(f + \lambda g) - \frac{d}{dx}\frac{\partial}{\partial y'}(f + \lambda g) = 0.$$

This equation is usually written in the form

$$\frac{\partial h}{\partial y} - \frac{d}{dx}\frac{\partial h}{\partial y'} = 0 \tag{I.36}$$

where $h = f + \lambda g$.

This is a differential equation of the second order containing an arbitrary parameter λ. (λ depends on λ_1, λ_2, which in turn depends on the choice of η_2, which is almost any permissible function.)

The two integration constants which are obtained in general by solving (I.36) are absorbed by the boundary conditions while λ has to be used in order to satisfy (I.31).

Let us state this result as:

Theorem I.3. *Let $f = f(x, y, y')$ and $g = g(x, y, y')$ be continuous functions with continuous first- and second-order derivatives with respect to each variable and let L be a given number.*

If $y = y(x)$ is continuous and has continuous first- and second-order derivatives, with

$$y(x_1) = y_1, \quad y(x_2) = y_2,$$

and is such that

$$I = \int_{x_1}^{x_2} f(x, y, y')\, dx$$

yields a minimum while

$$C = \int_{x_1}^{x_2} g(x, y, y')\, dx = L,$$

then it is necessary that $y(x)$ satisfy the differential equation

$$\frac{\partial h}{\partial y} - \frac{d}{dx}\frac{\partial h}{\partial y'} = 0$$

where $h(x, y, y') = f(x, y, y') + \lambda g(x, y, y')$ and λ is an arbitrary parameter.

The solution of Dido's problem is now very easy.

We had $f = y$ and $g = \sqrt{1 + y'^2}$. Therefore,

$$h = y + \lambda\sqrt{1 + y'^2}, \qquad \frac{\partial h}{\partial y} = 1, \quad \frac{\partial h}{\partial y'} = \frac{\lambda y'}{\sqrt{1 + y'^2}},$$

and (I.36) for Dido's problem assumes the form

$$1 - \lambda\frac{d}{dx}\left(\frac{y'}{\sqrt{1 + y'^2}}\right) = 0.$$

The solution of this equation is much simpler than one might think at first glance since

$$\frac{d}{dx}\left(\frac{y'}{\sqrt{1 + y'^2}}\right) = \frac{y''}{(1 + y'^2)^{3/2}} = \kappa$$

is the curvature of $y(x)$. So, our equation says nothing else but that

$$\kappa = \frac{1}{\lambda};$$

i.e.: The curvature of the solution has to be constant. As Dido correctly guessed, this is a circle.

From the boundary conditions we stated in subsection 1 of this section and the condition $L = \pi/2$, it follows that the solution is

$$y = \sqrt{\tfrac{1}{4} - (x - \tfrac{1}{2})^2}.$$

Problems I.23–31

23. Show that the problem of finding a curve which shall encompass a given area, while having the smallest possible length, is equivalent to the problem of Dido. (*Hint:* Compare the Euler-Lagrange equations of the two problems).

***24.** Find the curve joining the points (0, 0) and (1, 0) with a given length such that the y-coordinate of its centroid is a minimum.

Find the Euler-Lagrange equations of the isoperimetric problems 25–28:

25. $\displaystyle\int_0^1 y'^2\,dx \to \text{minimum},$

$$\int_0^1 y^2\,dx = 1, \qquad y(0) = 0, \quad y(1) = 0.$$

26. $\displaystyle\int_0^1 xy'^2\,dx \to \text{minimum},$

$$\int_0^1 xy^2\,dx = 1, \qquad |y(0)| < \infty, \quad y(1) = 0.$$

27. $\displaystyle\int_{-1}^1 (1 - x^2)y'^2\,dx \to \text{minimum},$

$$\int_{-1}^1 y^2\,dx = 1, \qquad |y(-1)| < \infty, \quad |y(1)| < \infty.$$

28. $\displaystyle\int_{x_1}^{x_2} (py'^2 - qy^2)\,dx \to \text{minimum},$

$$\int_{x_1}^{x_2} ry^2\,dx = 1, \qquad y(x_1) = y_1, \quad y(x_2) = y_2.$$

29. Show that the Euler-Lagrange equations of problems 25–27 are special cases of the Euler-Lagrange equation of problem 28.

30. $y = \sqrt{\tfrac{3}{2}}x$ is a solution of problem 27. Find the corresponding minimum values of the integral and compare it with the values obtained by considering a variation of $y = \sqrt{\tfrac{3}{2}}x$.

31. Find the Euler-Lagrange equation of the two dimensional isoperimetric problem

$$\iint_R f\left(x, y, z(x, y), \frac{\partial z}{\partial x}, \frac{\partial z}{\partial y}\right) dx\,dy \to \text{minimum},$$

$$\iint_R g\left(x, y, z(x, y), \frac{\partial z}{\partial x}, \frac{\partial z}{\partial y}\right) dx\,dy = L,$$

where $z = z(x, y)$ satisfies the boundary condition

$$z(x, y) = z_0(x, y) \quad \text{on } C, \qquad \text{where } C \text{ is the boundary of } R.$$

§4. NATURAL BOUNDARY CONDITIONS

1. A Problem of Zermelo in Modified Form

Let us consider the following problem:

Given a stream, the left bank of which is represented by the curve $y = \phi_1(x)$ and the right bank of which is given by $y = \phi_2(x)$.

Let the stream velocity at every point (x, y) be represented by the vector field

$$\mathbf{v}(x, y) = u(x, y)\mathbf{i} + v(x, y)\mathbf{j}.$$

A dog, capable of swimming with the constant speed c (in still water), wishes to cross the stream in a minimum amount of time. As one will easily recognize, the question here is not only along which curve $y = y(x)$ he should traverse the stream but also what points of embarkment and landing he should choose so that the shortest crossing time is ensured.

The actual velocity components of the dog's motion in the stream are obviously

$$\bar{u} = u(x, y) + c \cdot \cos \sigma,$$

$$\bar{v} = v(x, y) + c \cdot \sin \sigma,$$

where σ represents the direction in which the dog would swim were he in still water.

Then, with $\bar{u} = \dfrac{dx}{dt}$, $\bar{v} = \dfrac{dy}{dt}$,

$$\left(\frac{dx}{dt} - u\right)^2 + \left(\frac{dy}{dt} - v\right)^2 = c^2$$

or, in terms of a parameter τ

$$\left(\frac{dx}{d\tau} \cdot \frac{d\tau}{dt} - u\right)^2 + \left(\frac{dy}{d\tau} \cdot \frac{d\tau}{dt} - v\right)^2 = c^2.$$

Multiplication by $\left(\dfrac{dt}{d\tau}\right)^2$ and solution of the resulting quadratic equation for $\dfrac{dt}{d\tau}$ yields with $\tau = x$

$$\frac{dt}{dx} = \frac{-(u + vy') \pm \sqrt{c^2(1 + y'^2) - (uy' - v)^2}}{c^2 - u^2 - v^2}$$

and we have consequently, for the time it requires the dog to cross

$$t = \int_{x_1}^{x_2} \frac{-(u + vy') \pm \sqrt{c^2(1 + y'^2) - (uy' - v)^2}}{c^2 - u^2 - v^2}\, dx$$

where x_1 and x_2 are the x-coordinates of points on the curves $y = \phi_1(x)$ and $y = \phi_2(x)$ respectively.

Our problem is to find a function $y = y(x)$ that yields a minimum for this integral, which has a variable upper and a variable lower limit.

The solution of this problem in its full generality is of no further consequences for our treatment. Only the special case where both banks are parallel straight lines will be of interest to us, and this case will be dealt with in the next subsection.

However, we wish to remark that the solutions of this problem are sometimes referred to as "dog curves" and the author, being himself ignorant of canine habits, has been given to understand that dogs actually cross rivers along such mathematically predictable curves.

The reader who is interested in pursuing the problem of variable end points in greater detail than will be given here is referred to the appropriate literature.[5]

2. Natural Boundary Conditions for the One-Dimensional Problem

As in Chapter I, §1.2, we wish to minimize the integral

$$I = \int_{x_1}^{x_2} f(x, y, y')\, dx \tag{I.37}$$

by a proper choice of $y = y(x)$. However, we do not now prescribe the boundary values for $y(x)$ at x_1 and x_2 but let the end points of $y(x)$ slide freely along the two vertical lines $x = x_1$ and $x = x_2$.

Consequently we will consider a variation of the form

$$y = y(x, \epsilon) \quad \text{on } x_1 \le x \le x_2$$

such that

$$y(x, 0) = \bar{y}(x)$$

is the solution of our problem. However, in contrast to our previous procedure, we will not assume that all curves in this one-parameter family join the same end points. Hence, if we again let

$$\frac{\partial y(x, \epsilon)}{\partial \epsilon} = \eta(x, \epsilon),$$

we cannot conclude that $\eta(x, \epsilon)$ vanishes for $x = x_1$ and $x = x_2$.

If we compute the first variation of I as we have done in §1.2 of this chapter and carry out the same integration by parts, we arrive at

$$\left.\frac{dI(\epsilon)}{d\epsilon}\right|_{\epsilon=0} = \eta f_{y'}\Big|_{x_1}^{x_2} - \int_{x_1}^{x_2} \eta\left(\frac{\partial f}{\partial y} - \frac{d}{dx}\frac{\partial f}{\partial y'}\right) dx.$$

[5] See O. Bolza: *Lectures on the Calculus of Variations*, The University of Chicago Press, 1904, p. 102; G. A. Bliss: *Lectures on the Calculus of Variations*, The University of Chicago Press, 1946, p. 147.

Again, it is necessary that $\left.\frac{dI(\epsilon)}{d\epsilon}\right|_{\epsilon=0} = 0$ in order for $\bar{y}(x)$ to be a solution of our problem. Now if a solution exists at all, then it will join some point $P_1(x_1, y_1)$ on $x = x_1$ with some point $P_2(x_2, y_2)$ on $x = x_2$ and yield a minimum for I compared with the values of the integral which are obtained for other curves joining the same two points. Hence, we can conclude that the Euler-Lagrange equation has to be satisfied. This leaves us with

$$\left.\frac{dI(\epsilon)}{d\epsilon}\right|_{\epsilon=0} = \eta \left.\frac{\partial f}{\partial y'}\right|_{x_1}^{x_2} = 0$$

and since $\eta(x_1)$, $\eta(x_2)$ may be different from zero, we have (see problem I.35)

$$\left.\frac{\partial f}{\partial y'}\right|_{x=x_1} = 0, \quad \left.\frac{\partial f}{\partial y'}\right|_{x=x_2} = 0, \tag{I.38}$$

which one calls *natural boundary conditions.*

Thus, we can state that if $y = \bar{y}(x)$ minimizes I, then it is necessary that the Euler-Lagrange equation as well as the natural boundary conditions (I.38) be satisfied.

A very simple example will illustrate this situation:

We seek a curve $y = y(x)$ on $0 \leq x \leq 1$ such that

$$I = \int_0^1 \sqrt{1 + y'^2}\, dx$$

yields a minimum. The natural boundary condition yields in this case

$$\frac{y'(0)}{\sqrt{1 + y'^2(0)}} = 0, \quad \frac{y'(1)}{\sqrt{1 + y'^2(1)}} = 0$$

and it follows that $y'(0) = y'(1) = 0$.

Since $y'' = 0$ is the Euler-Lagrange equation of this problem, we see that the solution is a straight line which is orthogonal to the two vertical lines in 0 and 1.

3. Natural Boundary Condition for the Two-Dimensional Problem

We will now generalize the idea which we pursued in the preceding subsection to the two-dimensional case which was dealt with in §2.2 of this chapter.

We wish to minimize the integral

$$I = \iint_R f\left(x, y, z(x, y), \frac{\partial z}{\partial x}, \frac{\partial z}{\partial y}\right) dx\, dy \tag{I.39}$$

where R is a simply connected region which is bounded by a rectifiable curve C, through a proper choice of $z = z(x, y)$ on R, of which we do not require that it pass through a prescribed space curve. In analogy to the problem which we discussed in the preceding subsection, we let $z = z(x, y)$ slide freely along the walls of a right cylinder with the contour C.

As before, we consider a variation

$$z = z(x, y, \epsilon)$$

with

$$z(x, y, 0) = \bar{z}(x, y)$$

where $\bar{z}(x, y)$ is the solution of our problem. We repeat, however, that we do not require that all the surfaces in this variation pass through the same space curve. Consequently,

$$\frac{\partial z(x, y, \epsilon)}{\partial \epsilon} = \zeta(x, y, \epsilon)$$

does not necessarily vanish along C, and in the first variation of I

$$\left.\frac{dI(\epsilon)}{d\epsilon}\right|_{\epsilon=0} = \iint_R \zeta\left(\frac{\partial f}{dz} - \frac{\partial}{\partial x}\frac{\partial f}{\partial z_x} - \frac{\partial}{\partial y}\frac{\partial f}{\partial z_y}\right) dx\, dy - \oint_C \zeta\left(\frac{\partial f}{\partial z_y}\, dx - \frac{\partial f}{\partial z_x}\, dy\right)$$

the line integral along C does not necessarily vanish.

Again, we can conclude that the Euler-Lagrange equation

$$\frac{\partial f}{\partial z} - \frac{\partial}{\partial x}\frac{\partial f}{\partial z_x} - \frac{\partial}{\partial y}\frac{\partial f}{\partial z_y} = 0$$

has to be satisfied because $z = \bar{z}(x, y)$ has to yield a minimum for a fixed boundary. Hence, the vanishing condition for the first variation reduces to

$$\oint_C \zeta\left(\frac{\partial f}{\partial z_y}\, dx - \frac{\partial f}{\partial z_x}\, dy\right) = 0$$

for all continuous functions ζ with continuous first-order derivatives, and from this we conclude that

$$\frac{\partial f}{\partial z_y} - \frac{\partial f}{\partial z_x} \cdot \frac{dy}{dx} = 0 \quad \text{on } C. \tag{I.40}$$

(This last step follows by an application of the fundamental lemma in modified form; see problem I.34.)

(I.40) is the two-dimensional analog of conditions (I.38) of the preceding subsection, and again we call it a *natural boundary condition.*

Once more we can state that, if $z = \bar{z}(x, y)$ minimizes the integral (I.39), then the Euler-Lagrange equation as well as the natural boundary condition (I.40) have to be satisfied.

Let us illustrate (I.40) with an example, the physical significance of which will be discussed in Chapter II, §2.3.

Let

$$f = \left(\frac{\partial z}{\partial x}\right)^2 + \left(\frac{\partial z}{\partial y}\right)^2.$$

Then

$$\frac{\partial f}{\partial z_x} = 2\frac{\partial z}{\partial x},$$

$$\frac{\partial f}{\partial z_y} = 2\frac{\partial z}{\partial y}$$

and the natural boundary condition yields

$$\frac{\partial z}{\partial y} - \frac{\partial z}{\partial x} \cdot \frac{dy}{dx} = 0 \quad \text{on } C.$$

In order to give this condition a geometric interpretation, we represent C in parametric form

$$x = x(s),$$

$$y = y(s).$$

Then, $dy/dx = \dfrac{(dy/ds)}{(dx/ds)}$ on C and (I.40) can be written as

$$\frac{\partial z}{\partial y}\frac{dx}{ds} - \frac{\partial z}{\partial x}\frac{dy}{ds} = 0 \quad \text{on } C.$$

The tangent vector to C is given by

$$\mathbf{t} = \frac{dx}{ds}\mathbf{i} + \frac{dy}{ds}\mathbf{j}.$$

Hence,

$$\mathbf{n} = -\frac{dy}{ds}\mathbf{i} + \frac{dx}{ds}\mathbf{j}$$

is the normal to C (**AI.B1**).

If the parameter s represents the arc length on C, the derivative of z in the direction normal to C can be written as

$$\frac{dz}{dn} = -\frac{\partial z}{\partial x}\frac{dy}{ds} + \frac{\partial z}{\partial y}\frac{dx}{ds}$$

and we see that (I.40) amounts to

$$\frac{dz}{dn} = 0 \quad \text{on } C.$$

This means that in this case, the surface $z = z(x, y)$ has to be orthogonal to the walls of the right cylinder with the contour C.

Problems I.32–35

32. Formulate the dog problem of subsection 1 for the case where $x_1 = 0$, $x_2 = 1$, the river velocity is given by

$$\mathbf{v} = x(x - 1)\mathbf{j},$$

and $c = 1$.

33. What are the Euler-Lagrange equation and the natural boundary conditions for problem 32?

***34.** Prove that from

$$\oint_C \zeta(x, y)M(x, y)\, ds = 0$$

where C is a rectifiable curve, $M(x, y)$ is continuous, and s is the arc length on C, for all continuous functions $\zeta(x, y)$ with continuous first-order derivatives, it follows that

$$M(x, y) = 0 \quad \text{on } C.$$

35. Prove that from

$$\eta \frac{\partial f}{\partial y'}\bigg|_{x_1}^{x_2} = 0$$

for all functions η which are continuous and have a continuous derivative, it follows that

$$\frac{\partial f}{\partial y'}\bigg|_{x=x_1} = 0, \quad \frac{\partial f}{\partial y'}\bigg|_{x=x_2} = 0.$$

RECOMMENDED SUPPLEMENTARY READING

O. Bolza: *Lectures on the Calculus of Variations*, The University of Chicago Press, 1904.

O. Bolza: *Vorlesungen ueber Variationsrechnung*, Teubner, Leipzig, 1909.

G. A. Bliss: *Calculus of Variations*, The Open Court Publishing Company, Chicago, 1925.

G. A. Bliss: *Lectures on the Calculus of Variations*, The University of Chicago Press, 1946.

C. Lanczos: *The Variational Principles of Mechanics*, University of Toronto Press, 1946.

R. Weinstock: *Calculus of Variations with Applications to Physics and Engineering*, McGraw-Hill Book Co., New York, 1952.

F. B. Hildebrand: *Methods of Applied Mathematics*, Prentice Hall, New York, 1952 Chapter 2.

R. Courant and D. Hilbert: *Methods of Mathematical Physics*, Vol. I, Interscience Publishers, New York, 1953.

Ph. Frank and R. von Mises: *Die Differential- und Integralgleichungen der Mechanik und Physik*, Vol. I, F. Vieweg and Sohn, Braunschweig, 1930, Chapters 5 and 20.

II

REPRESENTATION OF SOME PHYSICAL PHENOMENA BY PARTIAL DIFFERENTIAL EQUATIONS

§1. THE VIBRATING STRING

1. Vibrations of a Stretched String—Vectorial Approach

We are going to derive in this subsection the equation for the vertical displacements of a perfectly flexible string stretched between two fixed points. As to choice of coordinate system, a discussion of the physical properties of the string, and the neglect of quantities which are small of higher order, we refer the reader to Chapter I, §2.1 (p. 15).

Since the string is stretched between two fixed points, there is a certain tension $\boldsymbol{\tau}$ exerted upon the end points. We assume the string to be perfectly flexible; i.e., only tensile forces can be transmitted in the direction of its tangent line. It follows that the tension acts with the same magnitude upon any part of the string in the direction of the tangent.

In order to obtain an equation for the vertical displacements, we have to establish according to the principle of vectorial mechanics the resultant of all external forces exerted upon any element of the string in the u-direction and equate this to the inertial force of the respective element in the u-direction. Considering Fig. 3, which represents a section of the string, we find that the u-component of the tension $\boldsymbol{\tau}$ at P_1, the beginning point of the considered element of length Δs, amounts to $-\tau \sin \alpha_1$ and at P_2, the end point of the considered element, amounts to $\tau \sin \alpha_2$ where $\tau = |\boldsymbol{\tau}|$. Thus, the resultant of all external forces exerted upon the element of length Δs in the u-direction $\boldsymbol{\tau}_{(u)}$ is given by

$$\boldsymbol{\tau}_{(u)} = \tau(\sin \alpha_2 - \sin \alpha_1). \tag{II.1}$$

Since

$$\sin \alpha_1 = \frac{du(x)}{ds} = \frac{du(x)}{dx} \cdot \frac{dx}{ds} = \frac{du(x)}{dx} \cdot \frac{1}{\sqrt{1 + (\partial u/\partial x)^2}},$$

where we write $u(x)$ instead of $u(x, t)$ and du/dx instead of $\partial u/\partial x$, (because t does not change its value in this analysis), we have according to

$$\sqrt{1 + \left(\frac{\partial u}{\partial x}\right)^2} = 1 + \frac{1}{2}\left(\frac{\partial u}{\partial x}\right)^2 + \cdots$$

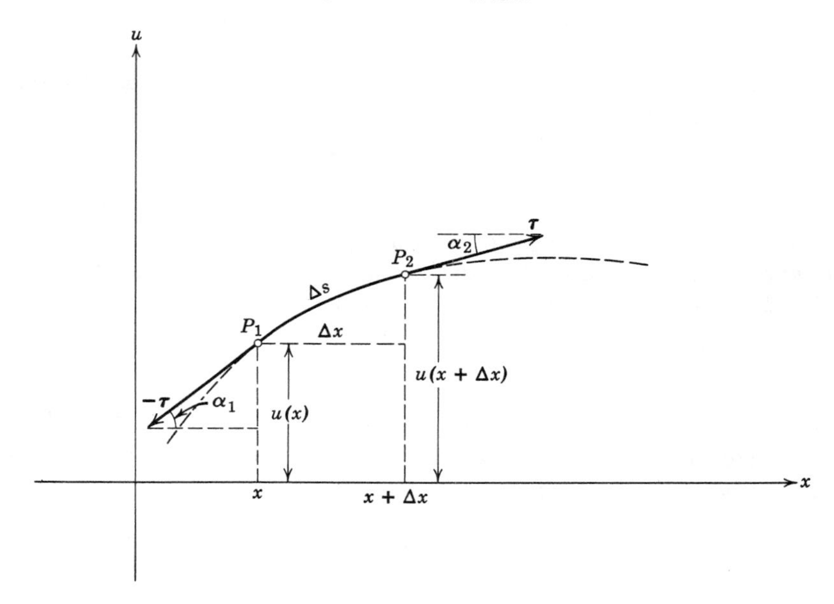

Fig. 3

in first approximation

$$\sin \alpha_1 = \frac{du(x)}{dx}.$$

Furthermore,

$$\sin \alpha_2 = \frac{du(x + \Delta x)}{ds} = \frac{du(x + \Delta x)}{dx} \cdot \frac{dx}{ds} \cong \frac{du(x + \Delta x)}{dx}.$$

If we apply the mean value theorem to $u'(x + \Delta x)$, we obtain

$$u'(x + \Delta x) = u'(x) + \frac{d}{dx}\, u'(\xi) \cdot \Delta x, \qquad \left(\frac{d}{dx}\, u'(\xi) = \frac{du'}{dx}\bigg|_{x=\xi}\right),$$

where

$$x \leq \xi \leq x + \Delta x$$

and consequently

$$\sin \alpha_2 = u'(x) + \frac{du'(\xi)}{dx} \Delta x.$$

Hence, we obtain for (II.1)

$$\boldsymbol{\tau}_{(u)} = \tau \frac{du'(\xi)}{dx} \Delta x. \tag{II.2}$$

The inertial force exerted by the same element of length Δs is given by

$$\rho \, \Delta s \frac{\partial^2 u}{\partial t^2} .$$

If we equate this quantity to (II.2) according to the principle of vectorial mechanics and divide by Δx, we obtain

$$\rho \frac{\Delta s}{\Delta x} \frac{\partial^2 u}{\partial t^2} = \tau \frac{d^2 u(\xi)}{dx^2} .$$

Taking the limit $\Delta x \to 0$ and accordingly $\xi \to x$ and approximating ds/dx by 1 in accordance with our agreement about the neglect of terms of second and higher order results in the equation

$$\boxed{\frac{\partial^2 u}{\partial t^2} = \frac{\tau}{\rho} \frac{\partial^2 u}{\partial x^2}} . \tag{II.3}$$

It is customary to denote the positive constant τ/ρ by α^2, which allows (II.3) to be written in the form

$$\frac{\partial^2 u}{\partial t^2} = \alpha^2 \frac{\partial^2 u}{\partial x^2} .$$

This is the form in which the string equation is usually stated in most of the standard texts on this subject.

The preceding discussion shows us that the displacement of a string, stretched between two fixed end points and satisfying the conditions stated in Chapter I, §2.1, obeys a partial differential equation of the second order which is linear in the derivatives and has constant coefficients.

This equation is, furthermore, homogeneous, which is due to the fact that no external forces except the tension are taken into account. This situation changes immediately if we consider the existence of other external forces, such as gravity. If we assume in general that a force of density $f(x, t)$ acts on the unit length of the string in the u-direction, we have to modify (II.2) by adding the term

$$f(x, t) \, \Delta s.$$

This force will contribute the additive term $(1/\rho)f(x, t)$ to the right side of (II.3) so that (II.3) assumes the more general form

$$\frac{\partial^2 u}{\partial t^2} = \frac{\tau}{\rho}\frac{\partial^2 u}{\partial x^2} + \frac{1}{\rho} f(x, t). \tag{II.4}$$

We obtained equation (II.3) earlier, in Chapter I, §2.2, by solving the string problem through the use of Hamilton's principle. Equation (II.4) too, can be obtained from a variational problem by taking into consideration that the potential energy of the string is increased jointly by the additional external force and the displacement, which means that we have to subtract the term

$$\int_0^1 f(x, t) \cdot u \, dx$$

from the expression for the potential energy V on p. 16. Thus, we obtain the variational problem

$$\frac{1}{2}\int_{t_1}^{t_2}\int_0^1 \left[\rho\left(\frac{\partial u}{\partial t}\right)^2 - \tau\left(\frac{\partial u}{\partial x}\right)^2 + 2fu\right] dx\, dt \rightarrow \text{minimum} \tag{II.5}$$

which has (II.4) as the Euler-Lagrange equation, as the reader can easily verify.

Problems II.1–5

1. Show that (II.4) is the Euler-Lagrange equation of the variational problem (II.5).

2. Write down the equation for the stretched string under tension and gravity.

3. Establish the expression for the kinetic energy of a vibrating elastic (laterally movable) rod.

4. Find the Euler-Lagrange equation for the vibrating elastic rod. (For potential energy see problem I.18.)

5. Consider only such deflections of the vibrating rod for which $\partial u/\partial x$ is small of first order and neglect all terms small of second and higher order. What is the Euler-Lagrange equation of the vibrating elastic rod under these simplified conditions?

2. An Attempt to Solve a Specific String Problem

In this subsection we are going to make a first attempt to solve equation (II.3) under specific conditions. To simplify matters we will perform the coordinate transformation $\sqrt{\tau}t \rightarrow t$, $\sqrt{\rho}x \rightarrow x$ such that the coefficient of $\partial^2 u/\partial x^2$ has the value 1 and (II.3) appears in the form

$$\frac{\partial^2 u}{\partial t^2} = \frac{\partial^2 u}{\partial x^2}. \tag{II.6}$$

Let the length of the string at rest be π. We choose the coordinate system so that the end points have the coordinates $(0, 0)$ and $(\pi, 0)$ and assume the string to have the initial position $\phi(x)$, where $\phi(x)$ may be any continuous function with a continuous derivative on the interval $0 \leq x \leq \pi$, and to be such that $\phi(x)$ vanishes at the end points. Suppose the string to be released from this position without initial velocity.

If we translate these conditions into mathematical language we obtain at the end points

$$\begin{aligned} u(0, t) &= 0, \\ u(\pi, t) &= 0 \end{aligned} \tag{II.7}$$

and two more conditions which describe the system at the time $t = 0$,

$$\begin{aligned} u(x, 0) &= \phi(x), \\ \left.\frac{\partial u(x, t)}{\partial t}\right|_{t=0} &= 0. \end{aligned} \tag{II.8}$$

One calls conditions of the type (II.7) *boundary conditions* and conditions of the type (II.8) *initial conditions*. (See the discussion in the following two subsections).

In order to obtain a solution of (II.6) under the conditions (II.7) and (II.8), we try to reduce the problem of solving a partial differential equation to one of solving ordinary differential equations, by assuming that the solution can be written in the form

$$u(x, t) = X(x) \cdot T(t). \tag{II.9}$$

Thus, the partial derivatives become ordinary derivatives of X with respect to x and of T with respect to t, and instead of one unknown function of two variables we have now two unknown functions of one variable each, which we have to determine so that their product satisfies (II.6) and as many of the conditions (II.7) and (II.8) as possible.

We call the method of representing the solution by a product of the type (II.9) *Bernoulli's separation method* in honor of the Swiss mathematician *Daniel Bernoulli*. The reader may think that we complicated the problem by making the assumption (II.9), but he will see in a moment that this is by no means the case. On the contrary: We will see that we actually reduced the partial differential equation (II.6) to two ordinary differential equations with very simple boundary conditions, as was our aim.

We can see immediately that the boundary conditions (II.7) simplify to

$$\begin{aligned} X(0) &= 0, \\ X(\pi) &= 0 \end{aligned} \tag{II.10}$$

and the second one of the initial conditions (II.8) becomes

$$T'(0) = 0. \tag{II.11}$$

We leave the first initial condition of (II.8) alone for the time being and take care of it later.

Substitution of (II.9) into (II.6) and division of the whole equation by $u = X \cdot T$ gives for a region in which u does not vanish

$$\frac{T''(t)}{T(t)} = \frac{X''(x)}{X(x)}. \tag{II.12}$$

Since the term on the left side is a function of t alone and the term on the right side is a function of x alone, (II.12) can hold only if

$$\frac{T''(t)}{T(t)} = \frac{X''(x)}{X(x)} = \lambda \tag{II.13}$$

where λ is an arbitrary constant.

This leads immediately to the following two ordinary differential equations of the second order:

$$X'' - \lambda X = 0, \tag{II.14}$$

$$T'' - \lambda T = 0. \tag{II.15}$$

The general solution of (II.14) is given by

$$X = a_1 e^{\sqrt{\lambda} \cdot x} + b_1 e^{-\sqrt{\lambda} \cdot x}.$$

If $\lambda > 0$, we see that we cannot possibly satisfy the boundary conditions (II.10) unless $a_1 = b_1 = 0$, which yields the trivial solution $u = 0$. However, this solution cannot possibly satisfy the first of the two initial conditions (II.8) unless $\phi(x) = 0$, in which case the whole problem becomes trivial. (It means physically that the string is at rest initially and since no disturbances occur, it will remain at rest.)

Let us, therefore, assume that $\lambda < 0$ and write

$$\lambda = -\mu^2.$$

Then, the general solution of (II.14) becomes

$$X = a_1 \cos \mu x + b_1 \sin \mu x.$$

From the first one of the conditions (II.10) it follows that

$$a_1 = 0$$

and from the second condition of (II.10) it follows that

$$\sin \mu\pi = 0,$$

which has $\mu = k$, where $k = 1, 2, 3, \cdots$, as a consequence. So we finally obtain for $X(x)$

$$X = b_1 \sin kx, \qquad k = 1, 2, 3, \cdots, \tag{II.16}$$

The general solution of (II.15) is achieved with $\lambda = -\mu^2 = 1, 4, 9, \cdots$:

$$T = a_2 \cos kt + b_2 \sin kt, \qquad \text{where } k = 1, 2, 3, \cdots,$$

with the derivative

$$T' = -a_2 \cdot k \sin kt + b_2 \cdot k \cos kt.$$

From (II.11) it follows that $b_2 = 0$, and consequently,

$$T = a_2 \cos kt, \qquad k = 1, 2, 3, \cdots. \tag{II.17}$$

(II.16) and (II.17) together, in view of (II.9), give for all integral values of k solutions,

$$u_k(x, t) = b_k \sin kx \cos kt \tag{II.18}$$

(where b_k is an arbitrary constant), which satisfy both boundary conditions and the second initial condition.

Now we have to discuss the first of the initial conditions (II.8). $\phi(x)$ is a given function which is continuous, has a continuous derivative, and vanishes at the end points $(0, 0)$ and $(\pi, 0)$, and that is all we know about it. The initial condition $u(x, 0) = \phi(x)$ requires that

$$u(x, 0) = b_k \sin kx = \phi(x).$$

It is clear that this condition could only be satisfied if $\phi(x)$ had the form $\phi(x) = C \sin kx$, which is certainly not the general case. We have to think now of some way to circumvent this obstacle in order to satisfy the first condition in (II.8). We have in (II.18) infinitely many solutions, all of which satisfy equation (II.6), the boundary conditions (II.7), and the second of the initial conditions (II.8). We can see immediately that any finite linear combination of solutions of the form (II.18) does the same. This furnishes us with a bright outlook for the future, since we can at least try to approximate the given function $\phi(x)$ by a trigonometric polynomial of the form

$$\sum_{k=1}^{n} b_k \sin kx \tag{II.19}$$

to a certain degree of accuracy. If such an approximation is possible at all—and we will see in Chapter IV, §1 that it is possible—this would be quite satisfactory from the point of view of practical computation, but it certainly does not satisfy us mathematically. Still there is no reason to lose hope. We have infinitely many solutions of the form (II.18) and we used a finite number of them only in our approximation (II.19). Why shouldn't we try to satisfy the first of the initial conditions (II.8) by considering an infinite linear combination of the solutions (II.18), that is, an infinite series of the form

$$\sum_{k=1}^{\infty} b_k \sin kx. \tag{II.20}$$

If the coefficients b_k can be determined such that this infinite series represents the function $\phi(x)$, we have solved our problem. We will see in Chapter IV that this is indeed possible under rather weak conditions imposed on $\phi(x)$. There we will discuss the problem of expanding functions into trigonometric series (so-called *Fourier series*).

As we pointed out earlier, it does not make much difference from a practical point of view whether we obtain an infinite series as a solution or a finite trigonometric polynomial as an approximate solution. Unless the infinite series can be evaluated in a closed form (as we will be able to do in case of the string problem), any infinite series degenerates anyway to a finite polynomial for all practical purposes at the moment one wishes to obtain numerical results. We will see in Chapter IV that the coefficients b_k of (II.19) will be the same ones as in (II.20), if we approximate $\phi(x)$ so that the integral of the square of the error becomes a minimum (a generalization of Gauss's least square method).

Before we close this section, we will list in short retrospect all the problems which arose in connection with our attempt to solve equation (II.6) and with which we have to deal in the following chapters.

Suppose we found a solution of (II.6) in the form $u(x, t) = X(x)T(t)$ which satisfies all the boundary and initial conditions. Does there exist another solution satisfying the same conditions, but representing a different type of vibration which cannot be written in the form $X(x)T(t)$? This is the uniqueness problem which will be dealt with in subsection 5 of this section.

Suppose we obtained infinitely many solutions of a linear partial differential equation. Is a linear combination of those infinitely many solutions again a solution? This problem will be dealt with in Chapter III (superposition principle).

Under what conditions can one expand a given function into an infinite trigonometric series? This question will lead us into the theory of Fourier expansions to be dealt with in Chapter IV.

Furthermore, the problem of finding such values of the constant λ for which the boundary-value problem (II.10), (II.14) has a solution—of which we could dispose of very easily in the preceding discussion—might not and will not always be that simple. The values λ for which a solution of the boundary value problem exists are called *eigenvalues*. A theoretical discussion concerning the eigenvalues of such boundary value problems will be given in Chapter V, §2 and a method for their numerical computation will be developed in Chapter VII.

Problems II.6–8

6. Reduce the following partial differential equations to ordinary differential equations by Bernoulli's separation method:

(a) $$\frac{\partial^2 u}{\partial x^2} + \frac{\partial^2 u}{\partial y^2} = 0,$$

(b) $$\frac{\partial u}{\partial t} = \frac{\partial^2 u}{\partial x^2} + \frac{\partial^2 u}{\partial y^2},$$

(c) $$\frac{\partial^2 u}{\partial t^2} = \frac{\partial^2 u}{\partial x^2} + \frac{1}{x}\frac{\partial u}{\partial x} + \frac{1}{x^2}\frac{\partial^2 u}{\partial y^2},$$

(d) $$x\frac{\partial^2}{\partial x^2}(xu) + \frac{1}{\sin y}\frac{\partial}{\partial y}\left(\sin y\frac{\partial u}{\partial y}\right) = 0.$$

7. Find all the values of λ for which the following boundary value problems have nontrivial solutions:

(a) $$y'' + \lambda y = 0, \qquad y(0) = 0, \quad y(1) = 0,$$

(b) $$y'' + \lambda y = 0, \qquad y(0) = 0, \quad y\left(\frac{3\pi}{2}\right) = 0.$$

8. Find constants a_0, a_1, b_1 such that

$$\int_0^{2\pi} \left(\phi(x) - \frac{a_0}{2} - a_1 \cos x - b_1 \sin x\right)^2 dx$$

is a minimum where $\phi(x)$ is a given function.

3. Boundary and Initial Conditions in Differential Equations

We used the terms "boundary conditions" and "initial conditions" in our discussion in the preceding subsection without bothering to define those terms. In order to clarify the nomenclature, we will shortly review how one uses these terms in general in connection with ordinary differential equations.

Given the ordinary differential equation of nth order

$$y^{(n)} = f(x, y, y', \cdots, y^{(n-1)})$$

where x shall represent a space coordinate. We know that the solution of such an equation contains in general n arbitrary integration constants

$$y = y(x, c_1, c_2, \cdots, c_n),$$

which means that a specific solution will be obtained by imposing a number of conditions. These conditions may concern the values of the solution and its derivatives up to the $(n - 1)$th order at a certain point—usually the beginning point of the interval within which the solution is desired; it may concern the values of the solution and of its derivatives at the beginning point and at the end point of the considered interval, as was always the case in the variational problems which we discussed in Chapter I.

Since x is considered here as space coordinate and the conditions imposed on the solution and its derivatives are in general conditions which have to be satisfied at the beginning and end point of the considered interval—the boundary of the interval—we call such conditions *boundary conditions.* In the case where the independent variable x represents the time, we will call the conditions imposed on the solution and its derivatives at the beginning point of the considered time interval *initial conditions* and those imposed at the end point *end conditions* (the latter are very unusual). Boundaries do not necessarily have to be finite points. The end point of a considered time interval may very well be the point ∞. The same holds true for space intervals, as the reader might well imagine.

The reader should be cautioned that the nomenclature introduced here is by no means standard. For example, the use of the term "boundary condition" is very frequent for all the different types of conditions we listed above. This use is certainly justified by the argument that also the beginning and end point of a time interval constitute a boundary of an interval, even though not in the strict geometrical sense.

The situation as discussed above changes rather radically at the moment one considers general solutions of partial differential equations. To make clear what one has to expect to happen, we will consider the partial differential equation

$$\frac{\partial^2 u}{\partial x\, \partial y} = 0.$$

We see immediately that

$$u = f(x) + g(y)$$

is a solution of this equation, where f and g are arbitrary, differentiable functions. We do not know, of course, whether this is the most general solution of this equation, but whether this is the case or not, we can see that we have to expect the appearance of arbitrary functions in solving partial differential equations.

We may consider another example as

$$\frac{\partial^2 u}{\partial x^2} - \frac{\partial^2 u}{\partial y^2} = 0$$

and we see again that

$$u = f(x + y) + g(x - y)$$

is a solution of this equation, whatever f and g are, as long as they are twice differentiable so that we have no difficulties in substituting this solution into the equation.

We could continue considering more and more examples always making the same observation, that is: A general solution of a partial differential equation contains arbitrary functions in analogy to the appearance of arbitrary constants in the theory of ordinary differential equations. (Note that we always carefully spoke of "a" general solution since we do not want to bother with an investigation as to whether we found "the" most general solution, which is quite unnecessary for our purposes.)

The fact that the solutions of partial differential equations contain arbitrary functions indicates that we have to impose rather severe conditions on a solution which is supposed to describe a specific process uniquely. If we have a partial differential equation in two independent variables, we can always interpret the solution as a two-dimensional surface in a three-dimensional space. Having in mind that the solution of an ordinary differential equation of the second order is uniquely determined by its values at the beginning and end points of a certain interval (boundary of the interval), we might guess that in the case of partial differential equations of the second order in two independent variables it will be required to know the values of the desired solution along the boundary of the region within which the solution is desired, that is, in general along a closed curve. Such a boundary is not necessarily a boundary in the strict geometrical sense. It could be, as we have seen in the example of the vibrating string in Chapter I, §2.1 a space-time boundary (e.g., a rectangle in an x, t–plane). Certain modifications of such a boundary condition (prescribed values along a closed curve) are certainly indicated. We only have to consider the string problem as formulated in subsection 2 of this section, where the boundary was the open polygon $t = 0$, $x = 0$, $x = \pi$ in the x, t–plane and where for lack of a condition on the open side, the values of the function and its derivative with respect to t were prescribed at $t = 0$. (Note the analogy with the case of an ordinary differential equation of the second order where value and derivative of the solution are prescribed at the beginning point instead of prescribing the values at beginning and end point.) Other modifications will be encountered in our subsequent discussions of the vibrating membrane (§2) and the potential- and heat-equation (§3) in this chapter.

In this treatment, we will use the terms "boundary condition" and "initial condition" as outlined above; however, we will not hesitate to deviate from this usage whenever it appears practical.

4. Boundary and Initial Value Problem for the String

We have seen in subsection 1 of this section that the displacements of a string stretched between two fixed points, say $P_1(x_1, 0)$ and $P_2(x_2, 0)$ under an external force in the u-direction of density $f(x, t)$ has to satisfy the equation

$$\frac{\partial^2 u}{\partial t^2} = \frac{\tau}{\rho} \cdot \frac{\partial^2 u}{\partial x^2} + \frac{1}{\rho} f(x, t).$$

This gives automatically the two boundary conditions

$$\begin{aligned} u(x_1, t) &= 0, \\ u(x_2, t) &= 0. \end{aligned} \tag{II.21}$$

It is clear that the behavior of the string from a certain time $t = t_0$ on will largely depend on the situation in which we find the string at this time; i.e.: we have to know the displacement of every point of the string at t_0, which furnishes the initial condition

$$u(x, t_0) = \phi(x) \tag{II.22}$$

where $\phi(x)$ is the displacement of the point x at the time t_0. Clearly,

$$\phi(x_1) = \phi(x_2) = 0$$

in order to be consistent with the boundary conditions (II.21).

We can see furthermore that what initial velocity is given to every point of the string will certainly make some difference in the behavior of the string, and this can be expressed as

$$\left. \frac{\partial u(x, t)}{\partial t} \right|_{t = t_0} = \psi(x) \tag{II.23}$$

where $\psi(x)$ is the initial velocity of the point x at the time t_0. Again,

$$\psi(x_1) = \psi(x_2) = 0$$

to be consistent with (II.21).

If $\psi(x) = 0$, then we have the case of the string released from rest as in subsection 2 of this section.

Considering now our problem together with the conditions (II.21), (II.22), and (II.23), we cannot help the feeling that this is about all we can impose on the problem without serious interference with its physical nature. The equation already takes care of the tension and the external force of density $f(x, t)$ and there obviously remains nothing else which could possibly influence the future behavior of the string. We feel definitely that we have formulated the problem as complete as we possibly can and feel justified in expecting a unique solution which will describe the actual physical process that will take place under the given circumstances. That this is indeed the case will be shown in the following subsection—and this is very fortunate because the playing of a string instrument would otherwise be quite a haphazard undertaking.

Problems II.9–16

9. Show that $u(x, y) = f(x + iy)$, where f is a twice-differentiable function and $i = \sqrt{-1}$, is a solution of

$$\frac{\partial^2 u}{\partial x^2} + \frac{\partial^2 u}{\partial y^2} = 0.$$

10. Show that the functions

$$u(x, y) = \log \sqrt{x^2 + y^2},$$
$$u(x, y) = x^2 - y^2,$$
$$u(x, y) = 2xy$$

are solutions of the partial differential equation in problem 9.

11. Show that $u(r, \theta) = \theta \log r$ is a solution of the equation

$$\frac{\partial^2 u}{\partial r^2} + \frac{1}{r}\frac{\partial u}{\partial r} + \frac{1}{r^2}\frac{\partial^2 u}{\partial \theta^2} = 0.$$

12. Show that $u = 1/r$ is a solution of the equation

$$\frac{\partial^2 u}{\partial x^2} + \frac{\partial^2 u}{\partial y^2} + \frac{\partial^2 u}{\partial z^2} = 0,$$

where $r = \sqrt{x^2 + y^2 + z^2}$.

13. Formulate the boundary and initial value problem for a stretched string of length π which is initially plucked at the midpoint; i.e., the point $x = \pi/2$ is raised through a distance h above the rest position and released.

14. Formulate the boundary and initial value problem for the stretched string of length l which is under no external forces except tension, if at the time 0 it is in rest position and is given an initial velocity which is zero at the end points and increases linearly toward the midpoint from both sides.

15. Apply Bernoulli's separation method to problems 13 and 14 and state all boundary and initial conditions for $X(x)$ and $T(t)$.

16. Formulate the boundary and initial value problem for the stretched string of length l under gravity, if it is released at $t = 0$ from the position $u = h(x - l)x$.

5. The Uniqueness of the Solution of the String Equation

We consider the boundary and initial value problem for the stretched, initially displaced and driven string under tension and an external force of density $f(x, t)$ as we formulated it in the preceding section.

If, in accordance with our previous notation, $\phi(x)$ is the initial displacement and $\psi(x)$ is the initial velocity, we have the problem of establishing the uniqueness of the solution of

$$\frac{\partial^2 u}{\partial t^2} = \frac{\tau}{\rho}\frac{\partial^2 u}{\partial x^2} + \frac{1}{\rho} f(x, t), \tag{II.4}$$

$$u(x_1, t) = 0, \qquad u(x_2, t) = 0, \tag{II.21}$$

$$u(x, t_0) = \phi(x), \tag{II.22}$$

$$\left.\frac{\partial u(x, t)}{\partial t}\right|_{t=t_0} = \psi(x) \tag{II.23}$$

and it is $\phi(x_1) = \phi(x_2) = \psi(x_1) = \psi(x_2) = 0$.

We now prove the following:

Theorem II.1. *Let R denote the region defined by $x_1 \leq x \leq x_2$, $t \geq t_0$ and let $u = u(x, t)$ be continuous and have continuous first- and second-order derivatives in R. If $u(x, t)$ satisfies the equation (II.4), the boundary conditions (II.21), and the initial conditions (II.22) and (II.23), then it is the only function in R with these properties.*

Proof (*by contradiction*). Let us assume that $v(x, t)$ is another function which is continuous in R and has continuous first and second order derivatives in R and suppose that $v(x, t)$ also is a solution of our boundary and initial value problem in R; i.e., $v(x, t)$ satisfies (II.4), (II.21), (II.22), and (II.23). Then,

$$\zeta(x, t) = u(x, t) - v(x, t)$$

is a solution of $\dfrac{\partial^2 \zeta}{\partial t^2} = \dfrac{\tau}{\rho}\dfrac{\partial^2 \zeta}{\partial x^2}$—since the equation (II.4) is linear—and $\zeta(x, t)$ satisfies the boundary conditions

$$\begin{aligned} \zeta(x_1, t) &= 0, \\ \zeta(x_2, t) &= 0, \end{aligned} \tag{II.24}$$

and the initial conditions

$$\zeta(x, t_0) = 0,$$

$$\left.\frac{\partial \zeta(x, t)}{\partial t}\right|_{t=t_0} = 0, \tag{II.25}$$

as one immediately sees by substitution.

In order to prove our theorem, we have to show that $\zeta(x, t) = 0$ identically in R. It is physically clear that this has to be the case since ζ stands for the displacement of a stretched string which is neither initially displaced nor driven from its rest position and is not subjected to external forces except the tension. Well, what should happen in such a case? Obviously, nothing is going to happen and the string stays in its rest position. We get a lead for our mathematical proof by considering the total energy (kinetic + potential energy) at any time t_1. From the physical point of view, if nothing happens, then the total energy is certainly zero at any time $t_1 \geq t_0$; i.e.,

$$\int_{x_1}^{x_2} \left[\rho \left(\frac{\partial \zeta}{\partial t} \right)^2 + \tau \left(\frac{\partial \zeta}{\partial x} \right)^2 \right]_{t=t_1} dx = 0 \tag{II.26}$$

and from this we will be able to deduce that $\zeta = 0$.

The first thing we have to do is show that (II.26) is true. We consider for this purpose the relation

$$\int_{t_0}^{t_1} \int_{x_1}^{x_2} \frac{\partial \zeta}{\partial t} \left[\rho \frac{\partial^2 \zeta}{\partial t^2} - \tau \frac{\partial^2 \zeta}{\partial x^2} \right] dx\, dt = 0 \tag{II.27}$$

which is certainly satisfied since ζ is a solution of (II.3). The integrand of (II.27) can be transformed as follows:

$$\frac{\partial \zeta}{\partial t} \left(\rho \frac{\partial^2 \zeta}{\partial t^2} - \tau \frac{\partial^2 \zeta}{\partial x^2} \right) = \frac{1}{2} \frac{\partial}{\partial t} \left[\rho \left(\frac{\partial \zeta}{\partial t} \right)^2 + \tau \left(\frac{\partial \zeta}{\partial x} \right)^2 \right] - \frac{\partial}{\partial x} \left(\tau \frac{\partial \zeta}{\partial t} \cdot \frac{\partial \zeta}{\partial x} \right),$$

as the reader can easily convince himself by carrying out the indicated operations on the right side of this identity. In view of this we can write (II.27) in the form

$$\int_{t_0}^{t_1} \int_{x_1}^{x_2} \left[\frac{1}{2} \frac{\partial}{\partial t} \left(\rho \left(\frac{\partial \zeta}{\partial t} \right)^2 + \tau \left(\frac{\partial \zeta}{\partial x} \right)^2 \right) - \frac{\partial}{\partial x} \left(\tau \frac{\partial \zeta}{\partial t} \cdot \frac{\partial \zeta}{\partial x} \right) \right] dx\, dt = 0. \tag{II.28}$$

This integral is extended over the region which is bounded by $t = t_0$, $t = t_1$, $x = x_1$, $x = x_2$ ($\bar{R}$). If we use Green's theorem (see **AI.C2**),

$$\iint_{\bar{R}} \left(\frac{\partial A}{\partial t} + \frac{\partial B}{\partial x} \right) dx\, dt = \oint_C B\, dt - A\, dx,$$

where C is the boundary of $\bar{R}$, we can transform (II.28) into a line integral, letting

$$A = -\frac{1}{2} \left[\rho \left(\frac{\partial \zeta}{\partial t} \right)^2 + \tau \left(\frac{\partial \zeta}{\partial x} \right)^2 \right],$$

$$B = \tau \frac{\partial \zeta}{\partial t} \cdot \frac{\partial \zeta}{\partial x},$$

and we obtain for (II.28)

$$\oint_C \left\{ \tau \frac{\partial \zeta}{\partial t} \cdot \frac{\partial \zeta}{\partial x} \, dt + \frac{1}{2} \left[\rho \left(\frac{\partial \zeta}{\partial t} \right)^2 + \tau \left(\frac{\partial \zeta}{\partial x} \right)^2 \right] dx \right\} = 0. \qquad \text{(II.29)}$$

We have $dx = 0$ along the vertical sides of $\bar{R}$ and $dt = 0$ on the horizontal sides of $\bar{R}$. Therefore, we simply obtain for (II.29)

$$\tau \int_{t_0}^{t_1} \frac{\partial \zeta}{\partial t} \frac{\partial \zeta}{\partial x} \bigg|_{x=x_2} dt - \int_{t_0}^{t_1} \tau \frac{\partial \zeta}{\partial t} \frac{\partial \zeta}{\partial x} \bigg|_{x=x_1} dt + \frac{1}{2} \int_{x_1}^{x_2} \left(\rho \left(\frac{\partial \zeta}{\partial t} \right)^2 + \tau \left(\frac{\partial \zeta}{\partial x} \right)^2 \right)_{t=t_0} dx$$
$$- \frac{1}{2} \int_{x_1}^{x_2} \left(\rho \left(\frac{\partial \zeta}{\partial t} \right)^2 + \tau \left(\frac{\partial \zeta}{\partial x} \right)^2 \right)_{t=t_1} dx = 0.$$

The first two integrals vanish because it follows from (II.24) that

$$\frac{\partial \zeta(x_1, t)}{\partial t} = \frac{\partial \zeta(x_2, t)}{\partial t} = 0.$$

The third integral vanishes in view of (II.25) and we see that

$$\int_{x_1}^{x_2} \left[\rho \left(\frac{\partial \zeta}{\partial t} \right)^2 + \tau \left(\frac{\partial \zeta}{\partial x} \right)^2 \right]_{t=t_1} dx = 0 \quad \text{for any } t_1 \geq t_0.$$

Since the integrand is a linear combination of two squares with positive coefficients, it follows that

$$\frac{\partial \zeta(x, t)}{\partial t} \bigg|_{t=t_1} = 0, \quad \frac{\partial \zeta(x, t)}{\partial x} \bigg|_{t=t_1} = 0$$

for all (t_1, x) in R. Hence

$$\zeta(x, t) = \text{constant} \quad \text{in } R.$$

Since $\zeta(x_1, t) = 0$ according to (II.24), we have $\zeta(x, t) = 0$ in R, which contradicts our assumption that u and v are two different solutions, q.e.d.

Problem II.17

17. Prove that

$$\frac{\partial \zeta}{\partial t} \left(\rho \frac{\partial^2 \zeta}{\partial t^2} - \tau \frac{\partial^2 \zeta}{\partial x^2} \right) = \frac{1}{2} \frac{\partial}{\partial t} \left[\rho \left(\frac{\partial \zeta}{\partial t} \right)^2 + \tau \left(\frac{\partial \zeta}{\partial x} \right)^2 \right] - \frac{\partial}{\partial x} \left(\tau \frac{\partial \zeta}{\partial t} \cdot \frac{\partial \zeta}{\partial x} \right).$$

§2. THE VIBRATING MEMBRANE

1. Vibrations of a Stretched Membrane

A membrane is an elastic skin which does not resist bending but does resist stretching. We will assume that such a membrane is stretched over a certain simply connected planar region R which is bounded by a rectifiable curve C. We choose the coordinate system so that C lies in the x, y–plane. The vertical displacement of the membrane at the point (x, y) at the time t shall be denoted by $u(x, y, t)$. We will assume that the tension per unit length has the constant value τ along the boundary C. If we represent the tension by a vector $\boldsymbol{\tau}$, then $|\boldsymbol{\tau}| = \tau$ and $\boldsymbol{\tau}$ is perpendicular to C but lies within the tangent plane to the membrane. If the membrane is in equilibrium, the tension will be constant over the entire membrane; i.e., along any arbitrary cut through the membrane, the tension will be transferred from one side of the cut to the other side without change of magnitude. Hence, the tension along any arbitrary cut through the membrane can be represented by a vector of the constant magnitude τ and a direction which is perpendicular to the cut and lies in the tangent plane to the membrane.

We assume that there are no external forces besides the tension present and proceed to establish an expression for the u-component of the external force due to the tension $\boldsymbol{\tau}$.

Let us consider for this purpose a surface element ΔS of the membrane which is quite arbitrarily cut out. ΔS shall be encompassed by the rectifiable curve γ and its projection into the x, y–plane, ΔR, shall have the area ΔA. Let

$$x = x(s),$$

$$y = y(s),$$

$$u = u(s)$$

be the parametric representation of the boundary γ, where s shall stand for the arc length on γ. If P is a point on γ, then the tension vector $\boldsymbol{\tau}$ at P is perpendicular to the tangent vector $\mathbf{t}$ to γ at P and lies in the tangent-plane to the membrane at P. (See Fig. 4.)

In order to find the u-component of $\boldsymbol{\tau}$ exerted upon γ, we proceed as follows: The membrane is represented by $u = u(x, y, t)$ at any time t. If we let $\Phi = u(x, y, t) - u$, we can represent the membrane by

$$\Phi(x, y, u, t) = 0$$

and

$$\mathbf{n} = \operatorname{grad} \Phi = \frac{\partial u}{\partial x}\mathbf{i} + \frac{\partial u}{\partial y}\mathbf{j} - \mathbf{k}$$

is a vector orthogonal to ΔS (see AI.B3). The tangent vector $\mathbf{t}$ to γ is given by

$$\mathbf{t} = \frac{dx}{ds}\mathbf{i} + \frac{dy}{ds}\mathbf{j} + \frac{du}{ds}\mathbf{k}$$

and in view of s being the arc length on γ, we have $|\mathbf{t}| = 1$ (see AI.B1). Let $(\ )_P$ denote "at P." Then

$$(\boldsymbol{\tau})_P = \tau \cdot \frac{\mathbf{n} \times \mathbf{t}}{|\mathbf{n}|}$$

(where $|\mathbf{n}| = \sqrt{1 + u_x^2 + u_y^2}$) is the tension vector in P, because it is perpendicular to $\mathbf{n}$ and thus lies in the tangent plane to ΔS, is perpendicular to $\mathbf{t}$, and

$$\left| \tau \cdot \frac{\mathbf{n} \times \mathbf{t}}{|\mathbf{n}|} \right| = \tau.$$

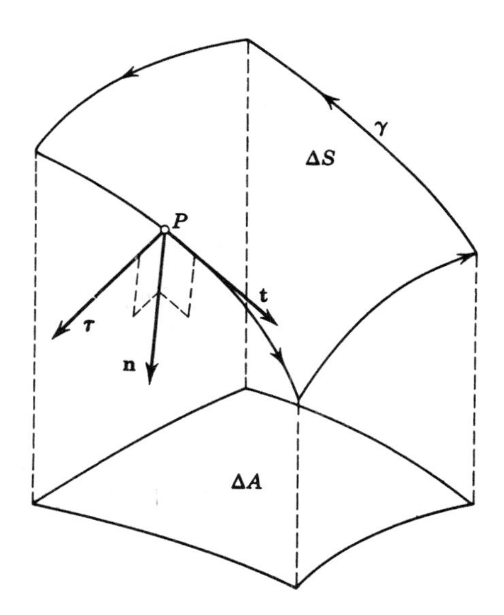

Fig. 4

Since

$$\tau \frac{\mathbf{n} \times \mathbf{t}}{|\mathbf{n}|} = \frac{\tau}{\sqrt{1 + u_x^2 + u_y^2}} \begin{vmatrix} \mathbf{i} & \mathbf{j} & \mathbf{k} \\ \dfrac{\partial u}{\partial x} & \dfrac{\partial u}{\partial y} & -1 \\ \dfrac{dx}{ds} & \dfrac{dy}{ds} & \dfrac{du}{ds} \end{vmatrix} \qquad \text{(see AI.A5)},$$

we obtain for the u-component of $\boldsymbol{\tau}$:

$$\boldsymbol{\tau}_{(u)} = \frac{\tau}{\sqrt{1 + u_x^2 + u_y^2}} \left(\frac{\partial u}{\partial x} \frac{dy}{ds} - \frac{\partial u}{\partial y} \cdot \frac{dx}{ds} \right).$$

This quantity represents the external force in the u-direction per unit length which is exerted upon the boundary γ of ΔS.

Again, we will only consider deformations for which $\partial u/\partial x$ and $\partial u/\partial y$ are small and consider terms of higher order neglectible compared with those of lower order.

Since

$$\sqrt{1 + u_x^2 + u_y^2} = 1 + \tfrac{1}{2}[u_x^2 + u_y^2] + \cdots$$

we can write in first approximation

$$\boldsymbol{\tau}_{(u)} = \tau\left(\frac{\partial u}{\partial x}\frac{dy}{ds} - \frac{\partial u}{\partial y}\frac{dx}{ds}\right).$$

In order to find the u-component of the external force exerted upon the entire boundary γ of ΔS, we have to take the line integral along γ:

$$\boldsymbol{\tau}_{(u)}(\gamma) = \tau\oint_\gamma \left(\frac{\partial u}{\partial x}\cdot\frac{dy}{ds} - \frac{\partial u}{\partial y}\cdot\frac{dx}{ds}\right) ds.$$

We invoke Green's theorem (see AI.C2) and obtain

$$\boldsymbol{\tau}_{(u)}(\Delta S) = \tau\iint_{\Delta R}\left(\frac{\partial^2 u}{\partial x^2} + \frac{\partial^2 u}{\partial y^2}\right) dx\, dy. \tag{II.30}$$

Since

$$\iint_{\Delta R} dx\, dy = \Delta A,$$

we obtain on applying the mean value theorem to (II.30) with $\xi, \eta \in \Delta R$

$$\boldsymbol{\tau}_{(u)}(\Delta S) = \tau\left(\frac{\partial^2 u}{\partial x^2} + \frac{\partial^2 u}{\partial y^2}\right)_{\xi,\eta} \Delta A, \tag{II.31}$$

where $(\ )_{\xi,\eta}$ stands for "taken at $x = \xi$ and $y = \eta$."

The principle of vectorial mechanics states that we may equate (II.31) to the u-component of the inertial force.

The surface element ΔS with the mass $\rho\Delta S$, where ρ is the constant mass density of the membrane, exerts the force

$$\rho\left(\frac{\partial^2 u}{\partial t^2}\right)_{\bar\xi,\bar\eta} \Delta S = \rho\left(\frac{\partial^2 u}{\partial t^2}\right)_{\bar\xi,\bar\eta} \iint_{\Delta R}\sqrt{1 + u_x^2 + u_y^2}\, dx\, dy$$

where $(\ )_{\bar\xi,\bar\eta}$ denotes "taken at some point $\bar\xi, \bar\eta$ in ΔR," which, as a first approximation, is equal to

$$\rho\left(\frac{\partial^2 u}{\partial t^2}\right)_{\bar\xi,\bar\eta} \Delta A. \tag{II.32}$$

Thus, we obtain, equating (II.31) and (II.32) according to the principle of vectorial mechanics

$$\rho\left(\frac{\partial^2 u}{\partial t^2}\right)_{\bar{\xi},\bar{\eta}} \Delta A = \tau\left(\frac{\partial^2 u}{\partial x^2} + \frac{\partial^2 u}{\partial y^2}\right)_{\xi,\eta} \Delta A.$$

If we cancel by ΔA and let ΔR shrink into one point (x, y), (ξ, η) as well as $(\bar{\xi}, \bar{\eta})$ will converge towards (x, y) and we obtain

$$\boxed{\frac{\partial^2 u}{\partial t^2} = \frac{\tau}{\rho}\left(\frac{\partial^2 u}{\partial x^2} + \frac{\partial^2 u}{\partial y^2}\right)}\,. \tag{II.33}$$

This is the equation for the vertical displacement of a vibrating stretched membrane under tensile forces, shortly called the *membrane equation*.

Again, we have obtained a linear partial differential equation of the second order with constant coefficients.

It is customary, as with the case of the vibrating string, to denote the positive constant τ/ρ by α^2, so that (II.33) takes the familiar form

$$\frac{\partial^2 u}{\partial t^2} = \alpha^2\left(\frac{\partial^2 u}{\partial x^2} + \frac{\partial^2 u}{\partial y^2}\right).$$

If the interior of the membrane is subjected to an external force in the u-direction of surface density $f(x, y, t)$, we have to modify (II.31) by adding the term

$$f(x, y, t)\,\Delta S \cong f(x, y, t)\,\Delta A,$$

thus obtaining the more general equation

$$\frac{\partial^2 u}{\partial t^2} = \alpha^2\left(\frac{\partial^2 u}{\partial x^2} + \frac{\partial^2 u}{\partial y^2}\right) + \frac{1}{\rho} f(x, y, t). \tag{II.34}$$

2. The Vibrations of a Membrane as a Variational Problem

The reader has certainly noticed that the derivation of the equation for the vibrating string, as carried out in Chapter I, §2.1 and 2 by using Hamilton's principle, was less complicated than setting up the equilibrium condition for the acting forces according to the principle of vectorial mechanics, as we have done in §1.1 of this chapter. The greater elegance and superiority of the variational method will become even more obvious in this section.

The potential energy of a stretched membrane—in analogy to the potential energy of the stretched string—depends jointly on the expansion of the membrane (change in area compared with the area at rest) and the tensile force, if no other external forces are in evidence. If an external force of surface density $f(x, y, t)$ acts in the u-direction upon the interior of the membrane in addition to the tension, then, this will contribute to the potential energy a term which is f times the displacement u per unit area.

Thus, the total potential energy of the membrane, subjected to the tension and the external force of density f, is

$$V = \tau\left(\iint_R dS - \iint_R dx\,dy\right) - \iint_R f \cdot u\,dx\,dy.$$

Since

$$\frac{dS}{dx\,dy} = \sqrt{1 + \left(\frac{\partial u}{\partial x}\right)^2 + \left(\frac{\partial u}{\partial y}\right)^2} = 1 + \frac{1}{2}\left[\left(\frac{\partial u}{\partial x}\right)^2 + \left(\frac{\partial u}{\partial y}\right)^2\right] + \cdots$$

we obtain by neglecting all terms of higher than second order

$$V = \frac{1}{2}\iint_R \left[\tau\left(\frac{\partial u}{\partial x}\right)^2 + \tau\left(\frac{\partial u}{\partial y}\right)^2 - 2fu\right] dx\,dy. \tag{II.35}$$

The kinetic energy of a surface element ΔS is

$$\Delta T = \frac{\rho\,\Delta S}{2}\left(\frac{\partial u}{\partial t}\right)^2$$

and as a first approximation

$$\Delta T = \frac{\rho\,\Delta x\,\Delta y}{2}\left(\frac{\partial u}{\partial t}\right)^2.$$

Hence, we obtain for the total kinetic energy of the membrane

$$T = \frac{\rho}{2}\iint_R \left(\frac{\partial u}{\partial t}\right)^2 dx\,dy. \tag{II.36}$$

Hamilton's principle requires that

$$\int_{t_0}^{t_1} (T - V)\,dt \to \text{minimum}$$

which in view of (II.35) and (II.36) appears in the form

$$\frac{1}{2}\int_{t_0}^{t_1}\iint_R \left[\rho\left(\frac{\partial u}{\partial t}\right)^2 - \tau\left(\frac{\partial u}{\partial x}\right)^2 - \tau\left(\frac{\partial u}{\partial y}\right)^2 + 2fu\right] dx\,dy\,dt \to \text{minimum}$$

through adequate choice of u.

The Euler-Lagrange equation of this variational problem is, according to (I.28),

$$\rho\frac{\partial^2 u}{\partial t^2} - \tau\left(\frac{\partial^2 u}{\partial x^2} + \frac{\partial^2 u}{\partial y^2}\right) - f = 0$$

which is identical with (II.34).

Problems II.18–23

18. Find the x-component and the y-component of the tension vector $\boldsymbol{\tau}$.

19. Transform

$$\oint_C y\,dx - x\,dy, \qquad C:\ x = \cos\phi,\quad y = \sin\phi;\quad 0 \le \phi < 2\pi,$$

into a double integral using Green's theorem and evaluate the double integral.

20. Transform

$$\iint_R (x^2 - y^2)\,dx\,dy, \qquad R:\ x^2 + y^2 \le 1,$$

into a line integral using Green's theorem and evaluate the line integral.

***21.** The potential energy of an elastic plate is proportional to

$$\iint_R (2H^2 - \lambda K)\,dx\,dy,$$

where λ is a constant, H is the mean curvature and K is the Gaussian curvature of the deformed plate. Neglect all terms of second and higher order in $\partial u/\partial x$ and $\partial u/\partial y$ and state Hamilton's principle for the vibrations of an elastic plate. (For kinetic energy see (II.36).)

22. Find the differential equation of a vibrating elastic plate (see problem I.22).

23. The equation for the equilibrium state of an elastic plate at rest is often referred to as the *plate equation.* Find the plate equation.

3. The Boundary and Initial Value Problem for the Membrane

We will start our discussion with the consideration of the stable state of the membrane, i.e., the case for which $\partial u/\partial t = 0$. We assume accordingly that the external force is time independent, hence equation (II.34) will simplify to

$$\frac{\partial^2 u}{\partial x^2} + \frac{\partial^2 u}{\partial y^2} + \frac{1}{\tau} f(x, y) = 0. \tag{II.37}$$

The membrane equation was obtained under the assumption that the membrane is stretched over a plane curve C, which furnishes the boundary condition

$$u(x(s), y(s)) = 0 \tag{II.38}$$

where

$$x = x(s),$$
$$y = y(s)$$

shall be the parametric representation of the curve C and s shall stand for the arc length on C.

We obtain the same differential equation with a similar boundary condition if we assume that the membrane is stretched over a closed rectifiable space curve L, which has a projection C in the x, y–plane and is such that C encompasses a simply connected region R.

In the first place, we obtain essentially the same expression for the potential energy. Instead of

$$V = \tau\left(\iint_R dS - \iint_R dx\,dy\right) - \iint_R f \cdot u\,dx\,dy$$

arrived at in the preceding subsection, we now get

$$V = \tau\left(\iint_R dS - \iint_R \overline{dS}\right) - \iint_R f(u - \bar{u})\,dx\,dy,$$

where the bars refer to the rest position of the membrane, stretched over L. If we expand the expressions $dS/dx\,dy$ and $\overline{dS}/dx\,dy$ according to the binomial law and neglect terms of higher order, the potential energy becomes

$$V = \frac{1}{2}\left[\tau\iint_R \left[\left(\frac{\partial u}{\partial x}\right)^2 + \left(\frac{\partial u}{\partial y}\right)^2\right] dx\,dy \right.$$
$$\left. - \tau\iint_R \left[\overline{\left(\frac{\partial u}{\partial x}\right)^2} + \overline{\left(\frac{\partial u}{\partial y}\right)^2}\right] dx\,dy - 2\iint_R f(u - \bar{u})\,dx\,dy\right],$$

where $\tau\iint_R (\bar{u}_x^2 + \bar{u}_y^2)\,dx\,dy - 2\iint_R f\bar{u}\,dx\,dy$ is a constant.

In the second place—since the stable state is characterized by a stationary value of the potential energy—we are led to consider the variational problem

$$\frac{1}{2}\iint_R \left\{\tau\left[\left(\frac{\partial u}{\partial x}\right)^2 + \left(\frac{\partial u}{\partial y}\right)^2\right] - 2fu\right\} dx\,dy \rightarrow \text{minimum}$$

with the boundary condition

$$u(x(s), y(s)) = \Phi(s), \tag{II.39}$$

if

$$x = x(s),$$
$$y = y(s),$$
$$u = \Phi(s)$$

is the parametric representation of L and obtain according to our general theory in Chapter I, §2.2, equation (II.37) as the Euler-Lagrange equation.

The problems (II.37), (II.38) respectively (II.37), (II.39) seem to be formulated completely since there apparently is nothing else which could possibly influence the membrane. A boundary value problem of this type is called a *Dirichlet problem* or *first boundary value problem*. Stated in words, Dirichlet's problem reads: Find the solution of a partial differential equation of the second order in two variables where the values of the solution are prescribed along a closed curve.

This is not the only possible type of boundary value problem we can formulate for the membrane—as already suggested by the name "*first* boundary value problem." The assumption that the membrane has to be stretched over a closed curve is actually not necessary to obtain equation (II.37).

This equation is also obtained if one lets the boundary of the membrane slide freely along the walls of a right cylinder with the trace C in the x, y–plane. We have in this case a variational problem as we discussed it in Chapter I, §4.3 and we obtain the natural boundary condition

$$\frac{\partial u}{\partial y} \cdot \frac{dx}{ds} - \frac{\partial u}{\partial x} \cdot \frac{dy}{ds} = 0 \quad \text{on } C.$$

This condition is identical with the one we secured for the example we discussed at the end of Chapter I, §4.3 if we let $z = u$. We saw on this previous occasion that the above condition is equivalent to

$$\frac{\partial u}{\partial n} = 0 \quad \text{on } C. \tag{II.40}$$

One calls (II.37), (II.40) the *second boundary value problem*.

If we consider now the more general problem of the vibrating membrane

$$\frac{\partial^2 u}{\partial t^2} = \alpha^2 \left(\frac{\partial^2 u}{\partial x^2} + \frac{\partial^2 u}{\partial y^2} \right) + \frac{1}{\rho} f(x, y, t), \tag{II.34}$$

we can understand that the solution of this equation cannot possibly be determined by (II.38), (II.39), or (II.40) since the behavior of the membrane will depend—in analogy to the string—on the position from which the membrane is released and the velocity which is imposed on it in the process of freeing it. This can be taken care of by imposing the two initial conditions

$$u(x, y, t_0) = \phi(x, y),$$

$$\left. \frac{\partial u(x, y, t)}{\partial t} \right|_{t = t_0} = \psi(x, y), \tag{II.41}$$

where $\phi(x, y)$ is the initial displacement of the point (x, y) at the time t_0 and $\psi(x, y)$ is the initial velocity of the point (x, y) at the time t_0. In order for these conditions to be consistent with the boundary conditions, we have to require in the case of the first boundary value problem that

$$\phi(x(s), y(s)) = \Phi(s), \qquad \psi(x(s), y(s)) = 0$$

and in the case of the second boundary value problem that

$$\frac{\partial \phi}{\partial n} = 0 \quad \text{on } C, \qquad \frac{\partial \psi}{\partial n} = 0 \quad \text{on } C.$$

4. An Attempt to Solve a Specific Membrane Problem

We choose the measure units so that τ/ρ has the value 1 and therefore (II.33) assumes the simpler form

$$\frac{\partial^2 u}{\partial t^2} = \frac{\partial^2 u}{\partial x^2} + \frac{\partial^2 u}{\partial y^2}. \tag{II.42}$$

We consider a circular membrane of radius 1, stretched over the circle

$$\left.\begin{aligned} x &= \cos \theta \\ y &= \sin \theta \end{aligned}\right\} 0 \leq \theta < 2\pi.$$

We displace the membrane initially by bringing it into the form $u = \phi(x, y)$ and release it from this position without initial velocity. If we introduce polar coordinates

$$\left.\begin{aligned} x &= r \cos \theta \\ y &= r \sin \theta \end{aligned}\right\}, \tag{II.43}$$

as is only natural in view of the axial symmetry of the boundary, we have to transform the right side of (II.42) according to (II.43) and obtain

$$\frac{\partial^2 u}{\partial x^2} + \frac{\partial^2 u}{\partial y^2} = \frac{\partial^2 u}{\partial r^2} + \frac{1}{r}\frac{\partial u}{\partial r} + \frac{1}{r^2}\frac{\partial^2 u}{\partial \theta^2}. \tag{II.44}$$

(see AI.B6).

Hence, (II.42) will appear in the form

$$\frac{\partial^2 u}{\partial t^2} = \frac{\partial^2 u}{\partial r^2} + \frac{1}{r}\frac{\partial u}{\partial r} + \frac{1}{r^2}\frac{\partial^2 u}{\partial \theta^2}. \tag{II.45}$$

The boundary condition that the membrane shall be stretched over the unit circle $r = 1$ is expressed by

$$u(1, \theta, t) = 0 \tag{II.46}$$

and the initial conditions stated above will take the form

$$u(r, \theta, 0) = \bar{\phi}(r, \theta) \qquad \text{where } \bar{\phi}(r, \theta) = \phi(x, y)_{r,\theta}, \tag{II.47}$$

$$\left.\frac{\partial u(r, \theta, t)}{\partial t}\right|_{t=0} = 0. \tag{II.48}$$

If we try again to obtain a solution by Bernoulli's method of separating the variables; i.e.: if we assume that the solution has the form

$$u(r, \theta, t) = R(r)\Theta(\theta)T(t), \tag{II.49}$$

the boundary and initial conditions will assume the form

$$R(1) = 0, \tag{II.50}$$

$$T(0)R(r)\Theta(\theta) = \bar{\phi}(r, \theta), \tag{II.51}$$

$$T'(0) = 0. \tag{II.52}$$

(II.49) substituted in (II.45) gives

$$R''\Theta T + \frac{1}{r} R'\Theta T + \frac{1}{r^2} R\Theta'' T = R\Theta T''. \tag{II.53}$$

Division by $R\Theta T$ leaves on the left side of (II.53) a function of r and θ only and on the right side a function of t only:

$$\frac{R''}{R} + \frac{1}{r}\frac{R'}{R} + \frac{1}{r^2}\frac{\Theta''}{\Theta} = \frac{T''}{T}.$$

This is consistent only if both sides equal an arbitrary positive or negative constant λ:

$$T'' - \lambda T = 0, \tag{II.54}$$

$$r^2\frac{R''}{R} + r\frac{R'}{R} - \lambda r^2 = -\frac{\Theta''}{\Theta}. \tag{II.55}$$

We can now repeat our argument with regard to (II.55) and equate either side of (II.55) to an arbitrary constant μ:

$$\Theta'' + \mu\Theta = 0, \tag{II.56}$$

$$r^2R'' + rR' - (\lambda r^2 + \mu)R = 0. \tag{II.57}$$

The general solution of (II.54) for $\lambda > 0$

$$T = c_1 e^{\sqrt{\lambda} t} + c_2 e^{-\sqrt{\lambda} t}$$

satisfies (II.52) only if $c_1 - c_2 = 0$, that is, we obtain the solution $T = c \sinh \sqrt{\lambda} t$. However, we will see in Chapter V, that we would encounter difficulties in trying to satisfy (II.50) with the solutions $R(r)$ of (II.57) for $\lambda > 0$. Let us assume therefore that $\lambda < 0$ and introduce a new parameter n by

$$-\lambda = n^2.$$

It follows that (II.54) has then the form

$$T'' + n^2 T = 0$$

with the general solution

$$T = c_1 \cos nt + c_2 \sin nt.$$

From (II.52) it follows that $c_2 = 0$ so that we finally arrive at

$$T = c_1 \cos nt. \tag{II.58}$$

Let us now consider the parameter μ in (II.56). If $\mu < 0$, we obtain as solutions

$$\Theta(\theta) = \bar{c}_1 e^{\sqrt{-\mu}\theta} + \bar{c}_2 e^{-\sqrt{-\mu}\theta}.$$

Since the solution $u(x, y, t)$ has to be at least continuous in (x, y) we have to reject the possibility of μ being negative for the following reason: Continuity of u in x, y requires that

$$\Theta(\theta + 2\pi) = \Theta(\theta) \tag{II.59}$$

which is certainly not satisfied by the expression above, unless $\bar{c}_1 = \bar{c}_2 = 0$ which yields the trivial solution only. Hence, we have to assume that $\mu > 0$ and let $\mu = m^2$.

Then we obtain as solution of (II.56)

$$\Theta(\theta) = \bar{c}_1 \cos m\theta + \bar{c}_2 \sin m\theta \tag{II.60}$$

and it follows from (II.59) that $m = 0, 1, 2, 3, \cdots$ (see problem II.24). The constants c_1, $\bar{c}_1$, $\bar{c}_2$ are still available to satisfy (II.51), which we ignored completely until now.

We still have equation (II.57) to discuss. We rewrite this equation in terms of the new constants n and m:

$$r^2 R'' + rR' + (n^2 r^2 - m^2)R = 0. \tag{II.61}$$

This is an ordinary differential equation of second order with a singularity at the point $r = 0$. As it is shown in the theory of ordinary differential equations (see Chapter VI, §2.2), the solution of an equation of this type is in general a nonelementary function that can, e.g., be calculated by means of a power series. Even though we will devote half of Chapter VI to the discussion of this equation we cannot resist making a few preliminary remarks about the solution right now in order to get an idea of what kind of mathematical problem we are going to face in the later developments.

The significance of the parameter n, contained in (II.61) will become clear as soon as we have performed the coordinate transformation

$$\rho = nr$$

which changes (II.61) into

$$\rho^2 \frac{d^2R}{d\rho^2} + \rho \frac{dR}{d\rho} + (\rho^2 - m^2)R = 0. \qquad \text{(II.62)}$$

Equation (II.62) is called the *Bessel equation* (we will make a few historic remarks about it in Chapter VI), and its solution for a certain value of the free parameter m is called the Bessel function of mth order.

We will denote the solutions of this equation by

$$R = J_m(\rho)$$

where m indicates the order of the function J. Therefore, the solution of (II.61) will have the form

$$R = J_m(nr). \qquad \text{(II.63)}$$

Since m can be any integer, we have obtained with this infinitely many solutions of (II.61). To satisfy (II.50),

$$J_m(n) = 0$$

has to hold. We will see in Chapter V, in a more general form, that this equation has for every m infinitely many solutions n, which we will call

$$n_{m,j}, \qquad j = 1, 2, 3, 4, \cdots$$

the *zeros of the Bessel Functions*. These numbers play here the role of eigenvalues and the reader can see already that the determination of eigenvalues is not always as simple as in §1.2 of this chapter.

Collecting the solutions (II.58), (II.60), and (II.63) and writing the free coefficients in suitable form, we obtain for u

$$u_{m,j}(r, \theta, t) = J_m(n_{mj}r)(A_{mj} \cos m\theta + B_{mj} \sin m\theta) \cos n_{mj}t. \qquad \text{(II.64)}$$

This is a two-parameter set of solutions and by taking the sum over all m and all j

$$\sum_{m=0}^{\infty} \sum_{j=1}^{\infty} J_m(n_{mj}r)(A_{mj} \cos m\theta + B_{mj} \sin m\theta) \cos n_{mj}t \qquad \text{(II.65)}$$

we hope again that we will be able to satisfy (II.51) by a proper choice of the arbitrary constants A_{mj} and B_{mj}.

In addition to the problems we formulated at the end of §1.2 of this chapter, we now have the following one:

Under what circumstances has (II.62) solutions, what are the properties of these solutions, and what can we find out about the zeros of these solutions? We are going to deal with this problem in Chapter V, §1 and Chapter VI, §2.

Problem II.24–26

***24.** Prove that m has to be an integer, if $\Theta = c_1 \cos m\theta + c_2 \sin m\theta$ shall satisfy (II.59).

25. Formulate the boundary and initial value problem for a membrane which is stretched over the unit circle $r = 1$ and its initial displacement is proportional to the distance from the boundary.

26. Formulate the boundary and initial value problem for the membrane which is stretched over the curve $r = 1$, $u = \sin s$ where s is the arc length measured along the circle $r = 1$, and initially driven with a velocity that is proportional to the projection of the distance from the boundary into the x, y–plane.

5. The Uniqueness of the Solution of the Membrane Equation

The vibrating, initially displaced or driven (or both) membrane under tension and an external force in the u-direction with the surface density $f(x, y, t)$ obeys the equation

$$\rho \frac{\partial^2 u}{\partial t^2} = \tau\left(\frac{\partial^2 u}{\partial x^2} + \frac{\partial^2 u}{\partial y^2}\right) + f(x, y, t) \qquad \text{(II.34)}$$

(see p. 53). This equation holds whether the membrane is stretched over a plane or space curve, or its boundary is permitted to move freely along the walls of a right cylinder, as we have seen in subsection 3 of this section. The boundary conditions are in the case of the stretched membrane

$$u(x(s), y(s), t) = \Phi(s), \qquad \text{(II.66)}$$

where

$$x = x(s),$$

$$y = y(s),$$

$$u = \Phi(s)$$

is the parametric representation of a closed space curve L. In the case of the free membrane we have the boundary condition

$$\frac{\partial u(x(s), y(s), t)}{\partial n} = 0 \qquad \text{(II.67)}$$

where

$$\left.\begin{aligned} x &= x(s) \\ y &= y(s) \end{aligned}\right\}$$

is the projection C of a right cylinder into the x, y–plane. As initial conditions, we have in general

$$u(x, y, t_0) = \phi(x, y), \tag{II.68}$$

$$\left.\frac{\partial u(x, y, t)}{\partial t}\right|_{t=t_0} = \psi(x, y), \tag{II.69}$$

where ϕ is the initial displacement and ψ the initial velocity. For the first boundary value problem with (II.66) we have of course $\phi(x(s), y(s)) = \Phi(s)$ and $\psi(x(s), y(s)) = 0$ and for the second boundary value problem with (II.67) we have $\partial\phi/\partial n = 0$ on C and $\partial\psi/\partial n = 0$ on C. We are now ready to prove

Theorem II.2. *Let the domain D in the (x, y, t)-space be defined by (x, y) in R, where R is the closed region bounded by the projection of L into the x, y–plane or by C, and $t_1 \geq t \geq t_0$ where t_1 is any value $\geq t_0$ and let $u = u(x, y, t)$ be continuous with continuous derivatives of first and second order on D. If $u = u(x, y, t)$ satisfies (II.34), the boundary condition (II.66) or (II.67) and the initial conditions (II.68) and (II.69), then it is the only function with this property.*

Proof (by contradiction). We proceed analogously to the proof we have given for theorem II.1. We assume that there exists another solution $v(x, y, t)$ of (II.34) satisfying the same boundary and initial conditions and consider the function

$$\zeta(x, y, t) = u(x, y, t) - v(x, y, t).$$

This function satisfies the homogeneous partial differential equation

$$\rho\frac{\partial^2\zeta}{\partial t^2} = \tau\left(\frac{\partial^2\zeta}{\partial x^2} + \frac{\partial^2\zeta}{\partial y^2}\right), \tag{II.70}$$

either one of the boundary conditions

$$\zeta(x(s), y(s), t) = 0, \tag{II.71}$$

or

$$\frac{\partial\zeta(x(s), y(s), t)}{\partial n} = 0 \tag{II.72}$$

and the initial conditions

$$\zeta(x, y, t_0) = 0, \tag{II.73}$$

$$\left.\frac{\partial\zeta(x, y, t)}{\partial t}\right|_{t=t_0} = 0. \tag{II.74}$$

Again it is plausible on physical grounds that under these conditions the membrane as described by ζ will stay in its rest position and therefore $\zeta = 0$.

To prove this, we consider the identity

$$\iiint\limits_D \frac{\partial \zeta}{\partial t}\left[\rho \frac{\partial^2 \zeta}{\partial t^2} - \tau \frac{\partial^2 \zeta}{\partial x^2} - \tau \frac{\partial^2 \zeta}{\partial y^2}\right] dt\, dx\, dy = 0$$

which has to be satisfied in view of ζ being a solution of (II.70). Transformation of the integrand analogous to the one performed in the preceding section furnishes the following form of this relation:

$$\iiint\limits_D \left[\frac{\partial}{\partial t}\frac{1}{2}(\rho\zeta_t^2 + \tau\zeta_x^2 + \tau\zeta_y^2) + \frac{\partial}{\partial x}(-\tau\zeta_t\zeta_x) + \frac{\partial}{\partial y}(-\tau\zeta_t\zeta_y)\right] dt\, dx\, dy = 0. \tag{II.75}$$

Let us apply the divergence theorem (see AI.C3)

$$\iiint\limits_D \operatorname{div} \mathbf{f}\, dt\, dx\, dy = \iint\limits_S \mathbf{f} \cdot \mathbf{n}^0\, dS,$$

where S is the surface enclosing D and $\mathbf{n}^0$ is the unit vector in the direction of the outward normal to S, to (II.75) with

$$\mathbf{f} = -\tau\zeta_t\zeta_x\mathbf{i} - \tau\zeta_t\zeta_y\mathbf{j} + \tfrac{1}{2}(\rho\zeta_t^2 + \tau\zeta_x^2 + \tau\zeta_y^2)\mathbf{k}$$

and let n_1, n_2, n_3 denote the components of $\mathbf{n}^0$:

$$\mathbf{n}^0 = n_1\mathbf{i} + n_2\mathbf{j} + n_3\mathbf{k}.$$

Then

$$\iiint\limits_D \left[\frac{\partial}{\partial t}\frac{1}{2}(\rho\zeta_t^2 + \tau\zeta_x^2 + \tau\zeta_y^2) - \frac{\partial}{\partial x}(\tau\zeta_t\zeta_x) - \frac{\partial}{\partial y}(\tau\zeta_t\zeta_y)\right] dt\, dx\, dy$$

$$= \iint\limits_S \left[-n_1\tau\zeta_t\zeta_x - n_2\tau\zeta_t\zeta_y + \frac{n_3}{2}(\rho\zeta_t^2 + \tau\zeta_x^2 + \tau\zeta_y^2)\right] dS \tag{II.76}$$

S consists of the bottom S_1 of the cylindrical domain D, the top S_2 of the cylindrical domain D and the lateral surface S_3.

We have on S_1

$$\mathbf{n}^0 = -\mathbf{k}, \quad dS_1 = dx\, dy,$$

on S_2

$$\mathbf{n}^0 = \mathbf{k}, \qquad dS_2 = dx\, dy,$$

and on S_3

$$\mathbf{n}^0 = \frac{dy}{ds}\mathbf{i} - \frac{dx}{ds}\mathbf{j}, \quad dS_3 = ds\, dt,$$

where $x = x(s)$, $y = y(s)$ is the parametric representation of the projection of L into the x, y–plane or the projection C of the right cylinder into the x, y–plane and s is the arc length on that curve. Hence, the surface integral on the right side of (II.76) can be written as

$$-\frac{1}{2}\iint_{S_1}(\rho\zeta_t^2 + \tau\zeta_x^2 + \tau\zeta_y^2)\Big|_{t=t_0} dx\,dy + \frac{1}{2}\iint_{S_2}(\rho\zeta_t^2 + \tau\zeta_x^2 + \tau\zeta_y^2)\Big|_{t=t_1} dx\,dy$$
$$+\iint_{S_3}\left(-\tau\zeta_t\zeta_x\frac{dy}{ds} + \tau\zeta_t\zeta_y\frac{dx}{ds}\right)_{\substack{x=x(s)\\y=y(s)}} ds\,dt.$$

The first integral vanishes in view of (II.73), which implies that $\zeta_x = \zeta_y = 0$ in $t = t_0$, and (II.74). If $\partial\zeta/\partial n$ denotes the derivative of ζ in the direction of the outward normal $\mathbf{n}^0$, we can write the integrand of the integral over S_3 in the following form

$$-\tau\zeta_t\left(\zeta_x\frac{dy}{ds} - \zeta_y\frac{dx}{ds}\right) = -\tau\zeta_t\frac{\partial\zeta}{\partial n}$$

(see Chapter I, §4.3, p.32). Thus, we obtain for the integral over S_3:

$$-\tau\iint_{S_3}\zeta_t\frac{\partial\zeta}{\partial n}\,ds\,dt.$$

If (II.71) is satisfied, we have $\zeta_t = 0$ on C and if (II.72) is satisfied, we have $\partial\zeta/\partial n = 0$ on C. Hence, the integral over S_3 vanishes in both cases and we obtain for (II.75)

$$\frac{1}{2}\iint_{S_2}\left[\rho\left(\frac{\partial\zeta}{\partial t}\right)^2 + \tau\left(\frac{\partial\zeta}{\partial x}\right)^2 + \tau\left(\frac{\partial\zeta}{\partial y}\right)^2\right]\Big|_{t=t_1} dx\,dy = 0.$$

It follows as in §1.5 of this chapter that

$$\frac{\partial\zeta}{\partial t} = 0, \quad \frac{\partial\zeta}{\partial x} = 0, \quad \frac{\partial\zeta}{\partial y} = 0 \qquad \text{in } D.$$

Hence

$$\zeta(x, y, t) = \text{constant}$$

and in view of (II.73) $\zeta(x, y, t) \equiv 0$, q.e.d.

Problem II.27

27. Show that

$$\zeta_t\left(\rho\frac{\partial^2\zeta}{\partial t^2} - \tau\frac{\partial^2\zeta}{\partial x^2} - \tau\frac{\partial^2\zeta}{\partial y^2}\right)$$
$$= \frac{\partial}{\partial t}\frac{1}{2}(\rho\zeta_t^2 + \tau\zeta_x^2 + \tau\zeta_y^2) + \frac{\partial}{\partial x}(-\tau\zeta_t\zeta_x) + \frac{\partial}{\partial y}(-\tau\zeta_t\zeta_y).$$

§3. THE EQUATION OF HEAT CONDUCTION AND THE POTENTIAL EQUATION

1. Heat Conduction without Convection

The considerations of this section will be based upon the fact that heat—which will be interpreted as some form of energy—flows within a certain medium from points of higher temperature to points of lower

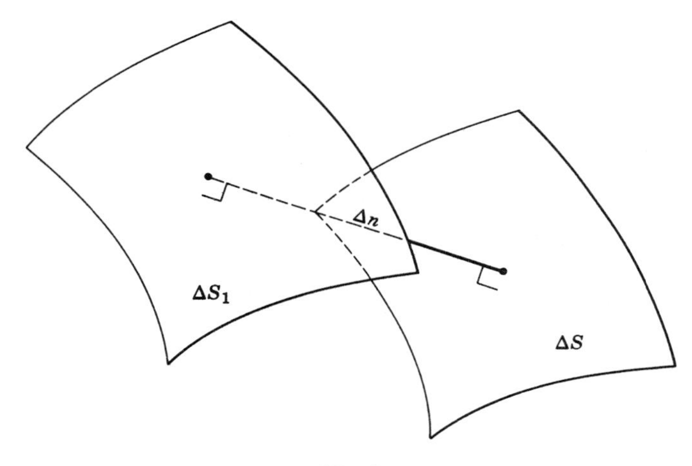

Fig. 5

temperature. This process takes place in such a way that the irregularly moving molecules transfer their kinetic energy by colliding with each other.

We will not consider the other possibility that heat transport can also be accomplished by a macroscopic flow of matter, as may be the case in fluids and gases (*convection*), or by radiation. Furthermore, we restrict our considerations to a *homogeneous* and *isotropic* medium. "Homogeneous" means that within the medium, physical properties are the same at all points and "isotropic" means that physical properties are independent of the direction selected.

In order to find the quantity of heat ΔQ penetrating an arbitrarily chosen surface element ΔS within the medium in unit time (*flux of heat*), we choose another surface element ΔS_1 of the same magnitude as ΔS, which is parallel to ΔS at an orthogonal distance Δn from ΔS (see Fig. 5.) We assume that ΔS is sufficiently small so that we can assume the temperatures $u = u(x, y, z, t)$ on ΔS and $u_1 = u(x + \Delta x, y + \Delta y, z + \Delta z, t)$ on ΔS_1 as constant over ΔS and ΔS_1 respectively.

From experimental thermodynamics and the kinetic theory of molecular motion it follows that heat will flow from the surface of higher temperature to the surface of lower temperature in the direction perpendicular to these surfaces. The magnitude of the flux of heat through the distance Δn is jointly proportional to the temperature difference between u and u_1, which we will call Δu and the area of the surface element ΔS:

$$\Delta Q \cdot \Delta n = k \cdot \Delta u \cdot \Delta S$$

where k, which is called the *thermal conductivity*, depends on the medium and in general on the temperature itself. This latter fact will be neglected and k assumed to be constant. (Thermal conductivity is flux of heat through a plate of unit area and unit thickness if the temperature difference between the two surfaces is 1 temperature unit.)

Hence

$$\Delta Q = k \cdot \frac{\Delta u}{\Delta n} \cdot \Delta S.$$

If we let $\Delta n \to 0$ we obtain the flux of heat through the surface element ΔS:

$$\Delta Q = k \frac{\partial u}{\partial n} \cdot \Delta S,$$

where $\partial u/\partial n$ is the derivative in the direction of the normal to ΔS. Thus we obtain for the flux through a volume element Δv which is enclosed by a surface S:

$$Q = k \iint_S \frac{\partial u}{\partial n}\, dS.$$

Since

$$\frac{\partial u}{\partial n} = \text{grad } u \cdot \mathbf{n}^0$$

(see AI.B8) where $\mathbf{n}^0$ is the unit vector in the direction perpendicular to S, we can write

$$Q = k \iint_S \text{grad } u \cdot \mathbf{n}^0\, dS$$

and apply the divergence theorem

$$\iint_S \mathbf{f} \cdot \mathbf{n}^0\, dS = \iiint_V \text{div}\, \mathbf{f}\, dV$$

(see AI.C3) with $\mathbf{f} = \text{grad } u$ and $V = \Delta v$:

$$Q = k \iint_S \text{grad } u \cdot \mathbf{n}^0\, dS = k \iiint_{\Delta v} \text{div (grad } u)\, dV.$$

We now apply the mean value theorem to the volume integral over Δv and obtain

$$Q = k \operatorname{div} \operatorname{grad} u(\xi_i, \eta_k, \zeta_l, t)\,\Delta v \tag{II.77}$$

where (ξ_i, η_k, ζ_l) is some point in Δv.

We see on the other hand that the temperature u in Δv is increased or decreased by the accumulation or loss of quantity of heat in Δv at the rate $\overline{(\partial u/\partial t)}$ per unit time, where $\overline{(\ \ )}$ indicates a mean value over Δv.

Hence, we can write the flux of heat into (or from) Δv as

$$\rho\sigma\overline{\left(\frac{\partial u}{\partial t}\right)}\,\Delta v, \tag{II.78}$$

where ρ is the density and σ the *specific heat* of the medium. (Specific heat is the amount of heat necessary to raise the temperature of a unit weight of the medium by one temperature unit in unit time.)

Equating (II.77) and (II.78), canceling Δv and letting the volume element Δv shrink to a point,

$$\rho\sigma\frac{\partial u}{\partial t} = k \operatorname{div} \operatorname{grad} u.$$

Since

$$\operatorname{grad} u = \frac{\partial u}{\partial x}\mathbf{i} + \frac{\partial u}{\partial y}\mathbf{j} + \frac{\partial u}{\partial z}\mathbf{k}$$

and

$$\operatorname{div} \mathbf{F} = \frac{\partial f_1}{\partial x} + \frac{\partial f_2}{\partial y} + \frac{\partial f_3}{\partial z}, \qquad \mathbf{F} = f_1\mathbf{i} + f_2\mathbf{j} + f_3\mathbf{k},$$

we have

$$\operatorname{div} \operatorname{grad} u = \frac{\partial^2 u}{\partial x^2} + \frac{\partial^2 u}{\partial y^2} + \frac{\partial^2 u}{\partial z^2}$$

and we can write the heat equation as follows

$$\boxed{\frac{\partial u}{\partial t} = \frac{k}{\rho\sigma}\left(\frac{\partial^2 u}{\partial x^2} + \frac{\partial^2 u}{\partial y^2} + \frac{\partial^2 u}{\partial z^2}\right)}\ . \tag{II.79}$$

(One frequently denotes the positive constant $k/\rho\sigma$ by α^2.)

This is again a linear partial differential equation of second order with constant coefficients.

Before we close this section, we will mention a few modifications of equation (II.79) for a more general circumstance than we assumed in the preceding derivation as well as considering some special cases.

If $f(x, y, z, t)$ is the rate at which heat is generated or absorbed within the medium at the point (x, y, z) at the time t (*sources* or *sinks*), we have to add to (II.77) the quantity

$$f(x, y, z, t)\,\Delta v$$

and hence (II.79) appears in the more general form

$$\frac{\partial u}{\partial t} = \alpha^2\left(\frac{\partial^2 u}{\partial x^2} + \frac{\partial^2 u}{\partial y^2} + \frac{\partial^2 u}{\partial z^2}\right) + \frac{1}{\sigma\rho} f(x, y, z, t). \tag{II.80}$$

If we consider the stationary (time-independent) temperature distribution with $\partial u/\partial t = 0$, we obtain *Poisson's equation* of potential theory

$$\frac{\partial^2 u}{\partial x^2} + \frac{\partial^2 u}{\partial y^2} + \frac{\partial^2 u}{\partial z^2} + \frac{1}{k} f(x, y, z) = 0. \tag{II.81}$$

A stationary temperature distribution in the absence of sources and sinks obeys simply the equation

$$\frac{\partial^2 u}{\partial x^2} + \frac{\partial^2 u}{\partial y^2} + \frac{\partial^2 u}{\partial z^2} = 0. \tag{II.82}$$

Further specializations are discussed in the problems.

Problems II.28–32

28. Let us consider a half space, bounded by a plane and assume that heat flows only in the direction perpendicular to this plane. Let the boundary plane coincide with the y, z–plane and state the heat equation.

29. In a thin plate the faces of which are insulated against loss or gain, no heat will flow perpendicularly to the plate. State the heat equation for this case.

30. What is the equation for a stationary temperature distribution in the case of a two-dimensional flow.

31. What integral equation has to hold for the flow of heat through a volume V bounded by a closed surface S?

32. State reasonable boundary and initial conditions for problem 28.

2. The Potential Equation

This section will be devoted to a short discussion of the equation

$$\frac{\partial^2 u}{\partial x^2} + \frac{\partial^2 u}{\partial y^2} = 0 \tag{II.83}$$

and its three-dimensional analog

$$\frac{\partial^2 u}{\partial x^2} + \frac{\partial^2 u}{\partial y^2} + \frac{\partial^2 u}{\partial z^2} = 0. \tag{II.84}$$

Equation (II.83) is called the *two-dimensional potential equation* or *potential equation in two variables* and (II.84) the *three-dimensional potential equation* or *potential equation in three variables.*

Equation (II.83) arises when one discusses the stable state of a membrane in absence of external forces, which is characterized by a minimization of the potential energy. The stationary temperature distribution in a two-dimensional field in the absence of sources and sinks also leads to equation (II.83) (see problem II.30), while the stationary temperature distribution in a three-dimensional space without sources and sinks is described by (II.84), as we have seen in the preceding subsection.

In the following passages, we will present still another interpretation of the potential equation.

Let the vector field

$$\mathbf{v} = u(x, y, z, t)\mathbf{i} + v(x, y, z, t)\mathbf{j} + w(x, y, z, t)\mathbf{k}$$

represent the velocity distribution in a medium (liquid) of density $\rho(x, y, z, t)$. If S is a closed rectifiable surface encompassing a simply connected domain D which lies entirely inside the medium, then

$$m = \iiint\limits_D \rho(x, y, z, t)\, dV$$

is the total mass of the medium at the time t in D and the rate of change of this mass m is given by

$$\frac{\partial m}{\partial t} = \iiint\limits_D \frac{\partial \rho}{\partial t}\, dV. \tag{II.85}$$

There are only two causes which could be responsible for a change of m in D, namely, matter may enter or leave (or both) D and in the second place, matter may be created or destroyed (or both) in D (sources or sinks).

If $\mathbf{n}^0$ is the unit vector in the direction of the outward normal to S, then $\mathbf{v} \cdot \mathbf{n}^0$ is the component of $\mathbf{v}$ in the direction perpendicular to S and $-\rho\mathbf{v} \cdot \mathbf{n}^0\, \Delta S$ is the amount of matter entering through ΔS per unit time. Hence, the total amount of matter entering through S into D per unit time is given by

$$-\iint\limits_S \rho\mathbf{v} \cdot \mathbf{n}^0\, dS. \tag{II.86}$$

Applying the divergence theorem (see AI.C3) to (II.86) renders

$$-\iint_S \rho\mathbf{v}\cdot\mathbf{n}^0\,dS = -\iiint_D \operatorname{div}(\rho\mathbf{v})\,dV. \tag{II.87}$$

Let $f(x, y, z, t)$ be the source (sink) density, i.e., the amount of matter created (destroyed) per unit volume at the time t; then the total amount of newly created (destroyed) matter in D is given by

$$\iiint_D f(x, y, z, t)\,dV. \tag{II.88}$$

Hence, we obtain in view of (II.85), (II.87), and (II.88)

$$\iiint_D \frac{\partial\rho}{\partial t}\,dV = -\iiint_D \operatorname{div}(\rho\mathbf{v})\,dV + \iiint_D f\,dV.$$

This equation has to hold for any D, however small it might be, which is entirely submerged in the medium. Applying the mean value theorem to the relation above and letting the diameter of D approach zero gives rise to the so-called *equation of continuity*

$$\boxed{\frac{\partial\rho}{\partial t} + \operatorname{div}(\rho\mathbf{v}) = f}\,. \tag{II.89}$$

an application of which will be discussed in Chapter VIII.

An important special case of this equation of continuity is obtained under the following restricting assumptions:

(a) Let the medium be *incompressible*; i.e., the density is constant and it follows that

$$\frac{\partial\rho}{\partial t} = 0 \quad \text{and} \quad \operatorname{div}(\rho\mathbf{v}) = \rho \operatorname{div}\mathbf{v}.$$

(b) Let the medium be free of sources and sinks; i.e., $f = 0$.

(c) Let the velocity field be *irrotational*; i.e., the line integral

$$\oint_L \mathbf{v}\cdot\mathbf{t}\,ds = 0$$

over any closed curve L in D is zero, where $\mathbf{t}$ is the tangent vector to L and $|\mathbf{t}| = 1$.

From this last condition it follows that there exists a function $u(x, y, z)$ such that

$$\mathbf{v} = -\text{grad } u$$

(see AI.C1).

Under the conditions a, b, and c, equation (II.89) reduces to

$$\text{div grad } u = 0. \tag{II.90}$$

Since grad $u = \nabla u$ and div $\mathbf{v} = \nabla \cdot \mathbf{v}$, we have for

$$\text{div grad} = \nabla \cdot \nabla = \frac{\partial^2}{\partial x^2} + \frac{\partial^2}{\partial y^2} + \frac{\partial^2}{\partial z^2}$$

(see AI.B5) and consequently (II.90) is the potential equation (II.84). Since div grad $= \nabla \cdot \nabla$, one often abbreviates this operator as ∇^2. Even more often one will find the symbol Δ (read "delta" or "Laplacian") for ∇^2 which we are going to use in the future.

The solutions of (II.83) and (II.84) are called *potential functions* or *harmonic functions* (or shortly *harmonics*). Occasionally, one refers to the solutions of (II.83) as *logarithmic potentials* because of the structure of one of its significant solutions (see problem II.34) and to the solutions of (II.84) as *Newtonian potentials* (see problem II.35).

Problems II.33–35

33. Let $\mathbf{v} = u\mathbf{i} + v\mathbf{j}$ be a two-dimensional irrotational velocity field. Show that in a medium of constant density which is free of sources and sinks there exists a velocity potential u, grad $u = \mathbf{v}$ such that $\Delta u = 0$.

34. Show that $u = \log r$, $r = \sqrt{x^2 + y^2}$, satisfies (II.83).

35. Show that Newton's gravitational potential $u = 1/r$, $r = \sqrt{x^2 + y^2 + z^2}$, satisfies (II.84).

3. The Initial and Boundary Value Problem for Heat Conduction

We found in subsection 1 of this section that the temperature distribution $u(x, y, z, t)$ in a homogeneous isotropic medium satisfies the equation

$$\frac{\partial u}{\partial t} = \alpha^2 \Delta u(x, y, z, t) \tag{II.79}$$

if there are neither sources nor sinks present.

This is a partial differential equation in four variables which is of the second order in x, y, z and of the first order in t.

Before we consider this general case, we will discuss the case of stationary heat distribution; i.e., the solution of the potential equation

$$\Delta u(x, y, z) = 0. \tag{II.84}$$

If we consider the temperature distribution in the interior of a three-dimensional region D, bounded by the closed surface S, it is physically clear that the temperature distribution on the boundary will determine the temperature distribution in the interior.

Let

$$x = x(v, w),$$
$$y = y(v, w),$$
$$z = z(v, w)$$

be the parametric representation of the closed surface S. Then

$$u(x(v, w), y(v, w), z(v, w)) = \Phi(v, w), \tag{II.91}$$

where $\Phi(v, w)$ is the given temperature distribution on S will determine the solution in the interior uniquely (we hope). (II.84), (II.91) constitutes the three-dimensional analog of Dirichlet's problem.

If we consider now the case where u is also time-dependent, (II.79), we have to be a little careful. If we know only the temperature distribution on the boundary which may or may not be a function of the time, we certainly do not know enough. The knowledge of the temperature distribution in the interior of the region D at a certain time t_0 will certainly be necessary information. This gives together with the boundary condition

$$u(x(v, w), y(v, w), z(v, w), t) = \Phi(v, w, t), \tag{II.92}$$

the initial condition

$$u(x, y, z, t_0) = \Psi(x, y, z) \tag{II.93}$$

where Φ is the given temperature distribution on S and Ψ is the temperature in the interior at the time t_0.

Also in dealing with heat conduction, we have an analogy to the second boundary value problem. Instead of prescribing the temperature on the closed surface S, we can prescribe the derivative of the temperature in the direction of the outward normal to S,

$$\frac{\partial u}{\partial n} = \mathrm{X}(x, y, z, t) \quad \text{on } S \tag{II.94}$$

as we can see from the following argument: We observed in subsection 1 of this section that the amount of heat flowing through a surface element ΔS per unit time is given by

$$k\frac{\partial u}{\partial n}\Delta S.$$

If we know the amount of heat flowing from outside the region D through a surface element ΔS at any time, we know how much heat the surface transmits into the interior of the region since this has to be the same amount for the simple reason that heat cannot possibly accumulate on a surface of mass zero. We can, therefore, find $\partial u/\partial n$ from the heat output of every point on the boundary and vice versa.

Problems II.36–38

36. State the boundary value problem for the stable state of a membrane under the influence of gravity and fixed along a circle of radius 1 in a plane parallel to the x, y–plane at a distance 1 from the x, y–plane.

37. State the boundary and initial value problem for the vibrations of a free membrane with the boundary sliding along the wall of the cylinder $x^2 + y^2 = 1$, if the membrane is released from the rest position with an initial velocity which is given by $\psi = 2\sqrt{x^2 + y^2} - x^2 - y^2$.

38. State the boundary and initial value problem for the temperature distribution in the interior of an infinite cylinder $x^2 + y^2 = 1$, if the temperature on the walls of the cylinder is given by

$$\Phi = \frac{t + 1}{t + z^2 + 1}$$

and the temperature in the interior at the time $t = 0$ is everywhere $1/(z^2 + 1)$.

4. Stationary Temperature Distribution Generated by a Spherical Stove

Let us assume that an infinite homogeneous, isotropic space is heated by a spherical stove with radius 1 whose surface is kept at a constant temperature $f(\phi)$ which is a function of the co-latitude ϕ alone. (The meaning of "constant" is of course "time independent" in this discussion.)

Since the space is supposed to be homogeneous and isotropic we can apply equation (II.82) to our problem:

$$\Delta u = 0.$$

If we choose our coordinate system so that the center of the stove lies in the coordinate origin, it is the proper thing to introduce spherical coordinates

$$\begin{aligned} x &= r \cos \theta \sin \phi, \\ y &= r \sin \theta \sin \phi, \\ z &= r \cos \phi, \end{aligned} \tag{II.95}$$

in which (II.82) assumes the form

$$\frac{1}{r^2} \cdot \frac{\partial}{\partial r}\left(r^2 \frac{\partial u}{\partial r}\right) + \frac{1}{r^2 \sin \phi} \frac{\partial}{\partial \phi}\left(\sin \phi \frac{\partial u}{\partial \phi}\right) + \frac{1}{r^2 \sin^2 \phi} \frac{\partial^2 u}{\partial \theta^2} = 0 \tag{II.96}$$

(see AI.B6).

The fact that the temperature on the surface of the stove is given as a function of the co-latitude introduces the boundary condition

$$u(1, \theta, \phi) = f(\phi). \tag{II.97}$$

In addition to this, we have to require

$$\left| \lim_{r\to\infty} u(r, \theta, \phi) \right| < \infty.$$

By using Bernoulli's separation method

$$u(r, \theta, \phi) = R(r)\Theta(\theta)\Phi(\phi),$$

we obtain for (II.97)

$$R(1)\Theta(\theta)\Phi(\phi) = f(\phi).$$

Hence, $\Theta(\theta) =$ constant and the term $\dfrac{\partial^2 u}{\partial \theta^2}$ in (II.96) will drop out. Thus we have to deal with the much simpler equation

$$\frac{1}{r^2}\frac{\partial}{\partial r}\left(r^2 \frac{\partial u}{\partial r}\right) + \frac{1}{r^2 \sin\phi}\frac{\partial}{\partial \phi}\left(\sin\phi \frac{\partial u}{\partial \phi}\right) = 0 \tag{II.98}$$

and we will assume now that the solution has the form

$$u(r, \phi) = R(r)\Phi(\phi). \tag{II.99}$$

Then, it follows from (II.97) that

$$R(1)\Phi(\phi) = f(\phi).$$

We require in addition that the solution be such that

$$\left| \lim_{r\to\infty} R(r) \right| < \infty. \tag{II.100}$$

If we consider the interior of the sphere instead of the exterior, condition (II.100) is to be replaced by

$$|R(0)| < \infty. \tag{II.101}$$

If we substitute (II.99) into (II.98) we obtain

$$\frac{1}{r^2}\frac{d}{dr}\left(r^2 \frac{dR}{dr}\right)\Phi + \frac{1}{r^2 \sin\phi}\cdot\frac{d}{d\phi}\left(\sin\phi \frac{d\Phi}{d\phi}\right)R = 0.$$

Division by $R\Phi$, multiplication by r^2, shifting terms which contain only functions of ϕ to the right side, and repeating the argument of §1.2 and §2.4 of this chapter, we obtain for any arbitrary parameter λ the two equations

$$\frac{d}{dr}\left(r^2 \frac{dR}{dr}\right) = \lambda R$$

and

$$\frac{1}{\Phi \sin \phi} \cdot \frac{d}{d\phi} (\sin \phi \cdot \Phi') = -\lambda.$$

Carrying out the indicated differentiations gives in the one case

$$r^2 R'' + 2rR' - \lambda R = 0. \tag{II.102}$$

In the case of the other equation we introduce a new variable ξ by

$$\xi = \cos \phi$$

and obtain

$$\frac{d}{d\xi}\left[(1 - \xi^2)\frac{d\Phi}{d\xi}\right] + \lambda \Phi = 0. \tag{II.103}$$

Equation (II.102) is a differential equation of second order of the *Euler-Cauchy* type and can be reduced to a linear differential equation of the second order with constant coefficients by means of the transformation

$$r = e^t.$$

By this we obtain the equation

$$\frac{d^2R}{dt^2} + \frac{dR}{dt} - \lambda R = 0$$

with the general solution

$$R = c_1 \exp\left(-\tfrac{1}{2} + \sqrt{\tfrac{1}{4} + \lambda}\right)t + c_2 \exp\left(-\tfrac{1}{2} - \sqrt{\tfrac{1}{4} + \lambda}\right)t$$

or in terms of the independent variable r

$$R = c_1 r^{-1/2 + \sqrt{1/4 + \lambda}} + c_2 r^{-1/2 - \sqrt{1/4 + \lambda}}.$$

If we designate the expression $-\frac{1}{2} + \sqrt{\frac{1}{4} + \lambda}$ by the letter n

$$n = -\tfrac{1}{2} + \sqrt{\tfrac{1}{4} + \lambda}$$

we have

$$\lambda = n(n + 1)$$

and

$$R = c_1 r^n + c_2 \cdot \frac{1}{r^{n+1}} \tag{II.104}$$

as the general solution of (II.102).

In order to satisfy (II.100), we have to choose $c_1 = 0$ and in the case of (II.101) we have to choose $c_2 = 0$.

Writing equation (II.103) in terms of n makes it appear as

$$\frac{d}{d\xi}\left[(1 - \xi^2)\frac{d\Phi}{d\xi}\right] + n(n + 1)\Phi = 0. \tag{II.105}$$

This equation is called *Legendre's equation*.

As we will show in Chapter VI, §1, the free parameter n has to be 0 or a positive integer in order to provide suitable solutions to our problem.

If we denote the solution of (II.105) for the moment with $\Phi_n(\xi)$ where the subscript indicates its dependence on the parameter n in (II.105), we can write the solution u to our problem in the form

$$u(r, \phi) = \frac{1}{r^{n+1}} \Phi_n(\cos \phi)$$

and try to satisfy the boundary condition (II.97) by a suitable choice of the coefficients A_n in

$$\sum_{n=0}^{\infty} \frac{A_n}{r^{n+1}} \Phi_n(\cos \phi).$$

Thus, we have now a new problem in addition to the ones we listed in previous sections, namely, to investigate Legendre's equation with particular emphasis on the possibility of expanding a given function in terms of its solutions.

5. The Uniqueness of the Solution of the Potential Equation

We consider the potential equation in three-dimensional space

$$\Delta u(x, y, z) = 0 \tag{II.84}$$

with the boundary condition

$$u(x(v, w), y(v, w), z(v, w)) = \Phi(v, w), \tag{II.106}$$

where

$$\left.\begin{aligned} x &= x(v, w) \\ y &= y(v, w) \\ z &= z(v, w) \end{aligned}\right\} S$$

is a closed surface enclosing a three-dimensional domain, or the boundary condition

$$\frac{\partial u}{\partial n} = \Psi(v, w) \quad \text{on } S. \tag{II.107}$$

This latter problem arises in dealing with the stationary heat distribution in space when the temperature distribution on the boundary is not given, as we have seen in subsection 3 of this section.

The analogous problem in two independent variables arises in dealing with the stationary heat distribution in a plane or in considering the stable state of a fixed or free membrane.

We will include both cases in the following theorem:

Theorem II.3. *If* $\begin{Bmatrix} u(x, y, z) \\ u(x, y) \end{Bmatrix}$ *is a solution of* $\Delta u = 0$ *in a bounded* $\begin{Bmatrix} \textit{three} \\ \textit{two} \end{Bmatrix}$ *dimensional region* $\begin{Bmatrix} D \\ R \end{Bmatrix}$ *which is encompassed by the rectifiable* $\begin{Bmatrix} \textit{surface } S \\ \textit{curve } C \end{Bmatrix}$, *is continuous and has continuous partial derivatives in* $\begin{Bmatrix} D \\ R \end{Bmatrix}$ *and on* $\begin{Bmatrix} S \\ C \end{Bmatrix}$ *and satisfies on* $\begin{Bmatrix} S \\ C \end{Bmatrix}$ *the conditions* $u = \Phi$ *or* $\partial u/\partial n = \Psi$, *then it is the only solution except for an additive constant in the case of the latter boundary value problem with* $\partial u/\partial n = \Psi$.

Proof (*by contradiction*). We give here the proof for the three-dimensional case only since the proof for the two-dimensional case is completely analogous and left for the problems. We consider again the possibility of having another solution v, which satisfies the same boundary conditions and the potential equation.

Then the function

$$\zeta = u - v$$

will satisfy $\Delta\zeta = 0$ and either the condition

$$\zeta = 0 \quad \text{on } S \tag{II.108}$$

or

$$\frac{\partial \zeta}{\partial n} = 0 \quad \text{on } S. \tag{II.109}$$

If we consider again the expression representing the total potential energy

$$\iiint_D (\zeta_x^2 + \zeta_y^2 + \zeta_z^2)\, dV = \iiint_D \text{grad}^2 \zeta\, dV,$$

we can apply Green's formula (see **AI.C4**)

$$\iiint_D [\text{grad}^2 u + u\, \Delta u]\, dV = \iint_S u \frac{\partial u}{\partial n}\, dS$$

and obtain in view of $\Delta\zeta = 0$

$$\iiint_D \text{grad}^2 \zeta\, dV = \iint_S \zeta \frac{\partial \zeta}{\partial n}\, dS.$$

Since ζ satisfies either condition (II.108) or (II.109) on S, the integral on the right side vanishes, therefore the integral on the left has to vanish and we conclude as in previous cases that

$$\zeta_x = \zeta_y = \zeta_z = 0$$

and therefore ζ = constant. In case (II.108) holds, this constant has to be zero and the solution is unique. In case (II.109) holds, we cannot draw this conclusion and ζ may be an arbitrary constant. This means that the solution is unique but for an arbitrary additive constant:

$$u = v + C.$$

This latter fact has a simple interpretation: the stable state of the free membrane is certainly independent of the distance of the membrane from the x, y–plane and we have to expect, therefore, a one-parameter family of solutions which are all parallel to each other.

This situation is also easy to understand in the case of the three-dimensional stationary heat distribution, if the flow through S is given, but the actual temperature distribution on S is not. There, the constant C represents the constant temperature ζ in D which is attained if S is insulated: $\partial\zeta/\partial n = 0$ on S.

Problem II.39

*39. Prove theorem II.3 for the case of the potential equation in two independent variables.

6. The Uniqueness of the Solution of the Heat Equation

We found in subsection 1 of this section that the temperature u in a homogeneous isotropic medium characterized by its density ρ, specific heat σ, and thermal conductivity k satisfies the equation

$$\frac{\partial u}{\partial t} = \frac{k}{\rho\sigma}\Delta u + \frac{1}{\rho\sigma}f(x, y, z, t) \tag{II.80}$$

where $f(x, y, z, t)$ shall represent a continuous source density.

For reasons of convenience we will now consider the rather general boundary value problem, where

$$\alpha u + \beta\frac{\partial u}{\partial n}\bigg|_{\text{on } S} = \phi(x, y, z, t) \tag{II.110}$$

is prescribed on the surface S, enclosing a certain volume V, in which we are interested, where α and β are constants which shall not vanish at the same time and shall both have the same sign.

For $\alpha = 0$ we have the case where the flow through S is given, for $\beta = 0$ the case where the temperature distribution on S is given.

As we have seen in subsection 3 of this section, we have to impose the initial condition

$$u(x, y, z, t_0) = \psi(x, y, z) \tag{II.111}$$

in order to formulate the problem completely.

We now state and prove the following theorem:

Theorem II.4. *If $u(x, y, z, t)$ is a solution of (II.80), is continuous in V and on S, has continuous derivatives of first and second order with respect to x, y, z in V and on S and a continuous derivative with respect to t for all $t \geq t_0$, and if it satisfies (II.110) and (II.111), then it is the only solution with this property.*

Proof (by contradiction). Again, we consider another possible solution v with the same properties, satisfying the same conditions.

Then $\zeta = u - v$ satisfies

$$\frac{\partial \zeta}{\partial t} = \frac{k}{\rho\sigma} \Delta\zeta, \tag{II.112}$$

the boundary condition

$$\alpha\zeta + \beta \frac{\partial \zeta}{\partial n} = 0 \quad \text{on } S, \tag{II.113}$$

and the initial condition

$$\zeta(x, y, z, t_0) = 0. \tag{II.114}$$

Since ζ is continuous in t, the function

$$Z(t) = \iiint_V \zeta^2 \, dV \geq 0 \tag{II.115}$$

is also continuous in t and, in view of (II.114),

$$Z(t_0) = 0.$$

If we differentiate Z with respect to t, we obtain, using (II.112),

$$Z'(t) = \iiint_V 2\zeta \frac{\partial \zeta}{\partial t} \, dV = \frac{k}{\rho\sigma} \iiint_V 2\zeta \, \Delta\zeta \, dV.$$

If we apply Green's formula (see **AI.C4**)

$$\iiint_V [\text{grad}^2 \zeta + \zeta \, \Delta\zeta] \, dV = \iint_S \zeta \frac{\partial \zeta}{\partial n} \, dS$$

to the expression occurring on the right side, we obtain

$$Z'(t) = \frac{2k}{\rho\sigma} \iint_S \zeta \frac{\partial \zeta}{\partial n} \, dS - \frac{2k}{\rho\sigma} \iiint_V \text{grad}^2 \zeta \, dV.$$

Since $\text{grad}^2 \zeta \geq 0$,

$$\iiint_V \text{grad}^2 \zeta \, dV \geq 0.$$

According to (II.113), for $\beta \neq 0$,

$$\frac{\partial \zeta}{\partial n} = -\frac{\alpha}{\beta}\zeta \quad \text{on } S, \text{ with } \frac{\alpha}{\beta} > 0.$$

Therefore

$$\iint_S \zeta \frac{\partial \zeta}{\partial n}\, dS = -\frac{\alpha}{\beta}\iint_S \zeta^2\, dS \leq 0$$

and hence

$$Z'(t) \leq 0 \quad \text{for all } t > t_0.$$

Z is zero at t_0 and is monotonically decreasing from then on; therefore, it would appear that

$$Z(t) \leq 0.$$

To bring this in agreement with (II.115), we must have $Z(t) = 0$ and from this it follows that $\zeta^2 = 0$ and $\zeta = 0$, q.e.d.

Problems II.40–41

40. Prove theorem II.4 for the case $\beta = 0$.
41. Prove the same as in problem 40 for the case $\alpha = 0$.

RECOMMENDED SUPPLEMENTARY READING

I. G. Petrovsky: *Lectures on Partial Differential Equations*, translated from the Russian by A. Shenitzer, Interscience Publishers, New York, 1954.

F. H. Miller: *Partial Differential Equations*, John Wiley and Sons, New York, 1941.

R. Courant and D. Hilbert: *Methods of Mathematical Physics*, Vol. I, Interscience Publishers, New York, 1953.

Ph. Frank and R. von Mises: *Differential- und Integral-Gleichungen der Mechanik und Physik*, Vol. II, F. Vieweg und Sohn, Braunschweig, 1935.

III

THEOREMS RELATED TO PARTIAL DIFFERENTIAL EQUATIONS AND THEIR SOLUTIONS

§1. GENERAL REMARKS ON THE EXISTENCE AND UNIQUENESS OF SOLUTIONS

1. The Problem of Minimal Surfaces

That the proof for the uniqueness of solutions in all the cases, which we dealt with in the preceding chapter, was quite simple should not give the reader the impression that this is always the case in the theory of partial differential equations. That such a proof is sometimes even impossible will be demonstrated with the following rather simple example.

We consider two closed curves which lie in planes parallel to each other. Let us represent those two curves by closed wires and perform the following experiment: The two wires which are by some means rigidly connected with each other are dipped in a soap solution and we observe a soap film, joining the two wires, upon removing them from the soap solution. The question is, what kind of a surface do we obtain? In order to investigate the nature of this surface to some extent, we formulate this problem mathematically:

We choose our coordinate system so that the two planes containing the boundary curves under consideration are perpendicular to the x-axis, at a distance d apart. Then the two curves can be represented by

$$C_1\colon\ \left.\begin{aligned} x &= 0 \\ y &= y_1(s) \\ z &= z_1(s) \end{aligned}\right\}, \qquad C_2\colon\ \left.\begin{aligned} x &= d \\ y &= y_2(s) \\ z &= z_2(s) \end{aligned}\right\}.$$

The soap film $z = z(x, y)$ which appears in our experiment will be such that its area (which is proportional to the potential energy due to surface tension[1]) is a relative minimum. That is, we have the following variational problem to consider:

$$\iint_R dS = \iint_R \sqrt{1 + z_x^2 + z_y^2}\, dx\, dy \to \text{minimum}.$$

The Euler-Lagrange equation of this problem is, according to (I.27),

$$\frac{(1 + z_y^2)z_{xx} - 2z_x z_y z_{xy} + (1 + z_x^2)z_{yy}}{(1 + z_x^2 + z_y^2)^{3/2}} = 0.$$

This states that the mean curvature of surfaces of this type—so-called *minimal surfaces*—is zero (see problem I.14). It is far too complicated and quite unnecessary for our purpose to pursue this problem any further in this generality. We can make our point quite clear if we consider as boundary curves C_1 and C_2, circles with radii r_1, r_2 and centers on the x-axis; i.e.:

$$C_1: \left.\begin{aligned} x &= 0 \\ y &= r_1 \cos\theta \\ z &= r_1 \sin\theta \end{aligned}\right\}, \qquad C_2: \left.\begin{aligned} x &= d \\ y &= r_2 \cos\theta \\ z &= r_2 \sin\theta \end{aligned}\right\}.$$

If we admit for consideration only surfaces of revolution (which is no restriction since the area of any other surface can be decreased by replacing every cross section perpendicular to the x-axis by a circle of the same area with its center on the x-axis), we have the variational problem

$$2\pi \int_0^d y\sqrt{1 + y'^2}\, dx \to \text{minimum}$$

or if we choose for matters of convenience a parametric representation for the generating curve $y = y(x)$ as

$$\left.\begin{aligned} x &= x(t) \\ y &= y(t) \end{aligned}\right\},$$

we have to minimize the integral

$$2\pi \int_{t_0}^{t_1} y\sqrt{\dot{x}^2 + \dot{y}^2}\, dt.$$

(For independence of this integral on the parameter t see problem III.1.)

[1] We neglect gravity and other forces that would interfere with the tendency of the surface tension to decrease the total area.

The Euler-Lagrange equations of the latter problem are, according to (I.17),

$$\dot{x}y = a\sqrt{\dot{x}^2 + \dot{y}^2}$$

$$\sqrt{\dot{x}^2 + \dot{y}^2} = \frac{d}{dt}\left(\frac{y\dot{y}}{\sqrt{\dot{x}^2 + \dot{y}^2}}\right)$$

where a is an integration constant (see problems I.11 and III.2).

The solutions of these equations are catenaries

$$x = at + b, \quad y = a \cosh t$$

as the reader can easily verify (see problem III.3).

With a certain amount of luck one can satisfy the boundary conditions and it would appear that the catenoid is a solution of this boundary value problem. Without going into this question more deeply, we will mention that it is not possible to satisfy the boundary conditions under certain circumstances.[2] In certain other circumstances one obtains two different catenaries, both of which satisfy the boundary conditions.[2]

We can see even by a much simpler argument that the solution of this problem cannot be unique: If we consider the two planes

$$\left.\begin{array}{l} x = 0 \\ y = t \end{array}\right\}, \quad \left.\begin{array}{l} x = d \\ y = t \end{array}\right\},$$

we can see that they satisfy the Euler-Lagrange equations for $a = 0$ and the boundary conditions and thus yield at least a second solution to our problem. This result is certainly not surprising to people who have performed soap-film experiments.

The reader might argue that the boundary value problem we just discussed differs essentially from those we considered in the preceding chapter insofar as we have here two boundaries and a doubly connected surface as a solution whereas we dealt with one boundary curve and simply connected surfaces as solutions at previous occasions. We can easily overcome this difficulty without changing anything as far as the non-uniqueness of the solution is concerned. If we cut out of the doubly connected catenoid a small strip bounded by two generating catenaries, we obtain a simply connected surface which has *eo ipso* a single boundary. Again one can see that both the catenoid short of the little strip we cut out as illustrated in Fig. 6a and the surface illustrated in Fig. 6b will furnish solutions to our problem.

[2] See G. A. Bliss, *Calculus of Variations*, The Open Court Publishing Company, Chicago, 1925.

To the reader who is interested in further pursuit of this problem, Courant's monograph[3] on Dirichlet's principle is highly recommended for more intense study and literature references. Upon studying this problem, as we formulated it in the beginning of this section, he will find out that very little is known as yet about the number and nature of its solutions.

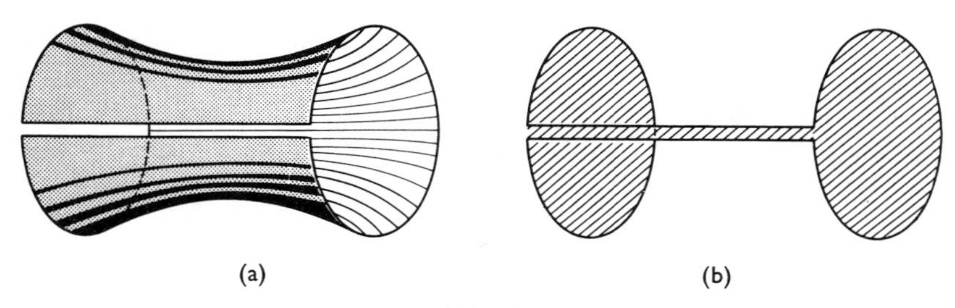

(a) (b)

Fig. 6

Problems III.1–3

***1.** Prove that the integral

$$\int_{t_0}^{t_1} y\sqrt{\dot{x}^2 + \dot{y}^2}\, dt$$

is invariant under a transformation of the parameter $t = t(s)$ if dt/ds is either positive or negative in the interval under consideration.

2. Find the Euler-Lagrange equations of the variational problem

$$\int_{t_0}^{t_1} y\sqrt{\dot{x}^2 + \dot{y}^2}\, dt \to \text{minimum}.$$

(*Hint:* Make use of the result in problem I.11.)

3. Show that $x = at + b$, $y = a \cosh t$ are solutions of the Euler-Lagrange equations in problem 2.

2. The Problem of Cauchy-Kowalewski

If we have an ordinary differential equation of the nth order

$$y^{(n)} = f(x, y, y', \cdots, y^{(n-1)}),$$

we can reduce this equation to a system of n differential equations of the first order

$$y_i' = f_i(x, y_1, y_2, \cdots, y_n), \qquad i = 1, 2, 3, \cdots, n$$

and prove under rather generous assumptions about the f_i that a system of n solutions $y_1(x), y_2(x), \cdots, y_n(x)$ with continuous derivatives through a given point exists and is unique. (See **A**III.B2,3.)

[3] R. Courant: *Dirichlet's Principle*, Interscience Publishers, New York, 1950, p. 180.

We have seen in the preceding subsection that the problem of uniqueness in the theory of partial differential equations is certainly not as simple as the one in the theory of ordinary differential equations. Unfortunately, the same has to be said about the existence problem; i.e., there exists no theorem of such generality in the theory of partial differential equations as in the theory of ordinary differential equations.

Nevertheless, existence theorems can be established in a number of cases, two of which we will shortly discuss.

Let us consider the system of two partial differential equations of the first order in the two unknown functions u and v[4]

$$\frac{\partial u}{\partial x} = \frac{\partial v}{\partial y},$$
$$\frac{\partial u}{\partial y} = -\frac{\partial v}{\partial x},$$

with the boundary conditions

$$u(0, y) = \frac{a}{a^2 + y^2}, \quad v(0, y) = \frac{y}{a^2 + y^2}.$$

The solutions of this problem are obviously

$$u = \frac{a - x}{(a - x)^2 + y^2}, \quad v = \frac{y}{(a - x)^2 + y^2},$$

as the reader can easily verify. These functions are not continuous at $x = a$, $y = 0$ and since we can choose a as close to the y-axis as we like, we can certainly choose it so that the discontinuity comes arbitrarily close to the y-axis. We can see from this that it cannot be said that the solutions are bounded in a neighborhood of the y-axis.

One can nevertheless give an existence proof for a system of this type which gives at the same time information about the properties of the solutions. Such a proof was first delivered by *Cauchy*, though in an incomplete form, and later completed by *Sonja Kowalewski* (1875).

With Cauchy, we consider the system

$$\frac{\partial u}{\partial x} = \phi\left(x, y, u, v, \frac{\partial u}{\partial y}, \frac{\partial v}{\partial y}\right),$$
$$\frac{\partial v}{\partial x} = \psi\left(x, y, u, v, \frac{\partial u}{\partial y}, \frac{\partial v}{\partial y}\right)$$

where the values of the solution on the y-axis $u(0, y)$, $v(0, y)$ shall be given.

[4] The reader will certainly recognize these equations as the Cauchy-Riemann equations, or, if he prefers, the integrability conditions for the velocity potential and the flux integral, and he will also know that the solutions of these equations satisfy the potential equation in two dimensions if they have continuous second-order derivatives.

Cauchy and Kowalewski[5] proved that if one can expand the functions ϕ, ψ into series with terms of the type

$$\alpha(x - x_0)^i(y - y_0)^k(u - u_0)^l(v - v_0)^m\left[\frac{\partial u}{\partial y} - \left(\frac{\partial u}{\partial y}\right)_0\right]^n\left[\frac{\partial v}{\partial y} - \left(\frac{\partial v}{\partial y}\right)_0\right]^p$$

and if the boundary functions are given along the y-axis by series of the type

$$u(0, y) = \sum_{k=0}^{\infty} a_k(y - y_0)^k,$$

$$v(0, y) = \sum_{k=0}^{\infty} b_k(y - y_0)^k,$$

then the solutions have the form

$$u = \sum_{n,m} a_{n,m}(x - x_0)^n(y - y_0)^m, \quad v = \sum_{n,m} b_{n,m}(x - x_0)^n(y - y_0)^m$$

and these series converge in a certain region.

We consider as another example the following partial differential equation of the second order

$$\frac{\partial^2 z}{\partial x\, \partial y} + a(x, y)\frac{\partial z}{\partial x} + b(x, y)\frac{\partial z}{\partial y} + c(x, y)z = 0$$

where the functions a, b, c shall be continuous in the rectangle

$$R: \quad \begin{cases} 0 \le x \le A, \\ 0 \le y \le B. \end{cases}$$

We impose the boundary conditions

$$z(x, 0) = \phi(x), \quad z(0, y) = \psi(y)$$

in $0 \le x \le A$ and $0 \le y \le B$ respectively, and assume that ϕ and ψ are continuous and have continuous derivatives and are such that $\phi(0) = \psi(0)$.

One can show by a method of successive approximations that the solution $z(x, y)$ exists at every point (x, y) in R and is uniquely determined by the boundary conditions.

We can see from this short discussion that the existence problem in the theory of partial differential equations does not seem to be as simple as the one in the theory of ordinary differential equations. For this reason, and since the equations we deal with in this treatment are of a very special type anyway, we will refrain from going into this problem at all. Furthermore, we give explicit solutions for the problems we are interested in and thus *a fortiori* prove the existence of solutions by actually finding the solutions themselves.

[5] See I. G. Petrovski: *Partial Differential Equations*, Interscience Publishers, New York, 1954, p. 14.

Problems III.4–5

4. Show that the solutions of $\frac{\partial u}{\partial x} = \frac{\partial v}{\partial y}$, $\frac{\partial v}{\partial x} = -\frac{\partial u}{\partial y}$ satisfy the potential equation $\Delta u = 0$, $\Delta v = 0$ if their second-order derivatives are continuous.

5. Show that $u = \frac{a - x}{(a - x)^2 + y^2}$, $v = \frac{y}{(a - x)^2 + y^2}$ are solutions of $\frac{\partial u}{\partial x} = \frac{\partial v}{\partial y}$, $\frac{\partial u}{\partial y} = -\frac{\partial v}{\partial x}$ and satisfy the boundary conditions $u(0, y) = \frac{a}{a^2 + y^2}$, $v(0, y) = \frac{y}{a^2 + y^2}$.

§2. INFINITE SERIES AS SOLUTIONS OF LINEAR HOMOGENEOUS PARTIAL DIFFERENTIAL EQUATIONS WITH HOMOGENEOUS BOUNDARY OR INITIAL CONDITIONS

1. Bernoulli's Separation Method

The procedure by which we are going to solve the problems we formulated in Chapter II has been indicated in Chapter II, §1.2, §2.4, and §3.4. We are not going to establish—as one would expect in analogy to the process usually followed in solving ordinary differential equations—the general solution of the given partial differential equation and then try to satisfy the given boundary or initial conditions. We rather try to find infinitely many particular solutions of the given equation which satisfy all but one or two of the given boundary or initial conditions and then attempt to satisfy the remaining conditions by proper choice of linear combinations of the infinitely many particular solutions.

We have seen that the bridge we have to cross in order to obtain particular solutions was the successful application of Bernoulli's separation method. This method worked all right in all the cases we considered, but this does not mean, of course, that it will always work. To clarify this situation, we will consider the most general form of a linear homogeneous partial differential equation of the second order in two independent variables:

$$a_{11}\frac{\partial^2 u}{\partial \xi^2} + 2a_{12}\frac{\partial^2 u}{\partial \xi\, \partial \eta} + a_{22}\frac{\partial^2 u}{\partial \eta^2} + a_{10}\frac{\partial u}{\partial \xi} + a_{01}\frac{\partial u}{\partial \eta} + a_{00}u = 0 \quad \text{(III.1)}$$

where the a_{ik} are functions of ξ and η.

We can easily prove the following:

Theorem III.1. *The partial differential equation (III.1) can be reduced to two ordinary differential equations of second order both containing an* *arbitrary parameter λ by Bernoulli's separation method if there exists a transformation*

$$\left.\begin{aligned} x &= x(\xi, \eta) \\ y &= y(\xi, \eta) \end{aligned}\right\}, \qquad \frac{\partial(x, y)}{\partial(\xi, \eta)} \neq 0, \tag{III.2}$$

so that the resulting equation

$$A_{11}\frac{\partial^2 u}{\partial x^2} + A_{22}\frac{\partial^2 u}{\partial y^2} + A_{10}\frac{\partial u}{\partial x} + A_{01}\frac{\partial u}{\partial y} + A_{00}u = 0 \tag{III.3}$$

does not contain a term of the type $\dfrac{\partial^2 u}{\partial x\, \partial y}$ *and if there exists a function* $B(x, y)$ *so that* $A_{11}/B = C_{11}$, $A_{10}/B = C_{10}$ *are functions of* x *only and* $A_{22}/B = C_{22}$, $A_{01}/B = C_{01}$ *are functions of* y *only, and if* A_{00}/B *can be split up into a function of* x *and a function of* y *as* $A_{00}/B = C_{00}^{(1)}(x) + C_{00}^{(2)}(y)$.

Proof. Let us substitute

$$u = X(x)\,Y(y)$$

into (III.3):

$$A_{11}X''Y + A_{22}XY'' + A_{10}X'Y + A_{01}XY' + A_{00}XY = 0.$$

If there exists a function $B(x, y)$ with the properties stated above, we divide the whole equation by $B \cdot X \cdot Y$ and obtain after shifting all terms which contain y to the right side

$$C_{11}(x)\frac{X''}{X} + C_{10}(x)\frac{X'}{X} + C_{00}^{(1)}(x) = -\left[C_{22}(y)\frac{Y''}{Y} + C_{01}(y)\frac{Y'}{Y} + C_{00}^{(2)}(y)\right].$$

The left side of this equation is a function of x only and the right side is a function of y only, and this is possible only if both sides equal a constant λ which gives the two ordinary differential equations of the second order

$$\begin{aligned} C_{11}(x)X'' + C_{10}(x)X' + C_{00}^{(1)}(x)X - \lambda X &= 0, \\ C_{22}(y)Y'' + C_{01}(y)Y' + C_{00}^{(2)}(y)Y + \lambda Y &= 0, \end{aligned} \tag{III.4}$$

q.e.d.

Let us illustrate this simple theorem by considering the situation as it presents itself in the case of the stable state of a circular, initially axially symmetrically displaced membrane under no external forces except tension. In this case, equation (II.45) will reduce to

$$\frac{\partial^2 u}{\partial r^2} + \frac{1}{r}\frac{\partial u}{\partial r} + \frac{1}{r^2}\frac{\partial^2 u}{\partial \theta^2} = 0$$

with $A_{11} = 1$, $A_{10} = 1/r$, $A_{22} = 1/r^2$.

This equation does not contain a term with $\partial^2 u/\partial r \partial \theta$ and we see immediately that we have to take $B = 1/r^2$ in order to obtain $C_{11} = r^2$, $C_{10} = r$, $C_{22} = 1$.

The condition that there exists a transformation (III.2) which causes the vanishing of the term $\partial^2 u/\partial x \partial y$ is always satisfied in the case of two independent variables inasmuch as such a transformation always exists, although it is not necessarily real.

One can show[6] that (III.1) can be transformed into one of the following *canonical forms*:

$$\frac{\partial^2 u}{\partial x^2} - \frac{\partial^2 u}{\partial y^2} + \text{first and zeroth order derivatives} = 0$$

in the case where $a_{12}^2 - a_{11}a_{22} > 0$. One calls such an equation *hyperbolic*.

$$\frac{\partial^2 u}{\partial x^2} + \text{first and zeroth order derivatives} = 0$$

in the case where $a_{12}^2 - a_{11}a_{22} = 0$, which is called a *parabolic* equation.

$$\frac{\partial^2 u}{\partial x^2} + \frac{\partial^2 u}{\partial y^2} + \text{first and zeroth order derivatives} = 0$$

in the case where $a_{12}^2 - a_{11}a_{22} < 0$, which one calls an *elliptic* equation.

In the case where the coefficients of (III.1) are functions of ξ and η, an equation may belong to different classes in different regions depending on the value of the discriminant.

In the light of this classification, the string equation (II.3) appears to be an equation of hyperbolic type, the equation of heat conduction (II.79) for the special case where no energy exchange takes place in the y and the z direction appears to be of parabolic type, and the equation for the stable state of the stretched membrane and the potential equation (II.83) in two dimensions are of elliptic type.

In the case of three and more variables, to which theorem III.1 can easily be generalized, the condition that a transformation which removes the terms containing the mixed derivative has to exist is not trivial, since it will be in general impossible to find a transformation which would reduce the given equation to a canonical form in any small neighborhood of a point $(x_1, x_2, x_3, \cdots)$.[7] Therefore, the successful application of

[6] See I. G. Petrovsky: *Lectures on Partial Differential Equations*, Interscience Publishers, New York, 1954, p. 45.

[7] See I. G. Petrovsky: *Lectures on Partial Differential Equations*, Interscience Publishers, New York, 1954, p. 43.

Bernoulli's method to equations in more than two variables will mainly depend on whether a term of the form $\partial^2 u/\partial x_i\, \partial x_k$ appears in the original form of the equation or not.

Problems III.6–10

6. State and prove theorem III.1 for the case of three independent variables.
7. Separate the variables in

$$\frac{\partial^2 u}{\partial x^2} + \frac{\partial^2 u}{\partial x\, \partial y} + \frac{\partial u}{\partial x} = 0.$$

This shows that the removal of the term $\partial^2 u/\partial x\, \partial y$ is not necessary.
8. Try to separate the variables in

$$a_{11}\frac{\partial^2 u}{\partial x^2} + 2a_{12}\frac{\partial^2 u}{\partial x\, \partial y} + a_{22}\frac{\partial^2 u}{\partial y^2} + a_{10}\frac{\partial u}{\partial x} + a_{01}\frac{\partial u}{\partial y} + a_{00}u = 0,$$

where all the $a_{ik} \neq 0$ without removing the term $\partial^2 u/\partial x\, \partial y$. (The a_{ik} are constants.)
9. Define $B(r, \phi)$ in the case of equation (II.98).
***10.** Find a transformation which reduces $\dfrac{\partial^2 u}{\partial \xi\, \partial \eta} = 0$ to canonical form. (*Hint:* Assume a general transformation of the type (III.2), transform the equation and try to satisfy the vanishing condition for the coefficient of $\dfrac{\partial^2 u}{\partial x\, \partial y}$ in the transformed equation.)

2. Solution of the Homogeneous Boundary Value Problem by Infinite Series

We reduced equation (III.1) in the preceding subsection to the two ordinary differential equations of the second order (III.4).

Both equations contain the arbitrary parameter λ and therefore their solutions—if there exist any—will also contain λ. By multiplying the solutions $X_\lambda(x)$ of the one equation with the solutions $Y_\lambda(y)$ of the other equation, we obtain infinitely many solutions of equation (III.1) or (III.3)

$$u_\lambda(x, y) = X_\lambda(x)\, Y_\lambda(y).$$

The course we took in Chapter II was that we tried to satisfy most of the given homogeneous boundary and initial conditions by these particular solutions by proper choice of the integration constants in $X(x)$ and $Y(y)$ and values $\lambda_1, \lambda_2, \lambda_3, \cdots$ for λ and hoped that it was possible to choose infinitely many constants A_i so that

$$\sum_{i=1}^{\infty} A_i u_{\lambda_i}(x, y) \tag{III.5}$$

satisfied the remaining nonhomogeneous condition(s). (The notation we use here suggests that λ may only assume discrete values as was the case with all the examples we considered in Chapter II. However, λ might very well be a continuous variable, in which case (III.5) would have the form $\int A(\lambda)u_\lambda(x, y)\,d\lambda$, where $A(\lambda)$ is a function which has to be chosen so that the remaining condition(s) will be satisfied.)

Most of the boundary and initial conditions of the homogeneous type which we have encountered this far have been of the form

$$u(x(s), y(s)) = 0, \quad \frac{\partial u(x(s), y(s))}{\partial x} = 0, \quad \frac{\partial u(x(s), y(s))}{\partial n} = 0.$$

In order that we may deal with these different types of conditions at once, we write them in the general form

$$\alpha u(x(s), y(s)) + \beta \frac{\partial u(x(s), y(s))}{\partial \xi} = 0 \tag{III.6}$$

where α and β are constants which are not both zero at the same time and where the derivative with respect to ξ denotes any directional derivative. Note that $\partial u/\partial \xi$ can be expressed as a linear combination of the derivatives of u with respect to x and with respect to y (see **AI.B8**).

We are now ready to give a satisfactory answer to a problem which we formulated at the end of Chapter II, §1.4; i.e.: Under what conditions is (III.5) a solution of (III.3) if the $u_{\lambda_i}(x, y)$ are solutions of (III.3) and satisfy the homogeneous boundary condition (III.6)?

Again, we restrict our considerations to the problem in two independent variables, as we have already formulated it; however, the reader can easily see that the number of variables will make no difference whatever. The only essential condition is that the equation itself and the boundary condition have to be linear and homogeneous.

Theorem III.2. (superposition principle). *If $u_{\lambda_i}(x, y)$ are solutions of (III.3), if they satisfy the boundary condition (III.6), and if*

$$\sum_{i=1}^{\infty} A_i \frac{\partial^2 u_{\lambda_i}}{\partial x^2}, \quad \sum_{i=1}^{\infty} A_i \frac{\partial^2 u_{\lambda_i}}{\partial y^2}, \quad \sum_{i=1}^{\infty} A_i \frac{\partial u_{\lambda_i}}{\partial x}, \quad \sum_{i=1}^{\infty} A_i \frac{\partial u_{\lambda_i}}{\partial y}$$

converge uniformly and $\sum_{i=1}^{\infty} A_i u_{\lambda_i}$ converges in a region R and on its boundary C, which is given by $x = x(s)$, $y = y(s)$ (the A_i are constants), then

$$u(x, y) = \sum_{i=1}^{\infty} A_i u_{\lambda_i}(x, y)$$

is a solution of (III.3) in R which satisfies the boundary condition (III.6) on C.

Remark: The conditions stated in this theorem are sufficient but by no means necessary, as we will see in §3 of Chapter IV.

Proof. According to our assumptions, $u_{\lambda_i}(x, y)$ is a solution of (III.3) and satisfies (III.6) for all integers i:

$$A_{11}\frac{\partial^2 u_{\lambda_i}}{\partial x^2} + A_{22}\frac{\partial^2 u_{\lambda_i}}{\partial y^2} + A_{10}\frac{\partial u_{\lambda_i}}{\partial x} + A_{01}\frac{\partial u_{\lambda_i}}{\partial y} + A_{00}u_{\lambda_i} = 0, \quad \text{(III.7)}$$

$$\alpha u_{\lambda_i}(x(s), y(s)) + \beta\frac{\partial u_{\lambda_i}(x(s), y(s))}{\partial \xi} = 0. \quad \text{(III.8)}$$

If we sum up the left sides of these equations after multiplying them by A_i we obtain

$$A_{11}\sum_{i=1}^{\infty} A_i\frac{\partial^2 u_{\lambda_i}}{\partial x^2} + A_{22}\sum_{i=1}^{\infty} A_i\frac{\partial^2 u_{\lambda_i}}{\partial y^2} + A_{10}\sum_{i=1}^{\infty} A_i\frac{\partial u_{\lambda_i}}{\partial x}$$

$$+ A_{01}\sum_{i=1}^{\infty} A_i\frac{\partial u_{\lambda_i}}{\partial y} + A_{00}\sum_{i=1}^{\infty} A_i u_{\lambda_i} = 0,$$

and

$$\alpha\sum_{i=1}^{\infty} A_i u_{\lambda_i}(x(s), y(s)) + \beta\sum_{i=1}^{\infty} A_i\frac{\partial u(x(s), y(s))}{\partial \xi} = 0.$$

Since

$$\frac{\partial u_{\lambda_i}}{\partial \xi} = \frac{\partial u_{\lambda_i}}{\partial x}\cdot\frac{dx}{d\xi} + \frac{\partial u_{\lambda_i}}{\partial y}\cdot\frac{dy}{d\xi}$$

(see **AI.B8**) and since $\sum_{i=1}^{\infty} A_i u_{\lambda_i}$ converges while the other sums involved converge uniformly, we can interchange summation and differentiation (see **AII.C5**) and obtain

$$A_{11}\frac{\partial^2}{\partial x^2}\sum_{i=1}^{\infty} A_i u_{\lambda_i} + A_{22}\frac{\partial^2}{\partial y^2}\sum_{i=1}^{\infty} A_i u_{\lambda_i} + A_{10}\frac{\partial}{\partial x}\sum_{i=1}^{\infty} A_i u_{\lambda_i}$$

$$+ A_{01}\frac{\partial}{\partial y}\sum_{i=1}^{\infty} A_i u_{\lambda_i} + A_{00}\sum_{i=1}^{\infty} A_i u_{\lambda_i} = 0$$

and

$$\alpha\sum_{i=1}^{\infty} A_i u_{\lambda_i}(x(s), y(s)) + \beta\frac{\partial}{\partial \xi}\sum_{i=1}^{\infty} A_i u_{\lambda_i}(x(s), y(s)) = 0;$$

i.e., $\sum_{i=1}^{\infty} A_i u_{\lambda_i}$ is a solution of (III.3) and satisfies the condition (III.6), q.e.d.

This leaves unanswered the problem of how to choose the A_i in order to satisfy the remaining, nonhomogeneous, boundary or initial condition(s). The solution to this problem will be found in Chapters IV, V, and VI.

Problem III.11

11. State and prove the superposition principle for the case of n independent variables.

RECOMMENDED SUPPLEMENTARY READING

I. G. Petrovsky: *Lectures on Partial Differential Equations*, Translated from the Russian by A. Shenitzer, Interscience Publishers, New York, 1954.

R. Courant: *Dirichlet's Principle*, Interscience Publishers, New York, 1950.

G. A. Bliss: *Calculus of Variations*, The Open Court Publishing Company, Chicago, 1925.

IV

FOURIER SERIES

§1. THE FORMAL APPROACH

1. Introductory Remarks

We tried in Chapter II, §1.2 to solve the problem of the initially displaced stretched string. We found out that a function which seems to satisfy the differential equation (II.6), the boundary conditions (II.7), and the second initial condition in (II.8) has the form

$$\sum_{k=1}^{\infty} b_k \sin kx \cos kt.$$

We know from our discussion in the preceding chapter that this is a solution if it satisfies the convergence conditions as stated in theorem III.2 (superposition principle), and this will depend entirely on the form of the coefficients b_k. As we have seen further the b_k have to be determined so that the first initial condition in (II.8)

$$u(x, 0) = \phi(x)$$

is satisfied where $\phi(x)$ is the displacement of the string at the distance x from the origin. This amounts to the problem of finding the b_k so that

$$\phi(x) = \sum_{k=1}^{\infty} b_k \sin kx.$$

If we do this and it turns out that the b_k permit $u(x, t)$ to satisfy the convergence conditions of theorem III.2, then we will have solved our problem and we will know moreover that according to theorem II.1 the only possible solution has been produced.

The fact that we obtained a series of sine functions only for the representation of $\phi(x)$ is purely accidental. If we had chosen our coordinate system differently, we might have obtained a series in cosine functions only or even a series in both sine and cosine functions.

The problem we will have to face in general will be the following one: Given a function $f(x)$ in a certain interval which we will assume for matters of convenience as $-\pi \leq x \leq \pi$. Is it possible to represent this function within this interval by a series of the type

$$\sum_{k=0}^{\infty} (a_k \cos kx + b_k \sin kx)$$

and if so, what form do the coefficients a_k and b_k have? Further: What convergence properties does this series have and how do these properties depend on the behavior of the function $f(x)$? (It seems obvious that there is a connection between $f(x)$ and the convergence of the series which is supposed to represent it.)

We call this the problem of expanding a function into a *Fourier series* in honor of the French mathematician *Jacques Fourier* who developed and studied this theory to a considerable extent in his classical work *La théorie analytique de la chaleur*[1] (Paris, 1822). However, such series had been used previously by *Daniel Bernoulli* in solving the string problem as we will point out in §3.3 of this chapter.

Before we go deeper into this problem in the form we just stated it, let us recall a remark made on page 40 about the approximation of a function $f(x)$ by a finite trigonometric polynomial to which our problem is automatically reduced the moment we want numerical results. We will deal with this finite problem in the following subsection.

2. Approximation of a Given Function by a Trigonometric Polynomial

Given the quadratically integrable function $f(x)$ on the interval $-\pi \leq x \leq \pi$. We will try to choose $2n + 1$ coefficients a_k, b_k such that the trigonometric polynomial

$$\frac{1}{2} a_0 + \sum_{k=1}^{n} (a_k \cos kx + b_k \sin kx) \tag{IV.1}$$

is the best approximation to $f(x)$ in the sense that the integral of the square of the error over the whole interval shall be a minimum.[2] That is, we have to determine the coefficients a_m, b_m so that

$$\Phi(a_0, a_1, \cdots, a_n, b_1, b_2, \cdots, b_n) = \int_{-\pi}^{\pi} \left[f(x) - \frac{1}{2} a_0 - \sum_{k=1}^{n} (a_k \cos kx + b_k \sin kx) \right]^2 dx \tag{IV.2}$$

[1] Analytic theory of heat.

[2] The reader will easily recognize this as a generalization of Gauss's least square method (see problems IV.1 and 2).

yields a minimum. This is an ordinary extreme value problem for a function of the $2n + 1$ variables a_k, b_k. A necessary condition for the minimum is that the partial derivatives of Φ with respect to the a_m, b_m vanish:

$$\frac{\partial \Phi}{\partial a_0} = -\int_{-\pi}^{\pi} \left[f(x) - \frac{1}{2}a_0 - \sum_{k=1}^{n}(a_k \cos kx + b_k \sin kx)\right] dx = 0, \quad \text{(IV.3)}$$

$$\frac{\partial \Phi}{\partial a_m} = -2\int_{-\pi}^{\pi} \left[f(x) - \frac{1}{2}a_0 - \sum_{k=1}^{n}(a_k \cos kx + b_k \sin kx)\right] \cos mx \, dx = 0, \quad \text{(IV.4)}$$

$$\frac{\partial \Phi}{\partial b_m} = -2\int_{-\pi}^{\pi} \left[f(x) - \frac{1}{2}a_0 - \sum_{k=1}^{n}(a_k \cos kx + b_k \sin kx)\right] \sin mx \, dx = 0. \quad \text{(IV.5)}$$

We can solve equation (IV.3) immediately for a_0, since the integrals of sine and cosine over the full period vanish and we obtain

$$a_0 = \frac{1}{\pi}\int_{-\pi}^{\pi} f(x)\, dx. \quad \text{(IV.6)}$$

We see from this that the absolute term $\frac{a_0}{2} = \frac{1}{2\pi}\int_{-\pi}^{\pi} f(x)\, dx$ is the mean value of $f(x)$ on $-\pi \leq x \leq \pi$. As to the equations (IV.4) and (IV.5), we have to observe the so-called *orthogonality relations* of the trigonometric functions which will be discussed in the next section.

Problems IV.1–2

1. Given a set of points (x_i, y_i) $(i = 1, 2, 3, \cdots, n)$ where $x_i \neq x_k$ for $i \neq k$. Gauss's method of least squares in its simplest application consists of determining the coefficients of $y = ax + b$ such that $\sum_{k=1}^{n} [y(x_k) - y_k]^2$ yields a minimum. Find the conditions on a and b.

***2.** Given a set of points as in problem 1 with $x_i = 2\pi i/n$. Find the conditions on a, b, c such that $y = a \cos x + b \sin x + c$ approximates the given points in the sense of Gauss's least square principle.

3. The Trigonometric Functions as an Orthogonal System

Let us consider all vectors [**A**] in an n-dimensional Euclidean space E_n. One calls [**A**] a *linear space* or a *vector space* because with any two elements **a**, **b** in [**A**] $\alpha\mathbf{a} + \beta\mathbf{b}$ is also in [**A**] where α, β are any real numbers. If **a** has the components $(a_1, a_2, \cdots, a_n)$ and **b** has components $(b_1, b_2, \cdots, b_n)$, one calls the expression

$$(\mathbf{a} \cdot \mathbf{b}) = a_1 b_1 + a_2 b_2 + \cdots + a_n b_n$$

the *dot product* or *inner product* of **a** and **b** and

$$|\mathbf{a}| = \sqrt{\mathbf{a} \cdot \mathbf{a}}$$

the *norm* of **a**. Clearly, the norm as defined here is the Euclidean length of the vector **a**. With the definition of a norm, a metric is introduced in our vector space and one calls such a space a *normed linear space* or a *normed vector space*. If for any two vectors **a**, **b** which are both different from the zero vector

$$(\mathbf{a} \cdot \mathbf{b}) = 0$$

one calls **a**, **b** *orthogonal*. If it is true for any two vectors from a set $\{\mathbf{a}_j\}$ that

$$(\mathbf{a}_i \cdot \mathbf{a}_k) = 0 \quad \text{for } i \neq k, \tag{IV.7}$$

one calls $\{\mathbf{a}_j\}$ an *orthogonal system* of vectors, and if in addition to (IV.7)

$$(\mathbf{a}_i \cdot \mathbf{a}_i) = 1 \quad \text{or} \quad |\mathbf{a}_i| = 1 \quad \text{for all } i \tag{IV.8}$$

holds, then $\{\mathbf{a}_j\}$ is called an *orthonormal system* of vectors. For example, the vectors $\mathbf{e}_1 = (1, 0, 0, \cdots, 0)$, $\mathbf{e}_2 = (0, 1, 0, \cdots, 0)$, $\cdots$, $\mathbf{e}_n = (0, 0, 0, \cdots, 0, 1)$ constitute an orthonormal system of vectors in E_n. As a matter of fact, the maximum number of elements in an orthonormal system of vectors in E_n is n. (See problem IV.3.)

This discussion merely serves the purpose of explaining the terminology in the subsequent treatment in terms of elementary geometric concepts. The reader who encounters difficulties in following this discussion in this abstract form will find it helpful to consider the case with $n = 2$ or $n = 3$, where the dot product $(\mathbf{a} \cdot \mathbf{b}) = |\mathbf{a}||\mathbf{b}| \cos(\mathbf{a}, \mathbf{b})$ has a simple geometric interpretation and where $(\mathbf{a} \cdot \mathbf{b}) = 0$ for $\mathbf{a} \neq 0$ and $\mathbf{b} \neq 0$ means that **a** is perpendicular to **b**.

Before generalizing the ideas as outlined above to even more general spaces, we wish to say a few more words about the dot product. Among other properties the dot product of two vectors satisfies the *Cauchy-Schwarz inequality*

$$|(\mathbf{a} \cdot \mathbf{b})| \leq |\mathbf{a}||\mathbf{b}|$$

which is quite obvious in the spaces E_2 and E_3 and can easily be proved for $n > 3$ (see problem IV.4).

The norm, as defined in terms of the dot product has the properties of a Euclidean length, which are

$$
\begin{aligned}
&|\mathbf{a}| > 0 \quad \text{for all } \mathbf{a} \neq 0 \quad \text{and} \quad |\mathbf{a}| = 0 \quad \text{if and only if } \mathbf{a} = \mathbf{0}, \\
&|\mathbf{a} + \mathbf{b}| \leq |\mathbf{a}| + |\mathbf{b}| \quad (\textit{triangle inequality}), \text{ and} \qquad \text{(IV.9)} \\
&|\beta \mathbf{a}| = |\beta||\mathbf{a}| \quad \text{for any real } \beta.
\end{aligned}
$$

We are now ready to generalize the introduced concepts to function spaces. Let us consider, for example, the set of all functions which are continuous on the interval $a \leq x \leq b$. Clearly, this set is a vector space, because any linear combination of any two elements in this space is again a function which is continuous on $a \leq x \leq b$. In order to make it a normed vector space, we introduce a metric which can be done via a sort of dot product as follows: We define the dot product of two functions f_1 and f_2 in our space by

$$(f_1 \cdot f_2) = \int_a^b f_1(x) f_2(x)\, dx.$$

Consequently, we define the norm of f as

$$|f| = \sqrt{(f \cdot f)}.$$

This norm satisfies all the conditions (IV.9) (see problem IV.6). The dot product as defined above satisfies the Cauchy-Schwarz inequality

$$(f_1 \cdot f_2) \leq |f_1||f_2|$$

as one can easily show (see problem IV.5).

Now, in analogy to (IV.7), we call a system of functions $\{f_j\}$ *orthogonal* on an interval $a \leq x \leq b$ if for any two functions f_i, f_k in $\{f_j\}$

$$(f_i \cdot f_k) = \int_a^b f_i(x) f_k(x)\, dx = 0 \qquad i \neq k \tag{IV.7a}$$

holds. If in addition to (IV.7a) in analogy to (IV.8)

$$|f_i|^2 = (f_i \cdot f_i) = \int_a^b f_i^2(x)\, dx = 1 \quad \text{for all } i \tag{IV.8a}$$

holds, we call $\{f_j\}$ an *orthonormal system* of functions on $a \leq x \leq b$.

If we consider now the system of functions which is of paramount interest to us, namely $\left\{\frac{1}{\sqrt{p}} \sin \frac{k\pi x}{p}, \frac{1}{\sqrt{p}} \cos \frac{k\pi x}{p}\right\}$, we see that it is orthonormal within any interval of length $2p$ in the sense of (IV.7a) and (IV.8a), since

$$\frac{1}{p}\int_{a-p}^{a+p} \sin\frac{n\pi x}{p}\sin\frac{m\pi x}{p}\,dx = \delta_{mn}, \tag{IV.10}$$

$$\frac{1}{p}\int_{a-p}^{a+p} \cos\frac{n\pi x}{p}\cos\frac{m\pi x}{p}\,dx = \delta_{mn}, \tag{IV.11}$$

$$\frac{1}{p}\int_{a-p}^{a+p} \sin\frac{n\pi x}{p}\cos\frac{m\pi x}{p}\,dx = 0 \tag{IV.12}$$

where a is any real number and $\delta_{nm} = \begin{Bmatrix} 1 \text{ if } n = m \\ 0 \text{ if } n \neq m \end{Bmatrix}$.

These relations can easily be verified by integrating the above expressions between $-p$ and p and making use of

Lemma IV.1. *If* $\Phi(x + 2p) = \Phi(x)$, *then* $\displaystyle\int_{a-p}^{a+p} \Phi(x)\,dx = \int_{-p}^{p} \Phi(x)\,dx.$

Proof. Let us first consider $\displaystyle\int_{\alpha}^{\beta} \Phi(x)\,dx$ and make the substitution $x + 2p = \xi$. Then we obtain in view of the $2p$-periodicity of $\Phi(x)$

$$\int_{\alpha}^{\beta} \Phi(x)\,dx = \int_{\alpha+2p}^{\beta+2p} \Phi(\xi)\,d\xi. \tag{IV.13}$$

Now, let $\alpha = a - p$ and $\beta = a + p$, then

$$\begin{aligned}\int_{a-p}^{a+p} \Phi(x)\,dx &= \int_{a-p}^{-p} \Phi(x)\,dx + \int_{-p}^{a+p} \Phi(x)\,dx \\ &= \int_{a+p}^{p} \Phi(x)\,dx + \int_{-p}^{a+p} \Phi(x)\,dx = \int_{-p}^{p} \Phi(x)\,dx\end{aligned}$$

since $\displaystyle\int_{a-p}^{-p} \Phi(x)\,dx = \int_{a+p}^{p} \Phi(x)\,dx$ in view of (IV.13), q.e.d.

Problems IV.3–10

3. Prove that n is the greatest number of elements in an orthogonal system of vectors in E_n. (*Hint:* Any $n + 1$ vectors in E_n are linearly dependent; i.e., there exists a linear composition equal to zero such that the coefficients are not all zero.)

4. Prove the Cauchy-Schwarz inequality for vectors

$$|(\mathbf{a}\cdot\mathbf{b})| \le |\mathbf{a}|\,|\mathbf{b}|.$$

(*Hint:* Consider the identity

$$\sum_{i=1}^{n} a_i^2 \sum_{i=1}^{n} b_i^2 - \left(\sum_{i=1}^{n} a_i b_i\right)^2 = \frac{1}{2}\sum_{k=1}^{n}\sum_{i=1}^{n}(a_i b_k - a_k b_i)^2.\Big)$$

5. Prove the Cauchy-Schwarz inequality for functions

$$(f \cdot g) \leq |f|\,|g|.$$

(*Hint:* Consider the identity

$$(f \cdot g)^2 = (f \cdot f)(g \cdot g) - \frac{1}{2}\int_a^b \int_a^b [f(x)g(\xi) - f(\xi)g(x)]^2 \, dx \, d\xi.\Big)$$

***6.** Show that $|f| = \sqrt{(f \cdot f)}$ satisfies the conditions (IV.9).

7. Show that $\int_{-\pi}^{\pi} \sin mx \cos nx \, dx = 0.$

8. Show that $\int_{-\pi}^{\pi} \sin^2 nx \, dx = \int_{-\pi}^{\pi} \cos^2 nx \, dx = \pi.$

9. Show that $\int_{-\pi}^{\pi} \sin nx \sin mx \, dx = 0, \int_{-\pi}^{\pi} \cos nx \cos mx \, dx = 0, n \neq m.$

10. Prove the orthogonality relations (IV.10), (IV.11), and (IV.12) by using the results of problems 7, 8, and 9.

4. Approximations by Trigonometric Polynomials, Continued

We are now in a position to solve equations (IV.4) and (IV.5) using the orthogonality relations of the trigonometric functions (IV.10), (IV.11), and (IV.12) for $p = \pi$, and we obtain

$$a_k = \frac{1}{\pi}\int_{-\pi}^{\pi} f(x) \cos kx \, dx, \tag{IV.14}$$

$$b_k = \frac{1}{\pi}\int_{-\pi}^{\pi} f(x) \sin kx \, dx. \tag{IV.15}$$

It is clear now why we have chosen to write the constant term in the form $a_0/2$, which might have appeared a little odd in the beginning, because (IV.6) appears now as a special case of (IV.14) for $k = 0$.

The constants a_k, b_k defined in (IV.14) and (IV.15) are called the *Fourier Coefficients* of $f(x)$. If we proceed now to the more general problem of trying to represent a function $f(x)$ by an infinite trigonometric series, we may proceed along the following lines:

We let n approach infinity in (IV.1) with coefficients determined according to (IV.14) and (IV.15) and investigate whether the series thus obtained converges and if so, whether it converges to $f(x)$. We will see in §2 of this chapter that we get an affirmative answer to this question under rather weak conditions on $f(x)$.

Another possibility would be—and we shall pursue this one in the next section—to assume that $f(x)$ can be represented by a series of this type and investigate from a formal standpoint what form the coefficients must have if such an expansion is permissible. We will discover that we obtain the same coefficients as arrived at for the finite trigonometric polynomial just discussed. The question of convergence remains, of course, open in performing this formal procedure. These theoretical questions will be dealt with in §2 of this chapter.

5. Formal Expansion of a Function into a Fourier Series

We assume that a function with the period 2π can be represented by a series of the form

$$f(x) = \frac{a_0}{2} + \sum_{k=1}^{\infty} (a_k \cos kx + b_k \sin kx). \tag{IV.16}$$

If this is considered to be true, then what form do the coefficients a_k, b_k have to have?

We assume that termwise integration in (IV.16) is permissible (see AII.C4) and find a_0 by integrating (IV.16) with respect to x over any interval of the length 2π

$$a_0 = \frac{1}{\pi} \int_{a-\pi}^{a+\pi} f(x)\, dx,$$

which is identical with (IV.6).

In order to obtain the remaining coefficients, we multiply (IV.16) step by step by all the $\cos nx$ ($n = 1, 2, 3, \cdots$) and integrate after each step over a period of the length 2π, again assuming that termwise integration is permissible. In view of the orthogonality relations (IV.10, 11, and 12), this will furnish

$$a_k = \frac{1}{\pi} \int_{a-\pi}^{a+\pi} f(x) \cos kx\, dx$$

which is identical with (IV.14). If we do now the same thing but using the sines, there results

$$b_k = \frac{1}{\pi} \int_{a-\pi}^{a+\pi} f(x) \sin kx\, dx$$

which is identical with (IV.15).

Whether (IV.16) with the coefficients (IV.14) and (IV.15) converges to $f(x)$ or not, any finite section of this series will nevertheless be the best approximation of this kind for $f(x)$ in the sense of the least square of the error as we know from subsection 2 of this section. That it does much more than that will become clear in the next section.

6. Formal Rules for the Evaluation of Fourier Coefficients

If the expansion we arrived at in the preceding subsection is of any value at all, it tells us that we can build up reasonably behaving functions $f(x)$ by setting infinitely many elements together. One "half" of these infinitely many elements (the cosines) are even functions and the other "half" (the sines) are odd functions. The following question arises: If $f(x)$ is itself an even function, wouldn't it be possible to build it up by using the even elements alone and if $f(x)$ is odd, by using the odd elements alone? This question appears to be very justified in view of the experience we had with the initially displaced string. The displacement function $f(x)$ can certainly be thought of as extended to the left side in such a way that it appears as an odd function and we indeed obtained a series of sine functions only to represent it. That such a thing is possible—or rather that it follows automatically from the formulas we derived for the coefficients—will be shown in the following rules.

Rule 1. *If $f(x)$ has the period 2π, $f(x + 2\pi) = f(x)$ and is an odd function,*

$$f(x) = -f(-x),$$

then the Fourier coefficients of the even terms all vanish; i.e., $a_k = 0$ for all k.

Proof. According to (IV.14)

$$a_k = \frac{1}{\pi}\int_{-\pi}^{\pi} f(x)\cos kx\,dx = \frac{1}{\pi}\left\{\int_{-\pi}^{0} f(x)\cos kx\,dx + \int_{0}^{\pi} f(x)\cos kx\,dx\right\}.$$

If we substitute $x = -\xi$ in the first term and switch the integration limits, we obtain

$$a_k = \frac{1}{\pi}\left\{-\int_{0}^{\pi} f(\xi)\cos k\xi\,d\xi + \int_{0}^{\pi} f(x)\cos kx\,dx\right\} = 0, \quad \text{q.e.d.}$$

Rule 2. *If $f(x)$ has the period 2π, $f(x + 2\pi) = f(x)$, and is an even function,*

$$f(x) = f(-x),$$

then the coefficients of the odd terms vanish; i.e., $b_k = 0$ for all k.

(For *proof* see problem IV.11.)

There is one more important rule which simplifies the computation of the coefficients considerably, namely,

Rule 3. *If $f(x)$ has the period 2π, $f(x + 2\pi) = f(x)$, and it is either even or odd, then the coefficients can be found by the formula*

$$a_k = \frac{2}{\pi} \int_0^{\pi} f(x) \cos kx \, dx$$

in case $f(x)$ is even and by

$$b_k = \frac{2}{\pi} \int_0^{\pi} f(x) \sin kx \, dx$$

in case $f(x)$ is odd.

Proof. We give the proof for even functions only and leave the proof for odd functions to the reader (see problem IV.12).

According to (IV.14)

$$a_k = \frac{1}{\pi} \left\{ \int_{-\pi}^{0} f(x) \cos kx \, dx + \int_0^{\pi} f(x) \cos kx \, dx \right\}.$$

With the substitution $x = -\xi$ in the first term and the observation that $f(-\xi) = f(\xi)$ we obtain

$$a_k = \frac{1}{\pi} \left\{ \int_0^{\pi} f(\xi) \cos k\xi \, d\xi + \int_0^{\pi} f(x) \cos kx \, dx \right\}, \quad \text{q.e.d.}$$

There are a great number of similar rules simplifying the computation of the Fourier coefficients for certain types of functions, which we are not going to discuss here. There is no doubt that such rules are of great advantage to people who do nothing but compute Fourier coefficients all day long; a more gainfully employed individual, however, would prefer to evaluate them the long way, if the occasion arises.

Problems IV.11–18

11. Give a proof for rule 2.
12. Give a proof for rule 3 in case $f(x)$ is odd.
13. Approximate the function

$$f(x) = \begin{Bmatrix} 1 & \text{for } -\frac{\pi}{2} \leq x < \frac{\pi}{2} \\ -1 & \text{for } \frac{\pi}{2} \leq x < \frac{3\pi}{2} \end{Bmatrix}, \quad f(x + 2\pi) = f(x)$$

by a trigonometric polynomial consisting of five terms.
14. Evaluate the error in problem 13 at $x = 0$ and at $x = \pi/3$.

15. Graph the function in problem 13 and the approximating polynomials consisting of the first term, the first three terms and all five terms, using different colors, in the same coordinate system.

16. Do the same as in problems 13, 14, and 15 with the function

$$f(x) = \begin{cases} x \quad \text{for } -\frac{\pi}{2} \le x < \frac{\pi}{2} \\ \pi - x \text{ for } \frac{\pi}{2} \le x < \frac{3\pi}{2} \end{cases}, \quad f(x + 2\pi) = f(x).$$

17. Expand the function $f(x) = x, \quad -\pi \le x < \pi, f(x + 2\pi) = f(x)$, into a Fourier series.

18. Do the same as in problems 13, 14, and 15 with the function in problem 17.

§2. THE THEORY OF FOURIER SERIES

1. Illustrative Examples

We will devote this subsection to the task of building up toward a theorem which will tell us under what conditions one can expand a periodic function into a Fourier series of the form established in the preceding section and how this series represents the expanded function.

In order to get some idea of what we have to expect, we will consider three examples which are very characteristic and which will serve our purpose quite well.

The Zig-Zag Function. In order to get a certain feeling about the way Fourier series converge (or not), we will discuss a simple example which will show us how the original function is approached by the partial sums— or the approximating trigonometric polynomials of the corresponding Fourier series. (We call specifically the sum of all terms up to and including the terms $a_n \cos nx$ and $b_n \sin nx$ the nth *partial sum.*) This will lead us eventually to the idea which is behind the customary convergence proofs for Fourier series.

Let us consider the following continuous function:

$$z(x) = \begin{cases} x \quad \text{for } -\frac{\pi}{2} \le x < \frac{\pi}{2} \\ \pi - x \quad \text{for } \frac{\pi}{2} \le x < \frac{3\pi}{2} \end{cases}, \quad f(x + 2\pi) = f(x). \qquad \text{(IV.17)}$$

This function resembles the sine function somewhat (see Fig. 7). Its characteristic feature is that it has corners wherever the sine has extreme values. For obvious reasons we call it the *zig-zag function.*

Since $z(x)$ is odd, we can apply rules 1 and 3 (subsection 6 of the preceding section) and obtain for the coefficients

$$b_k = \frac{2}{\pi}\int_0^{\pi/2} x \sin kx\, dx + 2\int_{\pi/2}^{\pi} \sin kx\, dx - \frac{2}{\pi}\int_{\pi/2}^{\pi} x \sin kx\, dx$$

$$= \begin{cases} 4/\pi \cdot (-1)^\nu/(2\nu+1)^2 & \text{for } k = 2\nu + 1, \\ 0 \quad \text{for } k = 2\nu. \end{cases}$$

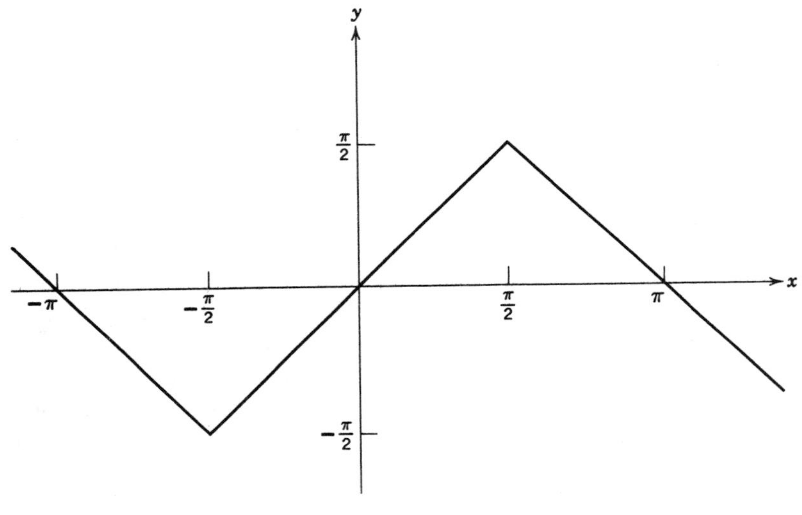

Fig. 7

Therefore, the Fourier expansion of (IV.17) has the form

$$z(x) = \frac{4}{\pi}\left(\sin x - \frac{1}{9}\sin 3x + \frac{1}{25}\sin 5x - + \cdots\right). \qquad \text{(IV.18)}$$

In order to freshen up our knowledge on the theory of convergence, we will prove that this series converges uniformly even though we can arrive at this result later on via a general theorem (theorem IV.2).

Since $|\sin kx| \leq 1$,

$$\left|\frac{(-1)^\nu}{(2\nu+1)^2}\sin(2\nu+1)x\right| \leq \frac{1}{(2\nu+1)^2}$$

holds. Since $\sum_{\nu=0}^{\infty}\frac{1}{(2\nu+1)^2}$ converges (see **AII.C2**), the uniform and absolute convergence of (IV.18) is assured in view of the comparison test (see **AII.B2**)

The best way to get an intuitive idea about how (IV.18) converges to the expanded function $z(x)$ is simply to sketch a few partial sums, as is done in Fig. 8. The bold line represents the original function, the broken line the first partial sum $s_1 = (4/\pi) \sin x$, and the dotted line the third partial sum $s_3 = (4/\pi)(\sin x - \frac{1}{9} \sin 3x)$.

We can see quite clearly how the partial sums come closer and closer to the original function and how they crowd more and more into its corners.

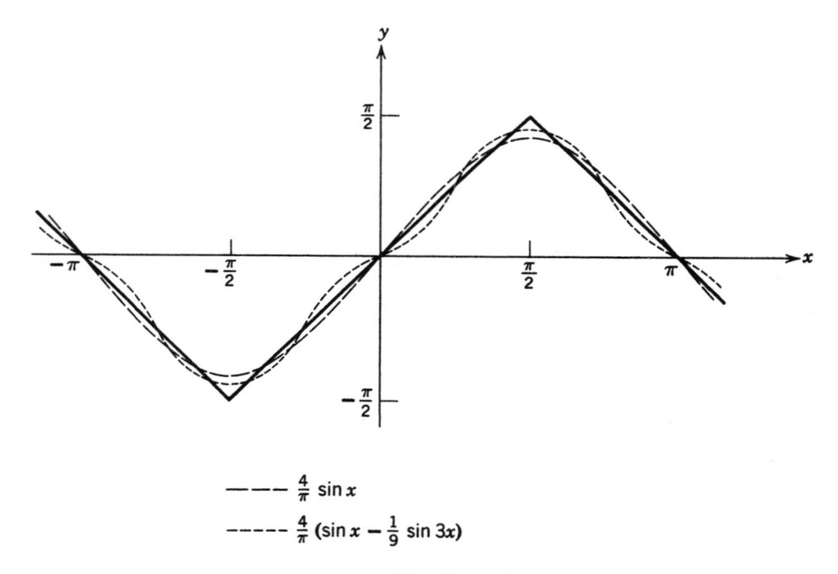

$\frac{4}{\pi}\sin x$

$\frac{4}{\pi}(\sin x - \frac{1}{9}\sin 3x)$

Fig. 8

The Saw-Tooth Function. We consider now a function with one discontinuity per period, namely

$$s(x) = x \quad \text{for } -\pi \leq x < \pi,\ s(x + 2\pi) = s(x). \qquad \text{(IV.19)}$$

This function, as shown in Fig. 9, is again an odd function and resembles the teeth of a saw, explaining its name *saw-tooth function.* Rules 1 and 3 again give us the coefficients

$$b_k = \frac{2}{\pi}\int_0^\pi x \sin kx\, dx = -2\frac{(-1)^k}{k}.$$

The Fourier expansion of (IV.19) is therefore

$$s(x) = 2\left(\sin x - \frac{\sin 2x}{2} + \frac{\sin 3x}{3} - \cdots\right). \qquad \text{(IV.20)}$$

We will not plunge into a discussion of the convergence of this series at the present time since this question is here more delicate than it was in the preceding example. We will, nevertheless, deal with this problem for a more general case in subsection 5 of this section.

In order to get a picture of how the approach to the original function takes place, we will sketch again some partial sums. Figure 10 shows in bold line the original function, in broken line the second partial sum

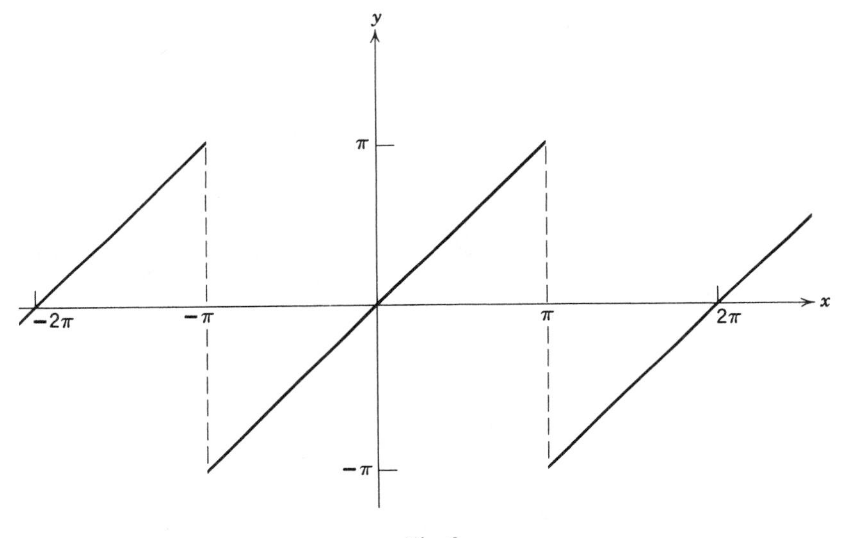

Fig. 9

$s_2 = 2(\sin x - \frac{1}{2}\sin 2x)$, in dotted line the third partial sum $s_3 = 2(\sin x - \frac{1}{2}\sin 2x + \frac{1}{3}\sin 3x)$, and finally in stroke-dotted line the fourth partial sum

$$s_4 = 2(\sin x - \tfrac{1}{2}\sin 2x + \tfrac{1}{3}\sin 3x - \tfrac{1}{4}\sin 4x).$$

It is obvious that these partial sums try to imitate the discontinuity at $x = \pi$ by going through the point $(\pi, 0)$ with increasing steepness.

The Interrupted Square Wave. We finally discuss a function with two discontinuities per period, namely

$$W(x) = \begin{cases} \dfrac{\pi}{2} & \text{for } 0 \le x < \dfrac{\pi}{2} \\ 0 & \text{for } \dfrac{\pi}{2} \le x < \dfrac{3\pi}{2} \\ -\dfrac{\pi}{2} & \text{for } \dfrac{3\pi}{2} \le x < 2\pi \end{cases}, \quad f(x + 2\pi) = f(x). \qquad \text{(IV.21)}$$

This function is shown in Fig. 11. Its shape suggests its name *interrupted square wave*. We see further that we have to deal again with an odd function and obtain for the coefficients

$$b_k = \frac{2}{\pi}\int_0^{\pi} f(x)\sin kx\,dx = \frac{2}{\pi}\int_0^{\pi/2}\frac{\pi}{2}\sin kx\,dx = -\left.\frac{\cos kx}{k}\right|_0^{\pi/2}$$

$$= \begin{cases} \dfrac{(-1)^{\nu+1}}{2\nu} + \dfrac{1}{2\nu} & \text{for} \quad k = 2\nu, \\[2ex] \dfrac{1}{2\nu+1} & \text{for} \quad k = 2\nu + 1. \end{cases}$$

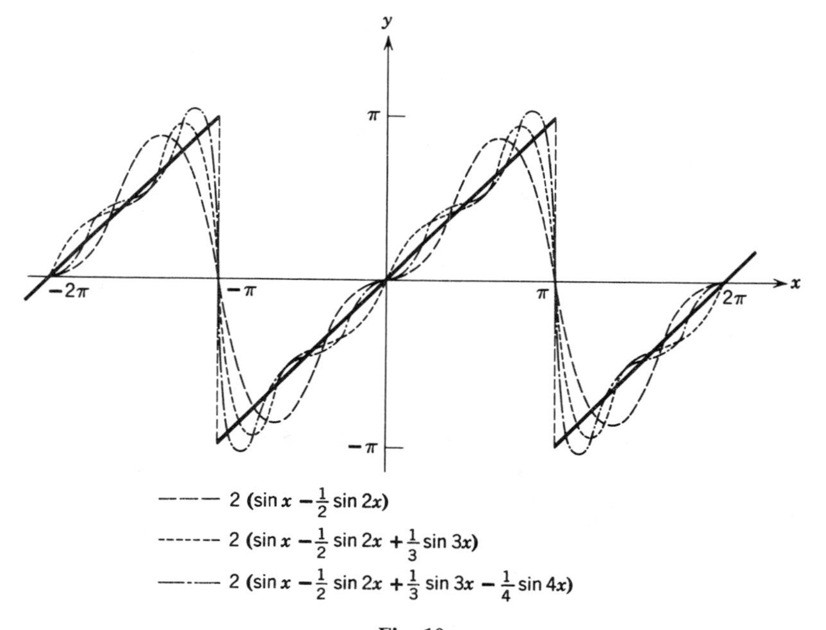

Fig. 10

The expansion therefore has the form

$$w(x) = \sin x + \sin 2x + \tfrac{1}{3}\sin 3x + \tfrac{1}{5}\sin 5x + \tfrac{1}{3}\sin 6x + \tfrac{1}{7}\sin 7x + \cdots. \tag{IV.22}$$

Figure 12 illustrates the third (full line), fifth (broken line), and sixth (dotted line) partial sums as they approach the original function.

In the last two examples—and we would have seen this much clearer if we had considered higher partial sums—the approximating curves make the comparatively largest oscillation before they "jump" over the discontinuities of the function. This interesting behavior which resembles somewhat the preliminary motions an exhibition diver goes through

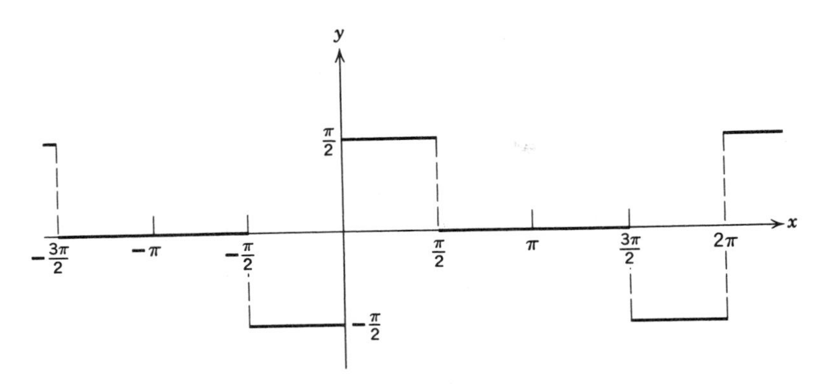

Fig. 11

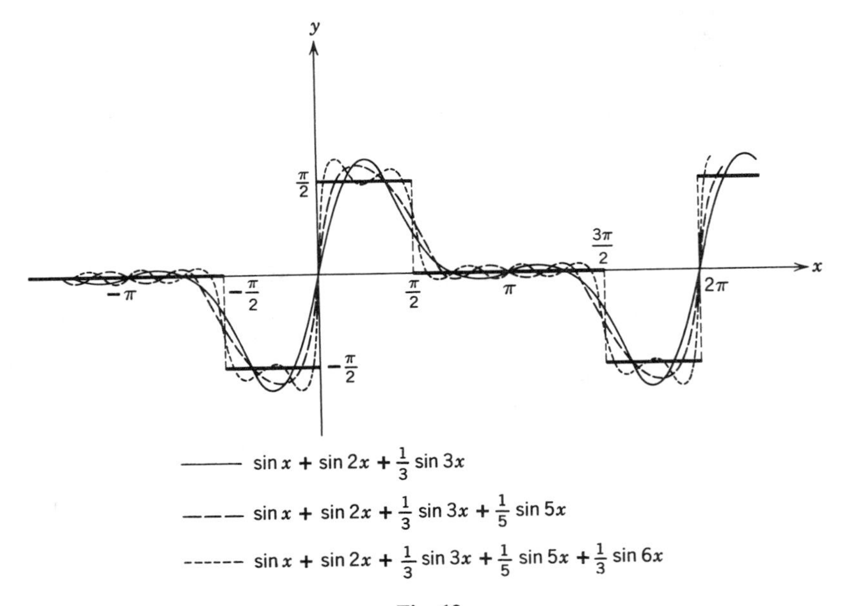

Fig. 12

before he jumps from the spring board is known as Gibbs' phenomenon in honor of the great American applied mathematician *J. W. Gibbs* (1839–1903), who made possibly the greatest contribution to its discovery and investigation. We cannot go into a discussion of this phenomenon in this treatment, and refer the reader for further study to the literature.[3] Apparently, from the examples we just discussed, the graphs representing the partial sums come closer to the original function within intervals in which the function is continuous than within intervals which contain a discontinuity of the function. In a point of discontinuity, of course, we cannot expect an approach to the function at all. In the example of the saw-tooth function, all partial sums passed through the point $(\pi, 0)$, i.e.: through the midpoint $\left(\pi, \frac{f(\pi + 0) + f(\pi - 0)}{2}\right)$ of the jump. This is not true in the example of the interrupted square wave at the points $(\pi/2) + k\pi$; however, the partial sums come closer and closer to the midpoint of the jump as the order of the partial sums increases. This makes us expect that the Fourier series of a discontinuous function—if it converges at all—converges at points of discontinuity to the mean value of the left- and right-hand limits of the function at this point.

Before we can deal with this situation, however, we have to investigate the possible convergence of Fourier series within intervals in which the function is continuous, a little further. We have seen already from the example of the zig-zag function that we can expect in such a case a rather "fast" and even uniform convergence, while the convergence of the series for the saw-tooth function and the interrupted square wave—if it is a matter of convergence at all—is rather slow and certainly not everywhere uniform.

2. Convergence of Fourier Series

We consider the continuous function $f(x)$ with period 2π, $f(x + 2\pi) = f(x)$. Let us expand this function according to our formal rules into a Fourier series

$$f(x) = \frac{a_0}{2} + \sum_{k=1}^{\infty} (a_k \cos kx + b_k \sin kx)$$

and substitute for the a_k and b_k their representations according to (IV.14) and (IV.15):

$$f(x) = \frac{1}{2\pi} \int_{-\pi}^{\pi} f(\xi)\, d\xi + \sum_{k=1}^{\infty} \frac{1}{\pi} \left\{ \int_{-\pi}^{\pi} f(\xi) \cos k\xi\, d\xi \cos kx + \int_{-\pi}^{\pi} f(\xi) \sin k\xi\, d\xi \sin kx \right\}. \quad \text{(IV.23)}$$

[3] See H. S. Carslaw: *Introduction to the Theory of Fourier Series and Integrals*, 3rd ed., Dover Publications, New York, 1930.

Then, the nth partial sum

$$s_n(x) = \frac{1}{2\pi}\int_{-\pi}^{\pi} f(\xi)\,d\xi + \frac{1}{\pi}\int_{-\pi}^{\pi} f(\xi)\sum_{k=1}^{n}(\cos k\xi\cos kx + \sin k\xi\sin kx)\,d\xi$$

can be transformed in view of

$$\cos k\xi\cos kx + \sin k\xi\sin kx = \cos k(\xi - x)$$

into

$$s_n(x) = \frac{1}{\pi}\int_{-\pi}^{\pi} f(\xi)\{\tfrac{1}{2} + \cos(\xi - x) + \cdots + \cos n(\xi - x)\}\,d\xi. \quad \text{(IV.24)}$$

It will be our goal to see whether the error $f(x) - s_n(x)$ approaches zero for every x, i.e.: whether the series (IV.23) converges to $f(x)$ or not. For this purpose we try to write (IV.24) in a simpler form, which can be done by using Euler's identity

$$\cos k(\xi - x) = \tfrac{1}{2}[e^{ik(\xi - x)} + e^{-ik(\xi - x)}].$$

If we sum from $k = 1$ to $k = n$, we obtain

$$\sum_{k=1}^{n}\cos k(\xi - x) = \frac{1}{2}\sum_{k=1}^{n}[e^{ik(\xi - x)} + e^{-ik(\xi - x)}].$$

Considering $e^{\pm ik(\xi - x)}$ as a term of a finite geometrical series, we get

$$\sum_{k=1}^{n}\cos k(\xi - x) = \frac{1}{2}\left[\frac{1 - e^{i(n+1)(\xi - x)}}{1 - e^{i(\xi - x)}} - 1 + \frac{1 - e^{-i(n+1)(\xi - x)}}{1 - e^{-i(\xi - x)}} - 1\right].$$

If we multiply numerator and denominator of the two fractions on the right side by $e^{-i(\xi - x)/2}$ and $e^{i(\xi - x)/2}$, respectively, we obtain in view of $(e^{i\alpha} - e^{-i\alpha})/2i = \sin\alpha$,

$$\sum_{k=1}^{n}\cos k(\xi - x) = \frac{1}{2}\cdot\frac{\sin(n + \frac{1}{2})(\xi - x)}{\sin\frac{1}{2}(\xi - x)} - \frac{1}{2}. \quad \text{(IV.25)}$$

If we substitute this into (IV.24) we obtain for the nth partial sum

$$s_n(x) = \frac{1}{2\pi}\int_{-\pi}^{\pi} f(\xi)\frac{\sin(n + \frac{1}{2})(\xi - x)}{\sin\frac{1}{2}(\xi - x)}\,d\xi. \quad \text{(IV.26)}$$

It must be shown that this expression approaches $f(x)$ as n becomes large. The basic idea of the proof can be explained by the following argument:

Let us consider the special case where $f(x) = 1$. Then

$$s_n(x) = \frac{1}{2\pi}\int_{-\pi}^{\pi}\frac{\sin(n + \frac{1}{2})(\xi - x)}{\sin\frac{1}{2}(\xi - x)}\,d\xi.$$

If the Fourier series is to converge to $f(x) = 1$, then $\lim_{n\to\infty} s_n(x) = 1$ and consequently

$$\lim_{n\to\infty} \int_{-\pi}^{\pi} \frac{\sin (n + \frac{1}{2})(\xi - x)}{\sin \frac{1}{2}(\xi - x)}\, d\xi = 2\pi.$$

We can show very easily, that the value of this integral is 2π regardless of n, provided n is an integer, since it follows from (IV.25) that

$$\frac{\sin (n + \frac{1}{2})(\xi - x)}{\sin \frac{1}{2}(\xi - x)} = 1 + 2\sum_{k=1}^{n} \cos k(\xi - x).$$

Hence

$$\int_{-\pi}^{\pi} \frac{\sin (n + \frac{1}{2})(\xi - x)}{\sin \frac{1}{2}(\xi - x)}\, d\xi = \int_{-\pi}^{\pi} \left(1 + 2\sum_{k=1}^{n} \cos k(\xi - x)\right) d\xi$$

$$= \left[\xi + \sum_{k=1}^{n} \frac{2 \sin k(\xi - x)}{k}\right]_{-\pi}^{\pi} = 2\pi$$

Thus, for all integers n

$$\int_{-\pi}^{\pi} \frac{\sin (n + \frac{1}{2})(\xi - x)}{\sin \frac{1}{2}(\xi - x)}\, d\xi = 2\pi. \qquad \text{(IV.27)}$$

Problems IV.19–21

19. Show that $\lim_{n\to\infty} \int_0^h \frac{\sin n\xi}{\xi}\, d\xi = \int_0^\infty \frac{\sin t}{t}\, dt, \qquad h > 0.$

***20.** Prove that $\int_0^\infty \frac{\sin t}{t}\, dt = \frac{\pi}{2}$. (*Hint:* Consider the function $F(x) = \int_0^\infty e^{-xt} \frac{\sin t}{t}\, dt$, calculate $F'(x)$, integrate again and compute $F(0)$.)

***21.** Transform

$$\int_{-h}^{h} \frac{\sin (n + \frac{1}{2})\xi\, d\xi}{\sin \frac{1}{2}\xi}$$

by using the identity

$$\sin (n + \tfrac{1}{2})\xi = \sin n\xi \cos \tfrac{1}{2}\xi + \cos n\xi \sin \tfrac{1}{2}\xi$$

and approximate $\sin \frac{1}{2}\xi$ by $\frac{1}{2}\xi$ and $\cos \frac{1}{2}\xi$ by 1, assuming that h is very small. Show that in case these approximations are justified

$$\lim_{n\to\infty} \int_{-h}^{h} \frac{\sin (n + \frac{1}{2})\xi}{\sin \frac{1}{2}\xi}\, d\xi = 2\pi$$

by making use of the result in problem 20.

3. The Convergence Proof

Before formulating the convergence theorem, we explain the concept of sectional (piecewise) continuity:

A function $\phi(x)$ is said to be *sectionally continuous* in the interval $a \leq x \leq b$ if there are only finitely many discontinuities $x_1, x_2, \cdots, x_n$ of $\phi(x)$ in $a \leq x \leq b$ and if

$$\lim_{x \to x_i - 0} \phi(x) = \phi(x_i - 0), \qquad \lim_{x \to x_i + 0} \phi(x) = \phi(x_i + 0)$$

exist for all $i = 1, 2, 3, \cdots n$. The notation $x \to x_i - 0$ and $x \to x_i + 0$ means that x approaches x_i from the left and from the right, respectively.

We are now ready to proceed to the proof of the following:

Theorem IV.1. *If the function $f(x)$ has the period 2π, i.e., $f(x + 2\pi) = f(x)$, is sectionally continuous in $-\pi \leq x \leq \pi$, and has a sectionally continuous derivative in $-\pi \leq x \leq \pi$ (functions which are sectionally continuous and have a sectionally continuous derivative are called* sectionally (piecewise) smooth), *then*

$$\frac{a_0}{2} + \sum_{k=1}^{\infty} (a_k \cos kx + b_k \sin kx) = \tfrac{1}{2}[f(x + 0) + f(x - 0)]$$

where

$$a_k = \frac{1}{\pi} \int_{-\pi}^{\pi} f(x) \cos kx \, dx, \qquad k = 0, 1, 2, 3, \cdots$$

$$b_k = \frac{1}{\pi} \int_{-\pi}^{\pi} f(x) \sin kx \, dx, \qquad k = 1, 2, 3, \cdots .$$

This means: At every point where the function is continuous, the series converges to the value of the function at this point and at every point where the function has a discontinuity, the series converges to the arithmetic mean of the left-hand and the right-hand limits of the function (*convergence in the mean*).

In case the period of the function is $2p$ instead of 2π, the analogous theorem holds:

Theorem IV.1a. *If $f(x)$ has the period $2p$ and is sectionally smooth on the interval $-p \leq x \leq p$, then*

$$\frac{a_0}{2} + \sum_{k=1}^{\infty} \left(a_k \cos \frac{k\pi x}{p} + b_k \sin \frac{k\pi x}{p} \right) = \tfrac{1}{2}[f(x + 0) + f(x - 0)],$$

where

$$a_k = \frac{1}{p} \int_{-p}^{p} f(x) \cos \frac{k\pi x}{p} \, dx, \qquad k = 0, 1, 2, 3, \cdots,$$

$$b_k = \frac{1}{p} \int_{-p}^{p} f(x) \sin \frac{k\pi x}{p} \, dx, \qquad k = 1, 2, 3, \cdots .$$

Proof of Theorem IV.1. (The proof of theorem IV.1a can be obtained from the proof which is given subsequently by making a few formal changes.) Using a result in subsection 2 of this section as a point of departure, we write the nth partial sum in the form

$$s_n(x) = \frac{1}{2\pi}\int_{-\pi}^{\pi} f(\xi)\frac{\sin(n+\frac{1}{2})(\xi - x)}{\sin\frac{1}{2}(\xi - x)}\,d\xi \qquad \text{(see (IV.26))}. \qquad \text{(IV.28)}$$

Let us assume without loss of generality that the point x under consideration lies in the interval $-\pi \leq x < \pi$. We now make the coordinate transformation $\xi - x = t$ and obtain in view of lemma IV.1

$$s_n(x) = \frac{1}{2\pi}\int_{-\pi}^{\pi} f(x+t)\frac{\sin(n+\frac{1}{2})t}{\sin\frac{1}{2}t}\,dt.$$

We interrupt the integration at $t = 0$ and transform the integrand of the first integral in

$$s_n(x) = \frac{1}{2\pi}\left\{\int_{-\pi}^{0} f(x+t)\frac{\sin(n+\frac{1}{2})t}{\sin\frac{1}{2}t}\,dt + \int_{0}^{\pi} f(x+t)\frac{\sin(n+\frac{1}{2})t}{\sin\frac{1}{2}t}\,dt\right\} \qquad \text{(IV.29)}$$

as follows

$$f(x+t) = f(x-0) - f(x-0) + f(x+t), \qquad t < 0.$$

If we denote the first integral in (IV.29) by I_1, we obtain

$$I_1 = f(x-0)\int_{-\pi}^{0}\frac{\sin(n+\frac{1}{2})t}{\sin\frac{1}{2}t}\,dt + \int_{-\pi}^{0}\frac{f(x+t)-f(x-0)}{\sin\frac{1}{2}t}\sin(n+\tfrac{1}{2})t\,dt.$$

In view of (IV.27) (see also problem IV.22), we have

$$f(x-0)\int_{-\pi}^{0}\frac{\sin(n+\frac{1}{2})t}{\sin\frac{1}{2}t}\,dt = \pi f(x-0)$$

and consequently

$$I_1 = \pi f(x-0) + \int_{-\pi}^{0}\frac{f(x+t)-f(x-0)}{\sin\frac{1}{2}t}\sin(n+\tfrac{1}{2})t\,dt. \qquad \text{(IV.30)}$$

The function

$$s(t) = \frac{f(x+t)-f(x-0)}{\sin\frac{1}{2}t}, \qquad t < 0$$

is sectionally continuous, since f is sectionally smooth and

$$\lim_{t\to 0-0}\frac{f(x+t)-f(x-0)}{\sin\frac{1}{2}t} = \lim_{t\to 0-0}\frac{f(x+t)-f(x-0)}{t}\cdot\frac{t}{\sin\frac{1}{2}t}$$

$$= 2\lim_{t\to 0-0}\frac{f(x+t)-f(x-0)}{t}$$

in view of $\lim_{t\to 0} t/(\sin\frac{1}{2}t) = 2$.

Since f is sectionally smooth, the left-hand and right-hand derivatives of f exist everywhere. Inasmuch as

$$\lim_{t\to 0-0} \frac{f(x+t)-f(x-0)}{t} = f'(x-0)$$

is the left-hand derivative of f at x, we have established the existence of $\lim_{t\to 0-0} s(t)$.

Before proceeding with our proof, we have to prove the following lemma:

Lemma IV.2 (Riemann-Lebesgue). *If $\sigma(t)$ is continuous in $a \leq x \leq b$, then* $\lim_{\lambda\to\infty} \int_a^b \sigma(t) \sin \lambda t \, dt = 0.$

Proof of Lemma IV.2. Let $t = \tau + \pi/\lambda$, then $\sin \lambda t = \sin \lambda(\tau + \pi/\lambda) = -\sin \lambda\tau$ and

$$I = \int_a^b \sigma(t) \sin \lambda t \, dt = -\int_{a-\pi/\lambda}^{b-\pi/\lambda} \sigma\left(\tau + \frac{\pi}{\lambda}\right) \sin \lambda\tau \, d\tau,$$

$$2I = \int_a^b \sigma(t) \sin \lambda t \, dt - \int_{a-\pi/\lambda}^{b-\pi/\lambda} \sigma\left(t + \frac{\pi}{\lambda}\right) \sin \lambda t \, dt$$

$$= -\int_{a-\pi/\lambda}^{a} \sigma\left(t + \frac{\pi}{\lambda}\right) \sin \lambda t \, dt + \int_a^{b-\pi/\lambda} \left(\sigma(t) - \sigma\left(t + \frac{\pi}{\lambda}\right)\right) \sin \lambda t \, dt$$

$$+ \int_{b-\pi/\lambda}^{b} \sigma(t) \sin \lambda t \, dt.$$

Since $\sigma(t)$ is continuous in $a \leq t \leq b$, there exists an M such that $|\sigma(t)| \leq M$. Hence

$$\left| \int_{a-\pi/\lambda}^{a} \sigma\left(t + \frac{\pi}{\lambda}\right) \sin \lambda t \, dt \right| = \left| \int_a^{a+\pi/\lambda} \sigma(t) \sin \lambda t \, dt \right| \leq \frac{\pi}{\lambda} M$$

and

$$\left| \int_{b-\pi/\lambda}^{b} \sigma(t) \sin \lambda t \, dt \right| \leq \frac{\pi}{\lambda} M$$

and consequently

$$2|I| \leq \frac{2\pi M}{\lambda} + \int_a^{b-\pi/\lambda} \left| \sigma(t) - \sigma\left(t + \frac{\pi}{\lambda}\right) \right| \sin \lambda t \, dt$$

$$\leq \frac{2\pi M}{\lambda} + \int_a^{b-\pi/\lambda} \left| \sigma(t) - \sigma\left(t + \frac{\pi}{\lambda}\right) \right| dt.$$

Since $\sigma(t)$ is continuous in $a \leq t \leq b$ and hence uniformly continuous, there exists for any arbitrary small $\epsilon > 0$ a δ_ϵ such that

$$\left| \sigma(t) - \sigma\left(t + \frac{\pi}{\lambda}\right) \right| < \epsilon \quad \text{for } \frac{\pi}{\lambda} < \delta_\epsilon \text{ and all } a \leq t \leq b - \frac{\pi}{\lambda}.$$

Now, we choose λ so large that $2\pi M/\lambda < \epsilon$ and $\pi/\lambda < \delta_\epsilon$. Then

$$|I| < \epsilon^* \quad \text{for } \lambda \text{ sufficiently large.}$$

Hence $\lim_{\lambda \to \infty} I = 0$, q.e.d.

Corollary to Lemma IV.2. *If $s(t)$ is sectionally continuous in $a \leq t \leq b$ then* $\lim_{\lambda \to \infty} \int_a^b s(t) \sin \lambda t \, dt = 0$.

The proof consists of a repeated application of the lemma IV.2 to every subinterval of $a \leq t \leq b$ in which $s(t)$ is continuous.

Continuation of the Proof for Theorem IV.1. It follows immediately from (IV.30) and the corollary to lemma IV.2 that

$$\lim_{n \to \infty} I_1 = \pi f(x - 0). \tag{IV.31}$$

If we denote the second integral in (IV.29) by I_2 and transform it according to

$$f(x + t) = f(x + 0) - f(x + 0) + f(x + t), \qquad t > 0,$$

we obtain by a procedure which is analogous to the one we carried out above

$$\lim_{n \to \infty} I_2 = \pi f(x + 0). \tag{IV.32}$$

Hence, we obtain in view of (IV.31) and (IV.32)

$$\lim_{n \to \infty} s_n(x) = \frac{1}{2\pi} \lim_{n \to \infty} (I_1 + I_2) = \tfrac{1}{2}[f(x + 0) + f(x - 0)], \quad \text{q.e.d.}$$

Before proceeding, we will make a few remarks concerning the assumption about $f(x)$ under which we proved theorem IV.1. We assumed that $f(x)$ has a sectionally continuous derivative; i.e., $f(x)$ has at every point a right-hand and a left-hand derivative and we recall that this assumption was essential in applying Riemann-Lebesgue's lemma, which in turn was essential for the proof offered of the convergence theorem. Dirichlet, who gave the first valid proof for this theorem, assumed that $f(x)$ is sectionally continuous and sectionally monotonic, i.e., that one can subdivide the interval into a finite number of subintervals such that in every one of them the function is either nondecreasing or nonincreasing. This implies that the function has only a finite number of extreme values in the interval. It might be of interest to note that continuity of $f(x)$ alone is not sufficient to prove the convergence of the corresponding Fourier series. Examples of this type, however, are not easily constructed and are necessarily of a rather "pathological" character.

Problems IV.22–25

22. Show that $\displaystyle\int_{-\pi}^{0} \frac{\sin(n+\frac{1}{2})t}{\sin\frac{1}{2}t}\,dt = \int_{0}^{\pi} \frac{\sin(n+\frac{1}{2})t}{\sin\frac{1}{2}t}\,dt = \pi.$

(See (IV.27).)

23. Prove the corollary to lemma IV.2.

24. Prove theorem IV.1 in the following modified form: Suppose $f(x)$ has the period 2π and is integrable on $-\pi \le x \le \pi$. Then the Fourier series converges to $\frac{1}{2}[f(x+0) + f(x-0)]$ at all points x at which the left-hand and the right-hand derivatives of $f(x)$ exist.

***25.** Give a proof of theorem IV.1 by writing the integrand in (IV.26) in terms of the sectionally smooth function

$$F(\xi) = f'(\xi)\frac{\xi - x}{2\sin\frac{1}{2}(\xi - x)}$$

and using the result of problem 20.

4. Bessel's Inequality

Let $f(x)$ be sectionally continuous in $-\pi \le x \le \pi$ and $f(x + 2\pi) = f(x)$. Clearly

$$\int_{-\pi}^{\pi}\left[f(x) - \frac{a_0}{2} - \sum_{k=1}^{n}(a_k\cos kx + b_k\sin kx)\right]^2 dx \ge 0, \qquad \text{(IV.33)}$$

where a_k and b_k denote the Fourier coefficients of $f(x)$. If we expand the integrand, we obtain

$$\int_{-\pi}^{\pi} f^2(x)\,dx + \frac{a_0^2\pi}{2} + \int_{-\pi}^{\pi}\left(\sum_{k=1}^{n} a_k\cos kx + b_k\sin kx\right)^2 dx - a_0\int_{-\pi}^{\pi} f(x)\,dx$$
$$- 2\int_{-\pi}^{\pi} f(x)\left(\sum_{k=1}^{n} a_k\cos kx + b_k\sin kx\right)dx$$
$$- a_0\int_{-\pi}^{\pi}\left(\sum_{k=1}^{n} a_k\cos kx + b_k\sin kx\right)dx \ge 0.$$

In view of the definitions of the Fourier coefficients (IV.14) and (IV.15) and the orthogonality relations for the trigonometric functions (IV.10), (IV.11), and (IV.12), we obtain (see also problem IV.26)

$$\int_{-\pi}^{\pi} f^2(x)\,dx - \pi\left[\frac{a_0^2}{2} + \sum_{k=1}^{n}(a_k^2 + b_k^2)\right] \ge 0$$

or

$$\frac{a_0^2}{2} + \sum_{k=1}^{n} (a_k^2 + b_k^2) \leq \frac{1}{\pi} \int_{-\pi}^{\pi} f^2(x)\, dx \tag{IV.34}$$

for all n.

Since the right side of (IV.34) is independent of n, we obtain, by passing to the limit,

$$\frac{a_0^2}{2} + \sum_{k=1}^{\infty} (a_k^2 + b_k^2) \leq \frac{1}{\pi} \int_{-\pi}^{\pi} f^2(x)\, dx, \tag{IV.35}$$

which is called *Bessel's inequality*.

If for any function $f(x)$ in a certain class C (e.g., the class of all sectionally smooth functions) it is possible to find for any arbitrary small $\epsilon > 0$ an N_ϵ such that

$$\int_{-\pi}^{\pi} \left[f(x) - \frac{a_0}{2} - \sum_{k=1}^{n} (a_k \cos kx + b_k \sin kx) \right]^2 dx < \epsilon$$

for all $n > N_\epsilon$, one calls the system of orthogonal functions $\{\cos kx, \sin kx\}$ *complete* with respect to C.

It can be shown easily, that if $\{\cos kx, \sin kx\}$ is complete with respect to a class C, then, for any $f(x)$ in C the so-called *completeness relation*

$$\frac{a_0^2}{2} + \sum_{k=1}^{\infty} (a_k^2 + b_k^2) = \frac{1}{\pi} \int_{-\pi}^{\pi} f^2(x)\, dx \tag{IV.36}$$

holds (see problem IV.27). We will discuss this aspect of the system of trigonometric functions $\{\cos kx, \sin kx\}$ in subsection 6 of this section.

Problems IV.26–27

26. Show that

$$\frac{a_0^2\pi}{2} - a_0 \int_{-\pi}^{\pi} f(x)\, dx = -\frac{a_0^2\pi}{2},$$

$$\int_{-\pi}^{\pi} \left(\sum_{k=1}^{n} a_k \cos kx + b_k \sin kx \right)^2 dx = \pi \sum_{k=1}^{n} (a_k^2 + b_k^2),$$

$$a_0 \int_{-\pi}^{\pi} \left(\sum_{k=1}^{n} a_k \cos kx + b_k \sin kx \right) dx = 0,$$

$$\int_{-\pi}^{\pi} f(x) \left(\sum_{k=1}^{n} a_k \cos kx + b_k \sin kx \right) dx = \pi \sum_{k=1}^{n} (a_k^2 + b_k^2).$$

27. Show that if the system of trigonometric function $\{\cos kx, \sin kx\}$ is complete with respect to a class C, then, for any $f(x)$ in C, the completeness relation (IV.36) is true.

5. Absolute and Uniform Convergence of Fourier Series

By using Bessel's inequality (IV.35) we can easily prove the following theorem regarding the uniform and absolute convergence of the Fourier series of a continuous function with a sectionally continuous derivative:

Theorem IV.2. *If $f(x)$ is continuous and has a sectionally continuous derivative in $-\pi \leq x \leq \pi$ and if $f(x)$ has the period 2π and if $f(-\pi) = f(\pi)$, then the Fourier series representing $f(x)$ converges everywhere uniformly and absolutely to $f(x)$.*

Proof. Let $u_n(x) = a_n \cos nx + b_n \sin nx$, where a_n, b_n are the Fourier coefficients of $f(x)$. We have to show according to Cauchy's convergence criterion for uniform convergence (see **AII.A2**) that for any arbitrary small $\epsilon > 0$ there exists an N_ϵ independent of x such that

$$|u_{n+1}(x) + \cdots + u_m(x)| < \epsilon \quad \text{for all } n, m > N_\epsilon.$$

Now,

$$u_{n+1}(x) + \cdots + u_m(x) = \sum_{k=n+1}^{m} (a_k \cos kx + b_k \sin kx).$$

In view of

$$a_k \cos kx + b_k \sin kx = \sqrt{a_k^2 + b_k^2} \cos (kx - \delta)$$

with $\delta = \arctan b_k/a_k$, we have

$$|u_{n+1}(x) + \cdots + u_m(x)| \leq \sum_{k=n+1}^{m} |a_k \cos kx + b_k \sin kx| \leq \sum_{k=n+1}^{m} \sqrt{a_k^2 + b_k^2}. \tag{IV.37}$$

Since $f'(x)$ is sectionally continuous, its Fourier coefficients α_k, β_k

$$\alpha_k = \frac{1}{\pi} \int_{-\pi}^{\pi} f'(x) \cos kx \, dx,$$

$$\beta_k = \frac{1}{\pi} \int_{-\pi}^{\pi} f'(x) \sin kx \, dx$$

satisfy Bessel's inequality (IV.35)

$$\frac{\alpha_0^2}{2} + \sum_{k=1}^{\infty} (\alpha_k^2 + \beta_k^2) \leq \frac{1}{\pi} \int_{-\pi}^{\pi} f'^2(x) \, dx. \tag{IV.38}$$

In view of the continuity of $f(x)$ in $-\pi \leq x \leq \pi$ and $f(-\pi) = f(\pi)$ there are simple relations between the a_k, b_k and the α_k, β_k as one can see by integrating the representations of a_k and b_k by parts:

$$a_k = \frac{1}{\pi}\int_{-\pi}^{\pi} f(x)\cos kx\,dx = -\frac{1}{k\pi}\int_{-\pi}^{\pi} f'(x)\sin kx\,dx = -\frac{\beta_k}{k},$$

$$b_k = \frac{1}{\pi}\int_{-\pi}^{\pi} f(x)\sin kx\,dx = \frac{1}{k\pi}\int_{-\pi}^{\pi} f'(x)\cos kx\,dx = \frac{\alpha_k}{k}.$$

Hence, we obtain for (IV.37)

$$|u_{n+1}(x) + \cdots + u_m(x)| \leq \sum_{k=n+1}^{m} \frac{1}{k}\sqrt{\alpha_k^2 + \beta_k^2}. \tag{IV.39}$$

If we apply the Cauchy-Schwarz inequality (see problem IV.4)

$$(\boldsymbol{\xi}\cdot\boldsymbol{\eta}) \leq |\boldsymbol{\xi}|\,|\boldsymbol{\eta}|$$

with

$$\boldsymbol{\xi} = \left(\frac{1}{n+1}, \frac{1}{n+2}, \cdots, \frac{1}{m}\right)$$

and

$$\boldsymbol{\eta} = (\sqrt{\alpha_{n+1}^2 + \beta_{n+1}^2}, \cdots, \sqrt{\alpha_m^2 + \beta_m^2}),$$

i.e.:

$$\sum_{k=n+1}^{m} \frac{1}{k}\sqrt{\alpha_k^2 + \beta_k^2} \leq \sqrt{\sum_{k=n+1}^{m} \frac{1}{k^2} \sum_{k=n+1}^{m} (\alpha_k^2 + \beta_k^2)},$$

to the right side of (IV.39), we obtain

$$|u_{n+1}(x) + \cdots + u_m(x)| \leq \sqrt{\sum_{k=n+1}^{m} \frac{1}{k^2}}\sqrt{\sum_{k=n+1}^{m} (\alpha_k^2 + \beta_k^2)}. \tag{IV.40}$$

Since $\sum_{k=1}^{\infty} \frac{1}{k^2}$ converges (see **AII.C2**) there exists for any $\epsilon > 0$ an N_ϵ independent of x such that

$$\sum_{k=n+1}^{m} \frac{1}{k^2} < \epsilon \quad \text{for all } n, m > N_\epsilon.$$

This fact together with (IV.38) yields

$$|u_{n+1}(x) + \cdots + u_m(x)| \leq \sqrt{\epsilon}\sqrt{\frac{1}{\pi}\int_{-\pi}^{\pi} f'^2(x)\,dx}$$

where $\int_{-\pi}^{\pi} f'^2(x)\,dx = Q < \infty$ since $f'(x)$ is sectionally continuous. Hence

$$|u_{n+1}(x) + \cdots + u_m(x)| \leq \sqrt{\epsilon}\sqrt{\frac{1}{\pi}Q} \quad \text{for all } n, m > N_\epsilon$$

where N_ϵ is independent of x. This establishes the uniform convergence. It also proves the absolute convergence (see **AII.A3**) because of (IV.37).

We are now going to extend this theorem; i.e., we will establish the uniform convergence of a Fourier series in every closed interval in which the function is continuous. A lead will be obtained from a discussion of the example of the saw-tooth function $s(x)$ (see subsection 1 of this section). The saw-tooth function as defined by

$$s(x) = x \quad \text{for } -\pi \leq x < \pi,\ s(x + 2\pi) = s(x),$$

is discontinuous at every odd multiple of π and has a sectionally continuous derivative. We found in subsection 1 of this section that the Fourier series of $s(x)$ is given by

$$s(x) = 2(\sin x - \tfrac{1}{2} \sin 2x + \tfrac{1}{3} \sin 3x - + \cdots).$$

Should this series converge uniformly everywhere, it would represent a continuous function (see **AII.C3**), which we know is not the case.

We will see, however, that this series converges uniformly in every closed interval which is contained in an interval $(2k - 1)\pi < x < (2k + 1)\pi$ whose boundaries are consecutive discontinuities of $s(x)$.

Without loss of generality we can assume that $k = 0$; i.e., we consider an interval contained in $-\pi < x < \pi$. For reasons of convenience we choose this interval as symmetric with respect to the origin; i.e.:

$$-a \leq x \leq a \qquad \text{where } a < \pi.$$

To accomplish our objective we apply a certain trick, which is not too artificial in the light of the following motivation. The saw-tooth function is generated by periodic continuation of the function $y = x$ in $-\pi \leq x < \pi$ and this leaves, as we know, discontinuities at all odd multiples of π. If we multiply this periodically continued function $s(x)$ by another function $\sigma(x)$ which we choose so that the product $s(x)\sigma(x)$ is continuous everywhere, is periodic, and has a sectionally continuous derivative, then the Fourier series for $s(x)\sigma(x)$ will converge uniformly in view of theorem IV.2. If the function $\sigma(x)$ can furthermore be chosen such that it assumes a positive minimum in $-a \leq x \leq a$

$$\min_{-a \leq x \leq a} \sigma(x) = M,$$

then the uniform convergence of the series for $s(x)$ in $-a \leq x \leq a$ can be established by utilizing the uniform convergence of the series for $s(x)\sigma(x)$ and the fact that

$$s(x)\sigma(x) \geq s(x)M \text{ in } -a \leq x \leq a.$$

All the requirements which we formulated for $\sigma(x)$ are clearly satisfied by the function

$$\sigma(x) = \cos \frac{x}{2}$$

because $\min\limits_{-a \le x \le a} \cos\frac{x}{2} = \cos\frac{a}{2} > 0$ and

$$\sigma(-\pi)s(-\pi) = \sigma(\pi)s(\pi) = 0.$$

The latter property guarantees that the function $f(x) = x \cos\frac{x}{2}$ for $-\pi \le x < \pi$ can be periodically continued into an everywhere-continuous function.

After these preparatory remarks we are ready to state and prove the following:

Lemma IV.3. *The Fourier series of the saw-tooth function as defined in subsection 1 of this section,*

$$s(x) = 2(\sin x - \tfrac{1}{2}\sin 2x + \tfrac{1}{3}\sin 3x - + \cdots)$$

converges uniformly in $-a \le x \le a$ *where* $a < \pi$.

Proof. We have to consider the expression (see **AII.A2**)

$$|u_{n+1}(x) + \cdots + u_m(x)| \quad \text{with } u_k(x) = \frac{(-1)^k}{k}\sin kx$$

and show that it can be made arbitrarily small for sufficiently large n, m independent of x.

$$|u_{n+1}(x) + \cdots + u_m(x)| = \left|\frac{\sin(n+1)x}{n+1} - \frac{\sin(n+2)x}{n+2} + \cdots \pm \frac{\sin mx}{m}\right|.$$

We multiply both sides by $\cos(x/2)$ and utilize the identity

$$2\sin\alpha\cos\beta = \sin(\alpha+\beta) + \sin(\alpha-\beta).$$

Then

$$\frac{1}{k}\sin kx \cos\frac{x}{2} = \frac{1}{2k}\left[\sin(k+\tfrac{1}{2})x + \sin(k-\tfrac{1}{2})x\right]$$

and consequently

$$|u_{n+1}(x) + \cdots + u_m(x)| \cos\frac{x}{2}$$

$$= \frac{1}{2}\left|\frac{\sin(n+\frac{3}{2})x}{n+1} - \frac{\sin(n+\frac{5}{2})x}{n+2} + \cdots \pm \frac{\sin(m+\frac{1}{2})x}{m}\right.$$

$$\left. + \frac{\sin(n+\frac{1}{2})x}{n+1} - \frac{\sin(n+\frac{3}{2})x}{n+2} + \cdots \pm \frac{\sin(m-\frac{1}{2})x}{m}\right|$$

$$= \frac{1}{2}\left|\frac{\sin(m+\frac{1}{2})x}{m} + \frac{\sin(n+\frac{1}{2})x}{n+1} + \frac{\sin(n+\frac{3}{2})x}{(n+1)(n+2)}\right.$$

$$\left. - \frac{\sin(n+\frac{5}{2})x}{(n+2)(n+3)} + \cdots \mp \frac{\sin(m-\frac{1}{2})x}{(m-1)m}\right|.$$

Because $\cos \frac{x}{2} \geq M$ in $-a \leq x \leq a$ and $|\sin t| \leq 1$, we obtain

$$|u_{n+1}(x) + \cdots + u_m(x)| \leq \frac{1}{M}\left(\frac{1}{m} + \frac{1}{n+1} + \frac{1}{(n+1)(n+2)} + \frac{1}{(n+2)(n+3)} + \cdots + \frac{1}{(m-1)m}\right)$$

$$< \frac{1}{M}\left(\frac{1}{m} + \frac{1}{n+1} + \frac{1}{n^2} + \frac{1}{(n+1)^2} + \cdots + \frac{1}{(m-2)^2}\right).$$

Since $\sum_{k=1}^{\infty} \frac{1}{k^2}$ converges (see **AII.C2**) we can find for any arbitrarily small $\epsilon > 0$ an N_ϵ independent of x such that

$$\frac{1}{m} + \frac{1}{n+1} + \sum_{k=n}^{m-2} \frac{1}{k^2} < \epsilon \quad \text{for all } n, m > N_\epsilon.$$

Hence

$$|u_{n+1}(x) + \cdots + u_m(x)| < \frac{\epsilon}{M} \quad \text{for all } n, m > N_\epsilon \text{ independent of } x, \quad \text{q.e.d.}$$

Lemma IV.3 will enable us to prove without difficulty the following extension of theorem IV.2:

Theorem IV.3. *If $f(x)$ has the period 2π and is sectionally smooth, then its Fourier series converges uniformly to $f(x)$ in every closed interval within which $f(x)$ is continuous.*

Proof. Since $f(x)$ is sectionally continuous, it has only finitely many discontinuities in $-\pi \leq x \leq \pi$. Let $-\pi \leq x_1 < x_2 < \cdots < x_n \leq \pi$ denote these discontinuities. In the light of the definition of sectional continuity, these discontinuities are finite jumps, at worst. Let j_k denote the height of the jump

$$j_k = f(x_k - 0) - f(x_k + 0)$$

at the point x_k and let $\phi_k(x)$ denote a function which has a jump of height j_k at x_k and is continuous everywhere else in $-\pi \leq x \leq \pi$. Then

$$F(x) = f(x) - \phi_1(x) - \phi_2(x) - \cdots - \phi_n(x)$$

is continuous everywhere in $-\pi \leq x \leq \pi$ except for removable discontinuities and has a continuous periodic continuation (see problem IV.29). Hence, $F(x)$ has a uniformly convergent Fourier series according to theorem IV.2.

Since we investigated the Fourier series of the saw-tooth function $s(x)$ extensively in lemma IV.3 we will proceed to represent the functions $\phi_k(x)$ in terms of $s(x)$.

$s(x)$ has a jump at $x = (2k + 1)\pi$ with the saltus 2π; hence $\dfrac{j_k}{2\pi} s(x + \pi - x_k)$ has a jump at $x = x_k$ with the saltus j_k.

Thus, we can write

$$\phi_k(x) = \frac{j_k}{2\pi} s(x + \pi - x_k)$$

and consequently

$$f(x) = F(x) + \frac{1}{2\pi} \sum_{k=1}^{n} j_k s(x + \pi - x_k).$$

The Fourier series for $F(x)$ converges uniformly everywhere and the Fourier series for

$$\frac{1}{2\pi} \sum_{k=1}^{n} j_k s(x + \pi - x_k)$$

converges uniformly in any closed interval $x_k < a \leq x \leq b < x_{k+1}$ according to lemma IV.3. Hence, the Fourier series for $f(x)$ which is obtained by termwise addition of the Fourier expansions for the $\phi_k(x)$ and the Fourier series of $F(x)$ (see problem IV.33) converges uniformly in any interval $x_k < a \leq x \leq b < x_{k+1}$, q.e.d.

Problems IV.28–33

28. Show that the Fourier series for the saw-tooth function $s(x)$ converges uniformly in any interval $a \leq x \leq b$, where $-\pi < a < b < \pi$.

29. Given two functions $f(x)$, $g(x)$, which are continuous everywhere except at $x = x_0$, where $f(x_0 - 0) - f(x_0 + 0) = j$, and $g(x_0 - 0) - g(x_0 + 0) = j$. Prove that $f(x) - g(x)$ is continuous everywhere, except for a removable discontinuity at x_0.

30. Expand

$$f(x) = \begin{cases} -1 + x & \text{for } -\pi \leq x < 0, \\ 1 - x & \text{for } 0 \leq x < \pi \end{cases}$$

into a Fourier series. $f(x)$ is discontinuous at $k\pi$ ($k = 0, 1, 2, \cdots$). By adding suitably chosen functions $\phi_k(x)$ obtain an everywhere-continuous function and its uniformly convergent Fourier series.

31. Find the Fourier expansions for the following functions

$$f_1(x) = x, \qquad -\pi \leq x < \pi \qquad \text{(saw-tooth function)}$$

$$f_2(x) = \begin{cases} x, & -\dfrac{\pi}{2} \leq x < \dfrac{\pi}{2} \\ -x + \pi, & \dfrac{\pi}{2} \leq x < \dfrac{3\pi}{2} \end{cases} \qquad \text{(zig-zag function)}$$

$$f_3(x) = \begin{cases} -x(x - \pi), & 0 \leq x < \pi \\ (x - \pi)(x - 2\pi), & \pi \leq x < 2\pi \end{cases}$$

$$f_4(x) = \begin{cases} x^2(x - \pi)^2, & 0 \leq x < \pi \\ (x - \pi)^2(x - 2\pi)^2, & \pi \leq x < 2\pi \end{cases}$$

where $f_k(x) = f_k(x + 2\pi)$.

Note that the Fourier coefficients a_n, b_n of $f_k(x)$ are of the form C/n^k. Note also that $f_k(x)$ has a discontinuous $(k-1)$th derivative (0th derivative of $f(x)$ is $f(x)$ itself) but all $(k-i)$th derivatives are continuous, where $k \geq i > 1$.

***32.** Given a function $f(x)$ which is continuous and has a sectionally continuous derivative. Prove that

$$|na_n| \leq M, \quad |nb_n| \leq M$$

for some M and all n. (*Hint:* Use integration by parts.)

33. Assume that the series $\sum_{k=1}^{\infty} u_n(x) = U(x)$ and $\sum_{k=1}^{\infty} v_n(x) = V(x)$ are uniformly convergent in $a \leq x \leq b$. Prove that $\sum_{k=1}^{\infty} [u_n(x) + v_n(x)]$ is uniformly convergent in $a \leq x \leq b$ and that its sum is $U(x) + V(x)$.

6. Completeness[4]

In subsection 4 of this section we defined completeness of the orthogonal system of trigonometric functions $\{\cos nx, \sin mx\}$ with respect to a class C of functions as follows:

$\{\cos nx, \sin mx\}$ is complete with respect to C if for every f in C there exists for any arbitrary small $\epsilon > 0$ an N_ϵ such that

$$\int_{-\pi}^{\pi} \left[f(x) - \frac{a_0}{2} - \sum_{k=1}^{m}(a_k \cos kx + b_k \sin kx)\right]^2 dx < \epsilon$$

for all $m > N_\epsilon$.

We are now in a position to prove the completeness of $\{\cos nx, \sin mx\}$ with respect to the class S_s of all sectionally smooth functions and subsequently for the class C_s of all sectionally continuous functions.

Theorem IV.4. *The orthogonal system of trigonometric functions* $\{\cos nx, \sin mx\}$ *is complete with respect to the class* S_s *of all sectionally smooth functions.*

Proof. If $f(x) \in S_s$ (where $\in$ stands for "is an element of"), then $f(x) \in C_s$, since $S_s \subset C_s$ (where $\subset$ stands for "is a subset of"). Hence, there exists an M such that (see problem IV.35)

$$\left|f(x) - \frac{a_0}{2} - \sum_{k=1}^{m}(a_k \cos kx + b_k \sin kx)\right| \leq M \quad \text{on} \quad -\pi \leq x \leq \pi.$$

Let $x_1, x_2, \cdots, x_{n-1}$ denote the discontinuities of $f(x)$ on $-\pi \leq x \leq \pi$ which are arranged such that $-\pi = x_0 \leq x_1 < x_2 < \cdots < x_{n-1} \leq x_n = \pi$. We now choose an arbitrarily small $\epsilon > 0$ and subsequently a $\delta = \epsilon/2nM^2$. The function $f(x)$ is continuous in I_k: $x_k + \delta \leq x \leq x_{k+1} - \delta$

[4] This subsection can be omitted without jeopardizing the continuity of the material.

for $k = 0, 1, 2, \cdots, n - 1$ and has a sectionally continuous derivative by hypothesis. Hence, its Fourier series converges uniformly in I_k according to theorem IV.3; i.e., for any arbitrarily small

$$\sqrt{\frac{\epsilon}{n(x_{k+1} - x_k)}}$$

there exists an N_{ϵ_k} such that

$$|E_m(x)| < \sqrt{\frac{\epsilon}{n(x_{k+1} - x_k)}} \quad \text{for all } m > N_{\epsilon_k}$$

where

$$E_m(x) = f(x) - \frac{a_0}{2} - \sum_{k=1}^{m} (a_k \cos kx + b_k \sin kx).$$

Thus:

$$\int_{x_k+\delta}^{x_{k+1}-\delta} E_m^2(x)\, dx < \frac{\epsilon}{n}$$

for all $m > N_{\epsilon_k}$.

Since

$$\int_{-\pi}^{\pi} E_m^2(x)\, dx = \sum_{k=0}^{n-1} \left[\int_{x_k}^{x_k+\delta} E_m^2(x)\, dx + \int_{x_{k+1}-\delta}^{x_{k+1}} E_m^2(x)\, dx + \int_{x_k+\delta}^{x_{k+1}-\delta} E_m^2(x)\, dx \right],$$

we obtain

$$\int_{-\pi}^{\pi} E_m^2(x)\, dx \leq 2nM^2\,\delta + \frac{n\epsilon}{n} = 2\epsilon$$

for all $m > \max N_{\epsilon_k}$, q.e.d.

From this theorem we can deduce:

Theorem IV.5. *The orthogonal system of trigonometric functions* $\{\cos nx, \sin mx\}$ *is complete with respect to the class* C_s *of all sectionally continuous functions.*[5]

Proof. We cannot make any use of the uniform convergence of the Fourier series of $f(x)$ in a direct way, because theorem IV.3 only establishes uniform convergence for Fourier series of continuous functions with a sectionally continuous derivative.

However, we can obtain a function with an everywhere-continuous derivative which approximates $f(x)$ uniformly in every interval I_k by invoking Weierstrass's approximation theorem, as follows:

[5] If a system of orthogonal functions is complete with respect to C_s, one frequently calls it simply *complete* without any further reference.

Let

$$F(x) = \begin{cases} f(x) \quad \text{for } x_k + \delta \le x \le x_{k+1} - \delta,\ k = 0, 1, 2, \cdots n - 1 \\ \dfrac{f(x_k + \delta) - f(x_k - \delta)}{2\delta}(x - x_k + \delta) + f(x_k - \delta) \\ \qquad\qquad \text{for } x_k - \delta < x < x_k + \delta,\ k = 1, 2, \cdots n - 1 \\ \dfrac{f(-\pi + \delta) - f(-\pi + 0)}{\delta}(x + \pi) + f(-\pi + 0) \\ \qquad\qquad \text{for } -\pi \le x < -\pi + \delta, \\ \dfrac{f(\pi) - f(\pi - \delta)}{\delta}(x - \pi) + f(\pi - 0) \quad \text{for } \pi - \delta < x \le \pi. \end{cases}$$

This function is pieced together from $f(x)$ in the I_k's and line segments joining the end points of $f(x)$ in I_{k-1} with the beginning points of $f(x)$ in I_k

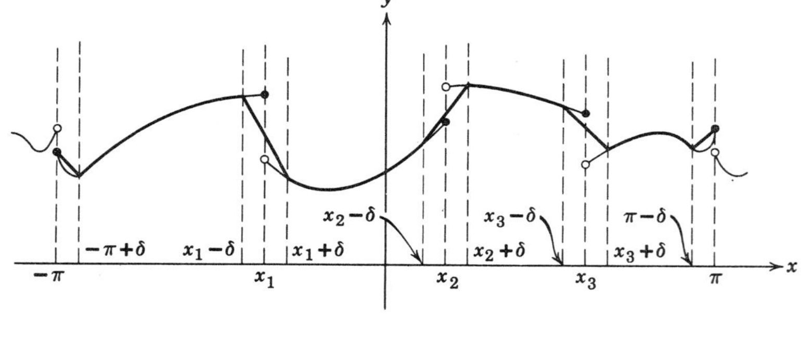

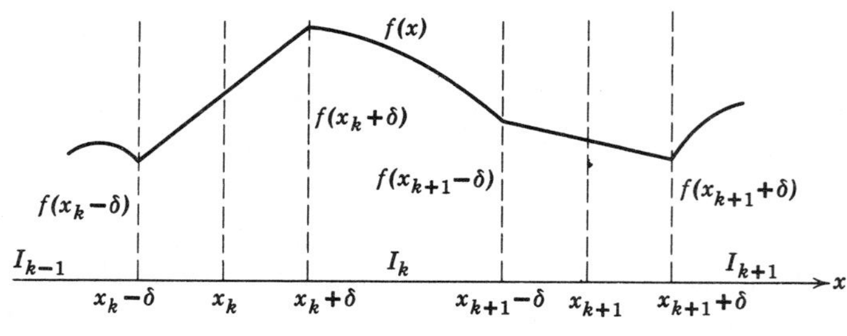

Fig. 13

(see Fig. 13). Clearly, this function $F(x)$ is continuous in $-\pi \le x \le \pi$. According to the theorem of Weierstrass[6] there exists for any arbitrarily

[6] See Apostol: *Mathematical Analysis*, Addison-Wesley, 1957, p. 481; and R. Courant and D. Hilbert: *Methods of Mathematical Physics*, Interscience Publishers, New York, 1953, p. 65.

small $\epsilon > 0$ a polynomial $P(x)$ such that

$$|F(x) - P(x)| < \epsilon \quad \text{for all } x \text{ in } -\pi \leq x \leq \pi.$$

Since

$$\int_{x_k+\delta}^{x_{k+1}-\delta} \left[f(x) - \frac{a_0}{2} - \sum_{k=1}^{m} a_k \cos kx + b_k \sin kx\right]^2 dx$$

$$= \int_{x_k+\delta}^{x_{k+1}-\delta} \left[(F(x) - P(x)) + \left(P(x) - \frac{a_0}{2} - \sum_{k=1}^{m} a_k \cos kx + b_k \sin kx\right)\right]^2 dx,$$

we can make this expression arbitrarily small (using the Cauchy-Schwarz inequality) if the a_k, b_k stand for the Fourier coefficients of $P(x)$, proceed in our proof as in the proof of theorem IV.4 (see problem IV.34), and note that

$$\int_{-\pi}^{\pi} \left[f(x) - \frac{c_0}{2} - \sum_{k=1}^{m} c_k \cos kx + d_k \sin kx\right]^2 dx$$

$$\leqslant \int_{-\pi}^{\pi} \left[f(x) - \frac{a_0}{2} - \sum_{k=1}^{m} a_k \cos kx + b_k \sin kx\right]^2 dx,$$

where the c_k and d_k are the Fourier coefficients of $f(x)$.

Problems IV.34–35

***34.** Carry out the details in the proof of theorem IV.5.

35. If $f(x)$ is continuous in $a \leq x \leq b$, there exists a K such that $|f(x)| \leq K$ in $a \leq x \leq b$. Show that if $g(x)$ is sectionally continuous in $a \leq x \leq b$, then, there exists an M such that $|g(x)| \leq M$ in $a \leq x \leq b$.

7. Integration and Differentiation of Fourier Series

In integrating Fourier series we will experience a pleasant surprise. In order to guarantee the convergence of a termwise integrated series to the integral of the expanded function, we have to assume in general that the series to be integrated converges uniformly (see **AII.C4**). In case of a Fourier series, however, we do not even have to assume the convergence of the series to be integrated. A simple example will demonstrate this: The series

$$\cos x - \cos 2x + \cos 3x - \cos 4x + \cdots$$

does not converge, whatever value x may assume, since

$$\lim_{k \to \infty} (-1)^{k+1} \cos ka \neq 0$$

where a may be any fixed value, while a necessary condition for the convergence of a series $\sum_{k=1}^{\infty} a_k$ is that $\lim_{k \to \infty} a_k = 0$ (see **AII.B1**).

However, if we integrate this series termwise:

$$\sin x - \frac{\sin 2x}{2} + \frac{\sin 3x}{3} - + \cdots,$$

we obtain a convergent series which even converges uniformly within any closed interval which is contained in $(k - 1)\pi < x < (k + 1)\pi$, as we have seen in subsection 5 (saw-tooth function).

This will help us to understand the following theorem better:

Theorem IV.6 (termwise integration of Fourier series). *If $f(x)$ is sectionally continuous in $-\pi \leq x \leq \pi$ and if*

$$\frac{a_0}{2} + \sum_{k=1}^{\infty} (a_k \cos kx + b_k \sin kx)$$

is its Fourier series, i.e., the series formally obtained by formulas (IV.14) and (IV.15), then—no matter whether this series converges or not—it is true that

$$\int_a^x f(x)\, dx = \frac{1}{2} \int_a^x a_0\, dx + \sum_{k=1}^{\infty} \int_a^x (a_k \cos kx + b_k \sin kx)\, dx, \quad \text{(IV.41)}$$

where $-\pi \leq a \leq x \leq \pi$, converges uniformly in x for any fixed value of a.

Proof. The function

$$F(x) = \int_{-\pi}^x (f(x) - \tfrac{1}{2}a_0)\, dx \quad \text{(IV.42)}$$

is certainly continuous and sectionally smooth since $f(x)$ is sectionally continuous and

$$\int_a^x f(x)\, dx = F(x) - F(a) + \frac{a_0}{2} \int_a^x dx. \quad \text{(IV.43)}$$

Furthermore $\int_{-\pi}^{\pi} \left(f(x) - \frac{a_0}{2}\right) dx = 0$ because $a_0 = \frac{1}{\pi} \int_{-\pi}^{\pi} f(x)\, dx.$

Hence

$$F(-\pi) = F(\pi) = 0. \quad \text{(IV.44)}$$

We can therefore apply theorem IV.2 and expand (IV.42) into a Fourier series which converges uniformly:

$$F(x) = \frac{A_0}{2} + \sum_{k=1}^{\infty} (A_k \cos kx + B_k \sin kx)$$

where

$$A_k = \frac{1}{\pi}\int_{-\pi}^{\pi} F(x)\cos kx\, dx,$$

$$B_k = \frac{1}{\pi}\int_{-\pi}^{\pi} F(x)\sin kx\, dx.$$

In view of (IV.44) integration by parts yields

$$A_k = -\frac{1}{k}\, b_k,$$

$$B_k = \frac{1}{k}\, a_k$$

(see also subsection 5) and we therefore obtain

$$F(x) - F(a) = \sum_{k=1}^{\infty}\left[-\frac{b_k}{k}(\cos kx - \cos ka) + \frac{a_k}{k}(\sin kx - \sin ka)\right]$$

and from this according to (IV.43)

$$\int_a^x f(x)\, dx = \frac{a_0}{2}\int_a^x dx + \sum_{k=1}^{\infty}\left[-\frac{b_k}{k}(\cos kx - \cos ka) + \frac{a_k}{k}(\sin kx - \sin ka)\right]$$

which is identical with (IV.41). The uniform convergence of this series follows from the uniform convergence of the series for $F(x)$ and problem IV.33.

Theorem IV.7 (termwise differentiation of Fourier series). *If $f(x)$ is continuous in $-\pi \le x \le \pi$ with $f(-\pi) = f(\pi)$ and if its derivative is sectionally smooth, then its Fourier series*

$$f(x) = \frac{a_0}{2} + \sum_{k=1}^{\infty}(a_k \cos kx + b_k \sin kx)$$

can be differentiated termwise wherever $f'(x)$ exists; i.e., it is true that

$$f'(x) = \sum_{k=1}^{\infty} k(-a_k \sin kx + b_k \cos kx), \qquad \textit{wherever } f'(x) \textit{ exists.}$$

Proof. If $f'(x)$ is sectionally smooth, it has according to theorem IV.1 a Fourier expansion which converges in the mean

$$\tfrac{1}{2}[f'(x+0) + f'(x-0)] = \frac{\alpha_0}{2} + \sum_{k=1}^{\infty}(\alpha_k \cos kx + \beta_k \sin kx) \quad \text{(IV.45)}$$

with

$$\alpha_k = \frac{1}{\pi}\int_{-\pi}^{\pi} f'(x)\cos kx\,dx,$$

$$\beta_k = \frac{1}{\pi}\int_{-\pi}^{\pi} f'(x)\sin kx\,dx.$$

If we integrate by parts and take into account that $f(-\pi) = f(\pi)$, we obtain

$$\alpha_k = kb_k,$$

$$\beta_k = -ka_k,$$

and, for the same reason, $a_0 = 0$.

Therefore, we obtain

$$\tfrac{1}{2}[f'(x+0) + f'(x-0)] = \sum_{k=1}^{\infty} k(-a_k \sin kx + b_k \cos kx)$$

and this series converges in the sense of theorem IV.1. Wherever $f'(x)$ has a discontinuity it converges to $\frac{1}{2}[f'(x+0) + f'(x-0)]$, which is for most purposes useless, but to the left and the right of this point, the series will converge to the left- and the right-hand derivative of $f(x)$. Wherever $f'(x)$ exists, it will converge to $f'(x)$.

Problems IV.36–37

***36.** The function $f(x)$ with the period 2π shall be continuous, have continuous derivatives of 1st, 2nd, $\cdots$, kth order, and the $k+1$th derivative shall have finitely many jumps in $-\pi \le x < \pi$. Prove that its Fourier coefficients a_n, b_n satisfy the following relations:

$$|n^{k+1}a_n| \le M, \quad |n^{k+1}b_n| \le M$$

for some M and all n. (*Hint:* See problem IV.32 and use theorem IV.6.)

37. Is termwise differentiation of the Fourier series for the zig-zag function $z(x)$ (see for definition subsection 1 of this section) permissible? If yes, how often?

§3. APPLICATIONS

1. The Plucked String

We are now going to make use of almost everything we discussed until now in solving the string problem under specific boundary and initial conditions. The problem we are going to discuss first is the so-called problem of the *plucked string*, that is, a stretched string which is displaced by a single force exerted upon a single point of the string and then released without initial velocity. We choose this problem as the first one since it is very simple from a formal point of view but becomes very interesting and to a certain extent complicated at the moment we gain some insight into the solution and its behavior.

We will assume that the string whose displacement satisfies equation (II.3)

$$\frac{\partial^2 u}{\partial t^2} = \alpha^2 \frac{\partial^2 u}{\partial x^2}, \qquad \alpha^2 = \frac{\tau}{\rho}$$

is stretched between the two points $(0, 0)$ and $(\pi, 0)$ so satisfying the boundary conditions

$$\begin{aligned} u(0, t) &= 0, \\ u(\pi, t) &= 0 \end{aligned} \tag{IV.46}$$

and initially displaced so that it has the form

$$u(x, 0) = f(x) = \begin{cases} hx & \text{for } 0 \le x < \dfrac{\pi}{2}, \\ \\ h(\pi - x) & \text{for } \dfrac{\pi}{2} \le x \le \pi, \end{cases} \tag{IV.47}$$

where h is a small number which shall assure that the displacements are such that $\partial u/\partial x$ is small, as we assumed in the derivation of (II.3). We will further assume that the string is released without initial velocity

$$\left.\frac{\partial u(x, t)}{\partial t}\right|_{t=0} = 0. \tag{IV.48}$$

Bernoulli's separation method yields for $u = X(x)T(t)$ the two equations

$$T'' + \lambda^2 T = 0,$$

$$X'' + \frac{\lambda^2}{\alpha^2} X = 0,$$

and thus the infinitely many solutions

$$u_\lambda(x, t) = (a_\lambda \cos \lambda t + b_\lambda \sin \lambda t)\left(\overline{a_\lambda} \cos \frac{\lambda}{\alpha} x + \overline{b_\lambda} \sin \frac{\lambda}{\alpha} x\right).$$

Taking the boundary conditions (IV.46) and the initial condition (IV.48) into account leaves us with

$$u_k(x, t) = B_k \cos \alpha kt \cdot \sin kx$$

where $k = 1, 2, 3, \cdots$ and from this, by applying the superposition principle, theorem III.2,

$$u(x, t) = \sum_{k=1}^{\infty} B_k \cos \alpha kt \sin kx.$$

In order to satisfy the remaining initial condition (IV.47) we have to determine the B_k so that

$$u(x, 0) = \sum_{k=1}^{\infty} B_k \sin kx = f(x)$$

where $f(x)$ is the function defined in (IV.47) and for the sake of its mathematical analysis is assumed to be periodically continued as an odd function: $f(x + 2\pi) = f(x), f(-x) = -f(x)$. We know already from Chapter IV, §2.1 (zig-zag function) that in this case

$$B_k = \begin{cases} \dfrac{4}{\pi} \cdot \dfrac{(-1)^\nu h}{(2\nu + 1)^2} & \text{for } k = 2\nu + 1, \\ 0 \quad \text{for } k = 2\nu, \end{cases}$$

which furnishes the solution

$$u(x, t) = \frac{4h}{\pi} \sum_{\nu=0}^{\infty} \frac{(-1)^\nu}{(2\nu + 1)^2} \cos \alpha(2\nu + 1)t \sin (2\nu + 1)x. \qquad \text{(IV.49)}$$

This series converges uniformly and what is more important, absolutely, as one can immediately see. Therefore, after substituting according to the identity $\cos \alpha \sin \beta = \frac{1}{2}[\sin (\alpha + \beta) - \sin (\alpha - \beta)]$, we can rearrange terms according to Riemann's rearrangement law (see **AII.C6**) and arrive at

$$u(x, t) = \frac{2h}{\pi} \sum_{\nu=0}^{\infty} \frac{(-1)^\nu}{(2\nu + 1)^2} \sin (2\nu + 1)(x + \alpha t) + \frac{2h}{\pi} \sum_{\nu=0}^{\infty} \frac{(-1)^\nu}{(2\nu + 1)^2} \sin (2\nu + 1)(x - \alpha t).$$

According to (IV.18), we may write

$$u(x, t) = \tfrac{1}{2}\{f(x + \alpha t) + f(x - \alpha t)\}, \qquad \text{(IV.50)}$$

where f is the function defined in (IV.47).

It would appear that we have obtained a simple solution without trouble. But the question arises: Is this really a solution? The question is justified on very good grounds. We know from the definition of $f(x)$ in (IV.47) that $f'(x)$ does not exist at $x = \pi/2$. Hence

$$\left.\frac{\partial u}{\partial x}\right|_{x \pm \alpha t = \pi/2} = \tfrac{1}{2}[f'(x + \alpha t) + f'(x - \alpha t)]_{x \pm \alpha t = \pi/2},$$

and

$$\left.\frac{\partial u}{\partial t}\right|_{x \pm \alpha t = \pi/2} = \tfrac{1}{2}[\alpha f'(x + \alpha t) - \alpha f'(x - \alpha t)]_{x \pm \alpha t = \pi/2}$$

do not exist. However, the left-hand and the right-hand derivatives exist in all four cases. This is clearly the reason that the series representations for $\partial u/\partial x$, $\partial u/\partial t$ are not termwise differentiable (see problems IV.36 and 37) and theorem III.2 (superposition principle) is not applicable. Since this theorem states sufficient conditions only, (IV.50) can still be a solution of our problem and indeed is, as one can directly verify:

$$\frac{\partial^2 u}{\partial x^2} = \tfrac{1}{2}\{f''(x + \alpha t) + f''(x - \alpha t)\} = 0,$$

$$\frac{\partial^2 u}{\partial t^2} = \tfrac{1}{2}\{\alpha^2 f''(x + \alpha t) + \alpha^2 f''(x - \alpha t)\} = 0,$$

whereby $\dfrac{\partial^2 u}{\partial x^2}, \dfrac{\partial^2 u}{\partial t^2}$ at $x \pm \alpha t = \pi/2$ have to be understood as the left-hand and the right-hand derivatives. Thus, equation (II.3) is satisfied and so are the boundary and initial conditions, as one can immediately see.

However, the fact that (IV.50) satisfies (II.3) and the boundary and initial conditions does not necessarily mean that it represents the vibrations of a stretched string under these conditions, since we assumed in our derivation of (II.3) implicitly that $\partial u/\partial x$ is everywhere continuous (see Chapter II, §1.1). Thus, our derivation of (II.3) does not necessarily apply to this case.

We emphasize the point made at the beginning of this section: while the problem appeared simple enough at first, difficulties have inevitably appeared and they seem to be rather serious. Now that we have recognized an apparently grave situation, let us allay further fears of the reader by stating that everything turns out all right after all as was proved by *Christoffel*.[7] We will devote the next subsection to an investigation of this matter.

[7] E. B. Christoffel: "Untersuchungen ueber die mit dem Fortbestehen linearer partieller Differentialgleichungen vertraeglichen Unstetigkeiten," *Annali di Matematica*, VIII, 1876.

2. The Validity of the Solution of the String Equation for a Displacement Function with a Sectionally Continuous Derivative

Let us first of all discuss the form of the solution we obtained. According to (IV.47) we have

$$f(x + \alpha t) = \begin{cases} h(x + \alpha t) & \text{for } -\alpha t \leq x < \dfrac{\pi}{2} - \alpha t, \\ h(\pi - x - \alpha t) & \text{for } \dfrac{\pi}{2} - \alpha t \leq x < \pi - \alpha t, \end{cases} \tag{IV.51}$$

and

$$f(x - \alpha t) = \begin{cases} h(x - \alpha t) & \text{for } \alpha t \leq x < \dfrac{\pi}{2} + \alpha t, \\ h(\pi - x + \alpha t) & \text{for } \dfrac{\pi}{2} + \alpha t \leq x < \pi + \alpha t \end{cases} \tag{IV.52}$$

with $f(v) = -f(-v), f(v + 2\pi) = f(v)$.

We see that our solution is composed of a zig-zag function (IV.51) which propagates with the velocity α to the left and another zig-zag function (IV.52) which propagates with the velocity α to the right. The displacement of the string at a certain time t at a certain point x is thus obtained by taking the arithmetic mean of the corresponding values of these functions (see (IV.50)).

Figure 14 illustrates the two components (in broken line) and the resulting displacement function (in solid line) for three values of t. We can see that the string has two sharp corners, which travel with velocity α to the left and the right.

Since those corners raised our doubts as to the validity of the solution as we presented it in the preceding section, we have to investigate them more closely. This investigation is based upon the direct application of Newton's law to each element of the string through which the corners travel.

Let ξ be the x-coordinate of the sharp corner C (see Fig. 15). While the displacement $u(x, t)$ is continuous in x and t, the same is not true for the derivative of u with respect to x and t at $x = \xi$. In the following discussion we denote the left-hand derivatives of u with respect to x or t at $x = \xi$ by $\dfrac{\partial u(\xi - 0, t)}{\partial x}, \dfrac{\partial u(\xi - 0, t)}{\partial t}$ and the right-hand derivatives of u by $\dfrac{\partial u(\xi + 0, t)}{\partial x}, \dfrac{\partial u(\xi + 0, t)}{\partial t}$.

The sharp corner lies on both branches (the ascending and the descending) of the displacement curve. Hence, the total change du of C in a time interval

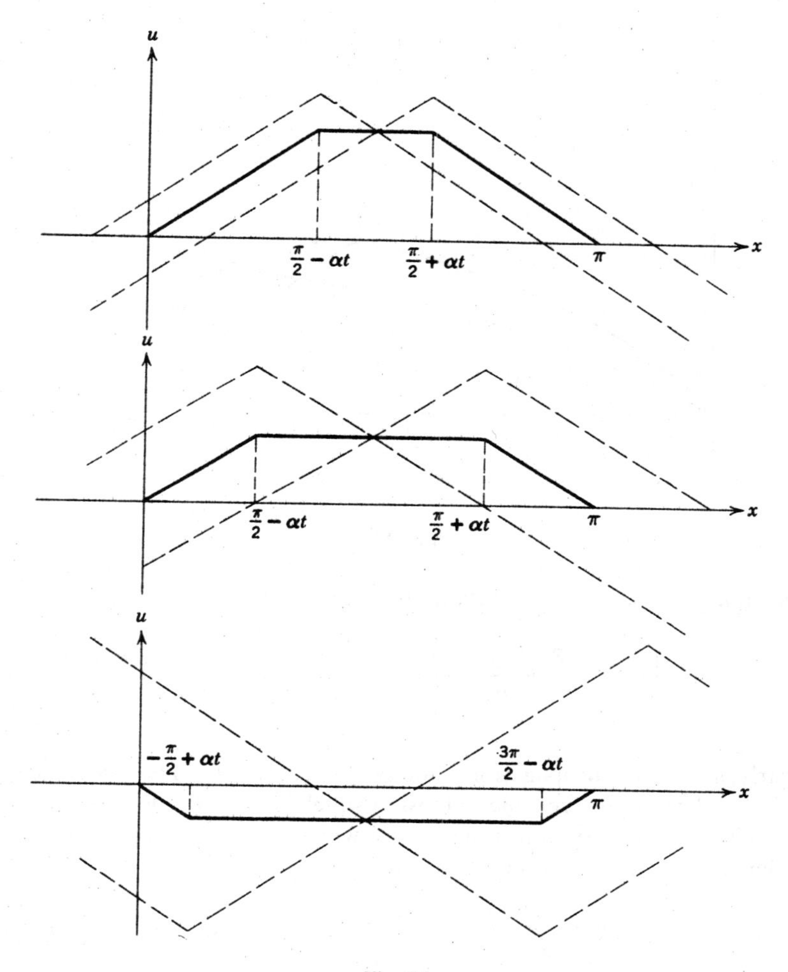

Fig. 14

dt in which C travels from $x = \xi$ to $x = \xi + d\xi$ has to be the same whether we compute it at $\xi - 0$ or at $\xi + 0$. We obtain at $x = \xi - 0$

$$(du)_{\xi-0} = \frac{\partial u(\xi - 0, t)}{\partial x} d\xi + \frac{\partial u(\xi - 0, t)}{\partial t} dt,$$

and at $x = \xi + 0$

$$(du)_{\xi+0} = \frac{\partial u(\xi + 0, t)}{\partial x} d\xi + \frac{\partial u(\xi + 0, t)}{\partial t} dt.$$

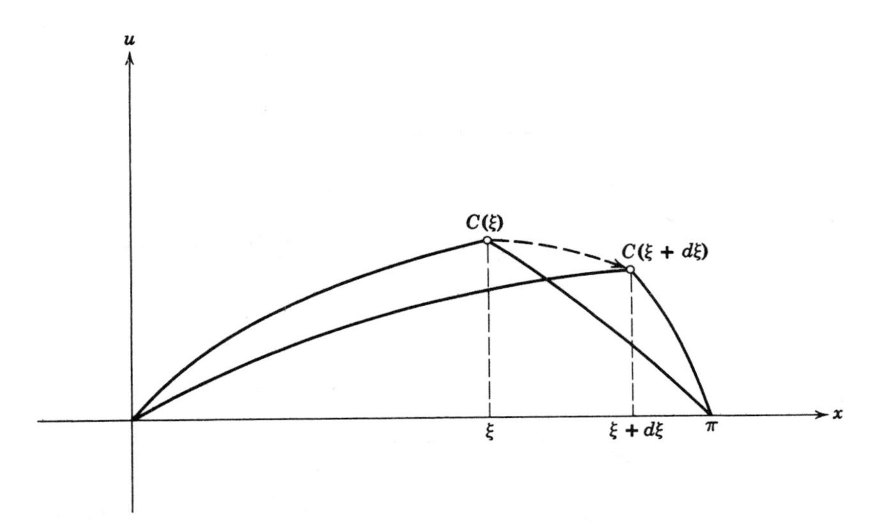

Fig. 15

Since $(du)_{\xi-0} = (du)_{\xi+0}$ we obtain after division by dt

$$\frac{\partial u(\xi - 0, t)}{\partial x} \cdot \frac{d\xi}{dt} + \frac{\partial u(\xi - 0, t)}{\partial t} = \frac{\partial u(\xi + 0, t)}{\partial x} \cdot \frac{d\xi}{dt} + \frac{\partial u(\xi + 0, t)}{\partial t}. \tag{IV.53}$$

Clearly, $d\xi/dt$ is the propagation velocity of the corner C. The u-component of the total external force due to tension which is exerted upon the element through which C travels in time dt and which has in first approximation the length $d\xi$ is in first approximation given by

$$F = \tau\left(\frac{\partial u(\xi + 0, t)}{\partial x} - \frac{\partial u(\xi - 0, t)}{\partial x}\right) \qquad \text{(see (II.1))}. \tag{IV.54}$$

The gain in momentum dM of the element $d\xi$ on the other hand is given in first approximation by

$$dM = \rho\, d\xi\left(\frac{\partial u(\xi - 0, t)}{\partial t} - \frac{\partial u(\xi + 0, t)}{\partial t}\right). \tag{IV.55}$$

(Observe that the corner C is considered to travel to the right.) According to Newton's law $dM = F\,dt$, we obtain from (IV.54) and (IV.55)

$$\tau\left(\frac{\partial u(\xi + 0, t)}{\partial x} - \frac{\partial u(\xi - 0, t)}{\partial x}\right) dt = \rho\, d\xi\left(\frac{\partial u(\xi - 0, t)}{\partial t} - \frac{\partial u(\xi + 0, t)}{\partial t}\right).$$

After division by dt, ρ, rearrangement of terms, and putting $\tau/\rho = \alpha^2$,

$$\alpha^2 \frac{\partial u(\xi + 0, t)}{\partial x} + \frac{d\xi}{dt}\frac{\partial u(\xi + 0, t)}{\partial t} = \alpha^2 \frac{\partial u(\xi - 0, t)}{\partial x} + \frac{d\xi}{dt}\frac{\partial u(\xi - 0, t)}{\partial t}. \tag{IV.56}$$

(IV.53) and (IV.56) are consistent if and only if $\alpha^2 = (d\xi/dt)^2$, i.e.:

$$\frac{d\xi}{dt} = \pm\alpha;$$

that is, the corner C has to propagate with a speed α to the right and to the left, respectively, as was the case in our problem of the plucked string. From (IV.53) and (IV.56) with $d\xi/dt = \pm\alpha$, it follows that

$$\frac{\dfrac{\partial u(\xi + 0, t)}{\partial t} - \dfrac{\partial u(\xi - 0, t)}{\partial t}}{\dfrac{\partial u(\xi + 0, t)}{\partial x} - \dfrac{\partial u(\xi - 0, t)}{\partial x}} = \pm\alpha; \tag{IV.57}$$

i.e., the ratio of the velocity jump at $x = \xi$ to the jump of the tangent line at $x = \xi$ has to equal the propagation speed of C. Let us check whether our solution of the plucked string problem satisfies this condition. We have according to (IV.50) with (IV.51) and (IV.52) at the sharp corner C at $x = (\pi/2) + \alpha t$ (see Fig. 14)

$$u\left(\frac{\pi}{2} + \alpha t - 0, t\right) = \frac{1}{2}\{h(x - \alpha t) + h(\pi - x - \alpha t)\} = \frac{h\pi}{2} - h\alpha t,$$

and

$$u\left(\frac{\pi}{2} + \alpha t + 0, t\right) = \frac{1}{2}\{h(\pi - x + \alpha t) + h(\pi - x - \alpha t)\} = -hx + h\pi.$$

Hence

$$\frac{\partial u(\xi + 0, t)}{\partial t} - \frac{\partial u(\xi - 0, t)}{\partial t} = h\alpha,$$

$$\frac{\partial u(\xi + 0, t)}{\partial x} - \frac{\partial u(\xi - 0, t)}{\partial x} = -h.$$

Thus we obtain for the quotient in (IV.57) the value $-\alpha$ as required. Likewise one can prove that (IV.57) is satisfied at the other corner at $x = (\pi/2) - \alpha t$ (see problem IV.38).

Problem IV.38

38. Verify (IV.57) for the sharp corner C at $x = (\pi/2) - \alpha t$ in the solution of the plucked string problem.

3. The General Solution of The String Problem—D'Alembert's Method

We are now going to solve the problem of the stretched string under the most general initial conditions. We require at $t = 0$

$$u(x, 0) = \phi(x), \tag{IV.58}$$

$$\left.\frac{\partial u(x, t)}{\partial t}\right|_{t=0} = \psi(x), \tag{IV.59}$$

where $\phi(x)$ and $\psi(x)$ may be any functions, provided they are sufficiently well behaved to guarantee the application of the superposition principle, the validity of the termwise integrations and differentiations of their series representations which are carried out in the following and are consistent with the boundary conditions

$$u(0, t) = u(\pi, t) = 0. \tag{IV.60}$$

The solution we obtain via Bernoulli's separation method and the superposition principle, and after imposing the boundary conditions (IV.60), has the form

$$u(x, t) = \sum_{k=1}^{\infty} (a_k \cos \alpha kt + b_k \sin \alpha kt) \sin kx \tag{IV.61}$$

and hence

$$\frac{\partial u}{\partial t} = \sum_{k=1}^{\infty} (-a_k \alpha k \sin \alpha kt + b_k \alpha k \cos \alpha kt) \sin kx.$$

In view of (IV.58) we have to determine the a_k so that

$$u(x, 0) = \phi(x) = \sum_{k=1}^{\infty} a_k \sin kx, \tag{IV.62}$$

and in view of (IV.59) the b_k so that

$$\left.\frac{\partial u(x, t)}{\partial t}\right|_{t=0} = \psi(x) = \sum_{k=1}^{\infty} b_k \alpha k \sin kx. \tag{IV.63}$$

We obtain for the a_k and b_k according to (IV.14), (IV.15), and rules 1 and 3 in subsection 6 of §1 of this chapter)

$$a_k = \frac{2}{\pi} \int_0^{\pi} \phi(x) \sin kx \, dx, \quad b_k = \frac{2}{\alpha k \pi} \int_0^{\pi} \psi(x) \sin kx \, dx.$$

(We remind the reader that $\phi(x)$ and $\psi(x)$ have to be considered as periodically continued beyond $0 \leq x \leq \pi$ so that they represent odd functions of period 2π.) Going through the same transformation process we performed in subsection 1 of this section we will be able to write this solution in a form analogous to (IV.50).

We transform the terms of (IV.61), using the identities

$$\cos\alpha\sin\beta = \tfrac{1}{2}[\sin(\alpha+\beta) - \sin(\alpha-\beta)],$$
$$\sin\alpha\sin\beta = \tfrac{1}{2}[\cos(\alpha-\beta) - \cos(\alpha+\beta)]$$

and rearrange the terms in the series, assuming that the series converges absolutely. Then we obtain

$$u(x, t) = \frac{1}{2}\sum_{k=1}^{\infty}[a_k \sin k(x + \alpha t) + a_k \sin k(x - \alpha t)]$$
$$+ \frac{1}{2}\sum_{k=1}^{\infty}[b_k \cos k(x - \alpha t) - b_k \cos k(x + \alpha t)]. \quad \text{(IV.64)}$$

If we compare the first term in this expression with (IV.62), we see that it is again

$$\tfrac{1}{2}\{\phi(x + \alpha t) + \phi(x - \alpha t)\}.$$

The second term is not quite ψ as we may expect, but if we differentiate this term with respect to t, assuming that termwise differentiation is permissible, we obtain

$$\frac{\partial}{\partial t}\left[\frac{1}{2}\sum_{k=1}^{\infty} b_k \cos k(x - \alpha t) - b_k \cos k(x + \alpha t)\right]$$

$$= \frac{1}{2}\sum_{k=1}^{\infty}(b_k\alpha k \sin k(x + \alpha t) + b_k\alpha k \sin k(x - \alpha t)) = \tfrac{1}{2}[\psi(x + \alpha t) + \psi(x - \alpha t)]$$

in view of (IV.63).

We may therefore write the last part of (IV.64) as

$$\frac{1}{2\alpha}\int_{x-\alpha t}^{x+\alpha t}\psi(s)\,ds$$

and consequently (IV.64) becomes

$$u(x, t) = \frac{1}{2}[\phi(x + \alpha t) + \phi(x - \alpha t)] + \frac{1}{2\alpha}\int_{x-\alpha t}^{x+\alpha t}\psi(s)\,ds. \quad \text{(IV.65)}$$

It is left to the reader to verify that (IV.65) satisfies the boundary conditions, initial conditions, and equation (II.3). This is the solution as presented by *Daniel Bernoulli*.[8]

(IV.65) was actually known as the solution of the string problem six years before *Bernoulli* came forward with his solution. *D'Alembert* derived this solution[9] by a method we will briefly discuss:

We try to transform the coordinates x, t in (II.3) with the idea of securing a partial differential equation in the new variables, say ξ, τ so that terms with the second derivative with respect to ξ and with the second derivative with respect to τ do not appear.

If

$$\left.\begin{aligned} \xi &= \xi(x, t) \\ \tau &= \tau(x, t) \end{aligned}\right\}, \quad \frac{\partial(\xi, \tau)}{\partial(x, y)} \neq 0$$

(see also Chapter III, §2.1 and problem III.10) stands for the transformation, we have in general

$$u_x = u_\xi \xi_x + u_\tau \tau_x,$$

$$u_{xx} = u_{\xi\xi}\xi_x^2 + 2u_{\xi\tau}\xi_x\tau_x + u_\xi \xi_{xx} + u_{\tau\tau}\tau_x^2 + u_\tau \tau_{xx}$$

and in analogy to this

$$u_{tt} = u_{\xi\xi}\xi_t^2 + 2u_{\xi\tau}\xi_t\tau_t + u_\xi \xi_{tt} + u_{\tau\tau}\tau_t^2 + u_\tau \tau_{tt}.$$

Under these substitutions equation (II.3) appears in the form

$$\frac{\partial^2 u}{\partial \xi^2}(\alpha^2 \xi_x^2 - \xi_t^2) + 2\frac{\partial^2 u}{\partial \xi\, \partial \tau}(\alpha^2 \xi_x \tau_x - \xi_t \tau_t) + \frac{\partial^2 u}{\partial \tau^2}(\alpha^2 \tau_x^2 - \tau_t^2)$$
$$+ u_\xi(\alpha^2 \xi_{xx} - \xi_{tt}) + u_\tau(\alpha^2 \tau_{xx} - \tau_{tt}) = 0.$$

To make the terms with $\partial^2 u/\partial \xi^2$ and $\partial^2 u/\partial \tau^2$ vanish, we have to choose ξ, τ so that

$$\alpha \frac{\partial \xi}{\partial x} = \pm \frac{\partial \xi}{\partial t},$$

$$\alpha \frac{\partial \tau}{\partial x} = \pm \frac{\partial \tau}{\partial t}.$$

We can see easily that it is sufficient to take

$$\xi = x + \alpha t,$$
$$\tau = x - \alpha t.$$

We can further see that this choice of ξ, τ makes also the terms with u_ξ and

[8] *Memoires de l'academie*, Berlin, 1753.

[9] *Memoires de l'academie*, Berlin, 1747.

u_τ disappear and furnishes for the coefficient of $\partial^2 u/\partial\xi\,\partial\tau$ the value 2. Hence, equation (II.3) in the new variables appears in the form

$$\frac{\partial u}{\partial\xi\,\partial\tau} = 0.$$

From this it follows that

$$\frac{\partial u}{\partial\xi} = f(\xi)$$

and

$$u(\xi, \tau) = \int f(\xi)\,d\xi + \Psi(\tau) = \Phi(\xi) + \Psi(\tau)$$

where Φ and Ψ are arbitrary differentiable functions.

In terms of our original variables x, t we have therefore

$$u(x, t) = \Phi(x + \alpha t) + \Psi(x - \alpha t).$$

If we now attempt to satisfy the initial conditions (IV.58) and (IV.59), we see that Φ and Ψ have to satisfy the conditions

$$\Phi(x) + \Psi(x) = \phi(x),$$
$$\alpha\Phi'(x) - \alpha\Psi'(x) = \psi(x).$$

The latter one can be written in the form

$$\Phi(x) - \Psi(x) = \frac{1}{\alpha}\int_0^x \psi(s)\,ds + C$$

which gives together with the first one

$$\Phi(x) = \frac{1}{2}\,\phi(x) + \frac{1}{2\alpha}\int_0^x \psi(s)\,ds + C,$$

$$\Psi(x) = \frac{1}{2}\,\phi(x) + \frac{1}{2\alpha}\int_x^0 \psi(s)\,ds - C,$$

and therefore

$$u(x, t) = \frac{1}{2}\,[\phi(x + \alpha t) + \phi(x - \alpha t)] + \frac{1}{2\alpha}\int_{x-\alpha t}^{x+\alpha t} \psi(s)\,ds,$$

which is identical with (IV.65).

Problems IV.39–46

39. Evaluate $\sum_{k=1}^{\infty}(a_k^2 + b_k^2)$ where a_k, b_k are the Fourier coefficients of the function

$$f(x) = \begin{cases} x & \text{for } 0 \le x < 1, \\ 1 & \text{for } 1 \le x < 3, \\ -x + 4 & \text{for } 3 \le x < 4, \end{cases}$$

$f(x + 4) = f(x)$. (*Hint:* Use the completeness relation.)

40. What conditions does the function in problem 39 satisfy as far as continuity and differentiability are concerned?

41. Establish a series for π^2 by integrating the series for the saw-tooth function termwise and substituting $x = \pi/2$.

42. Establish series for π^3, π^4, π^5 by the same process.

43. Find the solution of the string problem under the following initial and boundary conditions

$$u(0, t) = u(1, t) = 0$$

$$u(x, 0) = \begin{cases} 0.01x & \text{for } 0 \leq x < \frac{1}{2} \\ -0.01x + 0.01 & \text{for } \frac{1}{2} \leq x \leq 1 \end{cases}$$

$$\left.\frac{\partial u(x, t)}{\partial t}\right|_{t=0} = -0.1x^2 + 0.1x$$

44. At what time t does the plucked string coincide with the x-axis?

***45.** Find the temperature distribution between two parallel unbounded planes, 1 unit apart, if the planes are kept at the constant temperature 0 and the initial temperature is given by $f(x)$. (*Hint:* The temperature at any interior point obviously depends only on the distance from either plane (x or $1 - x$) and the time t.)

***46.** Solve the problem in 45 after changing the boundary conditions as follows: Instead of keeping the planes at the constant temperature 0, insulate them; i.e., insist that there is no flux of heat through these planes, or, equivalently, that $\partial u/\partial n = 0$ in the planes.

RECOMMENDED SUPPLEMENTARY READING

R. Courant: *Differential and Integral Calculus*, Vol. I, Interscience Publishers, New York, 1937, Chapter IX.

W. Rogosinski: *Fourier Series*, Chelsea Publishing Co., New York, 1950.

Felix Klein: *Elementary Mathematics from an Advanced Standpoint* (*Arithmetic, Algebra, Analysis*), The Macmillan Co., New York, 1932.

T. Apostol: *Mathematical Analysis*, Addison-Wesley Publishing Company, Reading, Mass., 1957, Chapter 15.

R. Courant and D. Hilbert: *Methods of Mathematical Physics*, Vol. I, Interscience Publishers, New York, 1953, Chapter II.

H. S. Carslaw: *Introduction to the Theory of Fourier Series and Integrals*, 3rd ed., Dover Publications, New York, 1930.

Ph. Frank and R. von Mises: *Differential und Integralgleichungen der Mechanik und Physik*, F. Vieweg and Sohn, Braunschweig, Germany 1935, Vol. I, Chapter IV, §4.

I. N. Sneddon: *Fourier Transforms*, McGraw-Hill Book Co., New York, 1951.

V

SELF-ADJOINT BOUNDARY VALUE PROBLEMS

§1. SELF-ADJOINT DIFFERENTIAL EQUATIONS OF THE SECOND ORDER

1. Self-adjoint Differential Equations

Let us consider a system of two linear homogeneous differential equations of the first order

$$\begin{aligned} \phi_1' &= a_{11}(x)\phi_1 + a_{12}(x)\phi_2 \\ \phi_2' &= a_{21}(x)\phi_1 + a_{22}(x)\phi_2 \end{aligned} \tag{V.1}$$

where the $a_{ik}(x)$ are real valued functions of x. One calls

$$A(x) = \begin{pmatrix} a_{11} & a_{12} \\ a_{21} & a_{22} \end{pmatrix}$$

the *coefficient matrix* of (V.1). A system of linear homogeneous differential equations of first order is called *adjoint* to (V.1) if its coefficient matrix B bears the following relationship to A:

$$B = -A^*$$

where A^* denotes the *transpose* of A.[1]

Since

$$A^* = \begin{pmatrix} a_{11} & a_{21} \\ a_{12} & a_{22} \end{pmatrix},$$

the system

$$\begin{aligned} \psi_1' &= -a_{11}\psi_1 - a_{21}\psi_2 \\ \psi_2' &= -a_{12}\psi_1 - a_{22}\psi_2 \end{aligned} \tag{V.2}$$

[1] All the definitions which are given here for a system of two linear homogeneous differential equations of first order can be immediately generalized to systems of n linear homogeneous differential equations of the first order (see problem V.1).

is, according to our definition, adjoint to (V.1). Clearly, (V.1) is by the same token adjoint to (V.2).

The name "adjoint" is derived from the following fact: If $\phi_1^{(1)}(x)$, $\phi_2^{(1)}(x)$ and $\phi_1^{(2)}(x)$, $\phi_2^{(2)}(x)$ are two linearly independent solutions of (V.1) (see AIII.B6), i.e.,

$$\Delta[\phi^{(1)}, \phi^{(2)}] = \begin{vmatrix} \phi_1^{(1)} & \phi_1^{(2)} \\ \phi_2^{(1)} & \phi_2^{(2)} \end{vmatrix} \neq 0,$$

then, $\psi_1^{(1)}(x) = \frac{1}{\Delta}\,\phi_2^{(2)}$, $\psi_2^{(1)}(x) = -\frac{1}{\Delta}\,\phi_1^{(2)}$ and $\psi_1^{(2)}(x) = -\frac{1}{\Delta}\,\phi_2^{(1)}$, $\psi_2^{(2)}(x) = \frac{1}{\Delta}\,\phi_1^{(1)}$ are two linearly independent solutions of (V.2). Thus, the *solution matrix* of (V.2)

$$\Psi(x) = \begin{pmatrix} \psi_1^{(1)} & \psi_1^{(2)} \\ \psi_2^{(1)} & \psi_2^{(2)} \end{pmatrix}$$

appears to be the *inverse transposed* or *adjoint* of the solution matrix

$$\Phi(x) = \begin{pmatrix} \phi_1^{(1)} & \phi_1^{(2)} \\ \phi_2^{(1)} & \phi_2^{(2)} \end{pmatrix}$$

of (V.1), i.e., $\Psi(x) = (\Phi^{-1}(x))^*$ (see problem V.2).

A differential equation of the second order can be reduced to a system of two differential equations of the first order. In particular, a linear homogeneous differential equation of the second order can be reduced to a system of two linear homogeneous differential equations of the first order. If we consider

$$y'' + a_1(x)y' + a_2(x)y = 0 \tag{V.3}$$

where the $a_i(x)$ are real valued functions of x and put $y = y_1$ and $y' = y_2$, then

$$\begin{aligned} y_1' &= y_2 \\ y_2' &= -a_2y_1 - a_1y_2 \end{aligned} \tag{V.4}$$

with the coefficient matrix

$$A = \begin{pmatrix} 0 & 1 \\ -a_2 & -a_1 \end{pmatrix}.$$

Since

$$A^* = \begin{pmatrix} 0 & -a_2 \\ 1 & -a_1 \end{pmatrix},$$

the system which is adjoint to (V.4) is

$$\begin{aligned} z_1' &= a_2z_2, \\ z_2' &= -z_1 + a_1z_2. \end{aligned} \tag{V.5}$$

Differentiation of the second equation and substitution of a_2z_2 for z_1' according to the first equation leads to

$$z'' - [a_1(x)z]' + a_2(x)z = 0, \tag{V.6}$$

where we have written z instead of z_2.

We call equations (V.3) and (V.6) *adjoint differential equations* because their corresponding systems of differential equations of first order are adjoint in the sense of our definition.

In case the coefficient of y'' in (V.3) is not 1 but some function $a_0(x) \not\equiv 0$,

$$a_0(x)y'' + a_1(x)y' + a_2(x)y = 0 \tag{V.7}$$

one calls in analogy to (V.6)

$$[a_0z]'' - [a_1z]' + a_2z = 0 \tag{V.8}$$

the adjoint differential equation to (V.7).

We note that the previous definition appears as a special case of this general definition for $a_0 =$ constant. However, if a_0 is some function of x, then the adjoint differential equation as defined in (V.8) differs from the one which is obtained from (V.6) after division by $a_0(x)$ (see problem V.3).

The operators

$$L = a_0\frac{d^2}{dx^2} + a_1\frac{d}{dx} + a_2, \tag{V.9}$$

$$M = \frac{d^2}{dx^2}(a_0\cdot) - \frac{d}{dx}(a_1\cdot) + a_2 \tag{V.10}$$

where the dot (.) denotes the place holder for the function to which M is to be applied, are called, accordingly, *adjoint differential operators.*

The following relation between adjoint differential expressions is of paramount importance for our subsequent discussions:

Lemma V.1 (Lagrange's identity).[2] *If L and M are adjoint differential operators of second order, then there exists a function $Q(x, y, y', z, z')$ such that*

$$zL[y] - yM[z] = \frac{d}{dx}Q(x, y, y', z, z').$$

Proof.

$$\begin{aligned} zL[y] - yM[z] &= a_0y''z + a_1y'z + a_2yz - y(a_0z)'' + y(a_1z)' - a_2yz \\ &= \frac{d}{dx}[a_0y'z - y(a_0z)' + y(a_1z)], \quad \text{q.e.d.} \end{aligned}$$

[2] Lemma V.I and its subsequent corollary are true for adjoint operators of any order, as one can easily see (see problem V.4).

Corollary to Lemma V.1. *If L and M are adjoint operators of second order, then*

$$\int_{x_1}^{x_2} (zL[y] - yM[z])\,dx = [a_0 y' z - y(a_0 z)' + y a_1 z]_{x_1}^{x_2}.$$

In our attempt in Chapter II to solve the problems of the vibrating string, the vibrating membrane, and a certain problem in heat conduction, we arrived via Bernoulli's separation method at the following ordinary differential equations of first order:

$$y'' + \lambda y = 0 \quad \text{(Chapter II, §1.2)}, \tag{V.11}$$

$$x^2 y'' + xy' + (\lambda x^2 - \mu^2) y = 0 \quad \text{(Bessel's equation, Chapter II, §2.4)}, \tag{V.12}$$

$$x^2 y'' + 2xy' + \lambda y = 0 \quad \text{(Euler-Cauchy equation, Chapter II, §3.4)}, \tag{V.13}$$

$$(1 - x^2) y'' - 2xy' + \lambda y = 0 \quad \text{(Legendre's equation, Chapter II, §3.4)}, \tag{V.14}$$

where we now consistently write y for the unknown function, x for the independent variable, and λ for the arbitrary parameter. (The parameter μ in (V.12) has to be considered as a fixed value which is determined by conditions which are not directly connected with the solution of Bessel's equation.)

Clearly, all these equations can be written in the form

$$L[y] = \frac{d}{dx}[p(x) y'] + q(x) y = 0 \tag{V.15}$$

with $p(x) = 1$, $q(x) = \lambda$ in case of (V.11); $p(x) = x$, $q(x) = (\lambda x^2 - \mu^2)/x$ in case of (V.12) after division by x; $p(x) = x^2$, $q(x) = \lambda$ in case of (V.13); and finally $p(x) = 1 - x^2$, $q(x) = \lambda$ in case of (V.14).

The significance of the fact that all these equations can be written in the form (V.15) lies in the following lemma:

Lemma V.2. *A differential operator of second order, L, is adjoint to itself, or* self-adjoint, *if and only if it has the form*

$$L = \frac{d}{dx}\left[p(x)\frac{d}{dx}\,\cdot\,\right] + q(x)$$

where $p(x)$ is differentiable.

The expression $L[y]$ is called a *self-adjoint differential expression* of the second order and $L[y] = 0$ a *self-adjoint differential equation* of the second order.[3]

Proof. If we expand L, we obtain

$$L = p'(x)\frac{d}{dx} + p(x)\frac{d^2}{dx^2} + q(x).$$

The adjoint operator M is, according to (V.10),

$$M = \frac{d^2}{dx^2}[p(x).] - \frac{d}{dx}[p'(x).] + q(x) = p(x)\frac{d^2}{dx^2} + p'(x)\frac{d}{dx} + q(x),$$

i.e., $L = M$.

On the other hand, if

$$L = a_0\frac{d^2}{dx^2} + a_1\frac{d}{dx} + a_2$$

and

$$M = \frac{d^2}{dx^2}(a_0.) - \frac{d}{dx}(a_1.) + a_2$$

are to be equal, it follows after one carries out all the indicated differentiations that

$$a_0' = a_1 = p'(x), \quad \text{q.e.d.}$$

According to lemma V.1, Lagrange's identity for self-adjoint differential expressions takes the form (see problem V.5)

$$zL[y] - yL[z] = \frac{d}{dx}[p(x)(y'z - yz')]. \tag{V.16}$$

Problems V.1–6

1. Given the system of n linear homogeneous differential equations of the first order

$$\begin{aligned} y_1' &= a_{11}y_1 + \cdots + a_{1n}y_n, \\ &\;\cdot \\ &\;\cdot \\ &\;\cdot \\ y_n' &= a_{n1}y_1 + \cdots + a_{nn}y_n. \end{aligned}$$

Write down the adjoint system.

***2.** Show that, if $\phi_1^{(1)}, \phi_2^{(1)}$ and $\phi_1^{(2)}, \phi_2^{(2)}$ are two linearly independent solutions of (V.1), then $\frac{1}{\Delta}\phi_2^{(2)}, -\frac{1}{\Delta}\phi_2^{(1)}$ and $-\frac{1}{\Delta}\phi_1^{(2)}, \frac{1}{\Delta}\phi_1^{(1)}$ are two linearly independent solutions of the adjoint system (V.2).

[3] Self-adjoint operators can be defined for all operators of even order by adjoint $L = L$. Differential operators of odd order, however, are called self-adjoint if adjoint $L = -L$, for obvious reasons.

3. Given $a_0(x)y'' + a_1(x)y' + a_2(x)y = 0$, where $a_0(x) \neq 0$. Divide through by $a_0(x)$ and find the adjoint equation according to (V.6). Carry out all the differentiations and show that the equation thus obtained is different from (V.8).

***4.** Show that Lagrange's identity is satisfied for two adjoint differential expressions of nth order, where the adjoint differential expression of nth order is defined in analogy to (V.8) such that the sign of the term with y is positive.

5. Show that for a self-adjoint operator L Lagrange's identity reduces to (V.16).

6. Given

$$y'' + a_1(x)y' + a_2(x)y = 0.$$

Show, that after multiplication of this equation by

$$p(x) = e^{\int a_1(x)\,dx},$$

it can be written in the form (V.15).

2. Examples

We will investigate in this section the solutions of self-adjoint differential equations especially with regard to their zeros. We have already seen in Chapter II that this is one of the vital points in connection with the solution of certain boundary value problems (see Chapter II, §2.4). To get a vague idea of what we have to expect in the general case—and this vague idea will turn out to be a rather good one—we will discuss shortly the simplest self-adjoint equation of the ones which are listed, namely (V.11)

$$y'' + \lambda y = 0.$$

Let us first discuss the case where $\lambda > 0$ and write $\lambda = \mu^2$.

Then,

$$y_1 = \sin \mu x, \quad y_2 = \cos \mu x$$

will constitute a system of two linearly independent solutions (see problems VI.1, 2.) We know that these functions for any fixed μ have infinitely many zeros which are, however, discretely distributed; i.e., in any finite interval there are only finitely many zeros, if any at all. The zeros of y_1 are located at $k\pi/\mu$, the zeros of y_2 at $\pi/2\mu + k\pi/\mu$, $(k = 0, \pm 1, \pm 2, \pm 3, \cdots)$. We see that between two consecutive zeros of the one solution there is one zero of the other solution, and as a matter of fact, of any other solution. This situation will change abruptly if we consider the case where $\lambda < 0$. We write $\lambda = -\mu^2$, and now

$$y_1 = e^{\mu x}, \quad y_2 = e^{-\mu x}$$

is a system of linearly independent solutions, neither of which has a zero at all. By proper choice of linear combinations of these two functions we can obtain another system of linearly independent solutions, such as

$$\sinh \mu x, \quad \cosh \mu x$$

one of which has a zero and the other one does not, or

$$\sinh \mu x, \quad \sinh \mu(x - a), \qquad a \neq 0$$

where both have one zero each.

It is, however, impossible to find a solution with more than one zero and it is therefore senseless to talk about zeros between consecutive zeros of a solution.

We will see in subsection 4 of this section that this situation is typical for the solutions of (V.15). Whether the first situation arises ($\lambda > 0$) or the second one ($\lambda < 0$) will depend on whether $p(x)$, $q(x)$ have equal sign or unequal sign (see theorems V.2, 3).

There is one other point we would like to make before proceeding to a general discussion:

Let λ_1, λ_2 be positive numbers such that $\lambda_1 < \lambda_2$. Then, with $\lambda_1 = \mu_1^2$ and $\lambda_2 = \mu_2^2$, the solutions of

$$y'' + \mu_1^2 y = 0,$$
$$y'' + \mu_2^2 y = 0$$

will be

$$y_1^{(1)} = \sin \mu_1 x, \quad y_2^{(1)} = \cos \mu_1 x$$

and

$$y_1^{(2)} = \sin \mu_2 x, \quad y_2^{(2)} = \cos \mu_2 x, \quad \text{respectively.}$$

The zeros of the solution of the first equation are

$$\frac{k\pi}{\mu_1}, \quad \frac{\pi}{2\mu_1} + \frac{k\pi}{\mu_1}$$

and those of the solutions of the second equation are

$$\frac{k\pi}{\mu_2}, \quad \frac{\pi}{2\mu_2} + \frac{k\pi}{\mu_2}$$

where k denotes an arbitrary integer, and we can see that between two consecutive zeros of a solution corresponding to the smaller λ_1, there is at least one zero of a solution corresponding to the greater λ_2 (see theorem V.5).

Problem V.7

7. Graph one solution of $y'' + 4y = 0$ in red pencil and one solution of $y'' + 9y = 0$ in blue pencil in the same coordinate system. Observe that the blue solution has at least one zero between two consecutive zeros of the red solution.

3. The x, ϕ–Diagram

We will assume that $p(x)$ and $q(x)$ in

$$\frac{d}{dx}[p(x)y'] + q(x)y = 0 \tag{V.15}$$

are continuous and that $p(x)$ is differentiable in a certain closed interval $a \leq x \leq b$. We will assume furthermore that

$$p(x) > 0 \quad \text{in } a \leq x \leq b.$$

(In case $p(x) < 0$ in $a \leq x \leq b$, we multiply the equation (V.15) by -1. The case where $p(x) = 0$ at certain points of the interval has to be excluded for reasons which will soon become obvious.)

If we introduce a new function z according to

$$z(x) = p(x)y'(x) \tag{V.17}$$

we can write (V.15) as a system of two differential equations of the first order

$$\begin{aligned} \frac{dy}{dx} &= \frac{1}{p(x)} z, \\ \frac{dz}{dx} &= -q(x)y, \end{aligned} \tag{V.18}$$

and apply to this system the standard existence and uniqueness theorems for ordinary differential equations (see AIII.B4). Under not too severe conditions concerning $p(x)$ and $q(x)$ we can establish the existence and uniqueness of the solutions of (V.18), which we will keep in mind during the following discussion.

It is of considerable help in investigating the nature of the solutions of (V.15) and especially their zeros to introduce polar coordinates.[4]

$$\begin{aligned} y &= r(x) \cos \phi(x), \\ z &= r(x) \sin \phi(x). \end{aligned} \tag{V.19}$$

Thus we obtain for (V.18)

$$r' \cos \phi - r \sin \phi \cdot \phi' = \frac{1}{p} r \sin \phi,$$

$$r' \sin \phi + r \cos \phi \cdot \phi' = -qr \cos \phi.$$

[4] The method which we follow in the discussions of this chapter was introduced by H. Pruefer, *Mathematische Annalen* 95 (1926).

Solving for r' and ϕ' yields

$$\frac{d\phi}{dx} = -\left(\frac{1}{p(x)}\sin^2\phi + q(x)\cos^2\phi\right), \tag{V.20}$$

$$\frac{dr}{dx} = \left(\frac{1}{p(x)} - q(x)\right) r \sin\phi\cos\phi. \tag{V.21}$$

Equation (V.20) does not contain $r(x)$; i.e., we can solve for ϕ independently of (V.21), substitute this solution ϕ into (V.21), separate the variables, and solve for $r(x)$:

$$r(x) = e^{1/2\int(1/p - q)\sin 2\phi\, dx} \tag{V.22}$$

It is sufficient for our purposes to concentrate our investigations on the solutions ϕ of (V.20). We can see, first of all, that if $\phi(x)$ is a solution of (V.20), then $\phi(x) + k\pi$ will also be a solution since, according to (V.19),

$$r\cos(\phi + k\pi) = (-1)^k r\cos\phi = (-1)^k y,$$
$$r\sin(\phi + k\pi) = (-1)^k r\sin\phi = (-1)^k z = (-1)^k py';$$

i.e., if k is odd, $\phi + k\pi$ will give the solution $-y(x)$ instead of $y(x)$ and if k is even, we obtain $y(x)$ itself. Since $-y$ is, in case of a homogeneous differential equation, a solution if and only if y is a solution, we can consider $\phi(x) + k\pi$ for all $k = 0, \pm 1, \pm 2, \cdots$ as one and the same solution. Hence, we can always assume that at a given point x_0, $\phi(x)$ satisfies the inequality

$$\frac{\pi}{2} + (k-1)\pi < \phi(x_0) \leq \frac{\pi}{2} + k\pi$$

for any given integer k.

Let us next prove the following lemma, which will turn out to be essential for all the subsequent investigations:

Lemma V.3. *The solution $y(x)$ of (V.15) has a zero at $x = x_1$ if and only if $\phi(x_1) = (\pi/2) + k\pi$. (We call such points x_1 the $\pi/2$-*intercepts *of $\phi(x)$.)*

Proof. If $\phi(x_1) = (\pi/2) + k\pi$, it follows from (V.19) that $y(x_1) = 0$.

Let us now assume that $y(x_1) = 0$ to prove the necessity of the condition. It follows then from (V.19) that

$$r(x_1)\cos\phi(x_1) = 0.$$

Since $r(x_1) \neq 0$ for x_1 in $a \leq x \leq b$ in view of (V.22), it follows that $\cos\phi(x_1) = 0$ and from this $\phi(x_1) = (\pi/2) + k\pi$.

Problem V.8

8. Find a ϕ-curve of a solution of $y'' + y = 0$ and list all $\pi/2$-intercepts and find a ϕ-curve of a solution of $y'' - y = 0$ and list all $\pi/2$-intercepts.

4. The Zeros of the Solutions of a Self-Adjoint Differential Equation

By means of the x, ϕ–diagram and the lemma which we just proved in the preceding subsection, we can easily establish the simplicity of the zeros of the solutions of (V.15).

Theorem V.1. *The zeros of the solutions of* (*V.15*) *are simple*; *i.e.*: *If* $y(x_1) = 0$, *then* $y'(x_1) \neq 0$.

Proof. $y(x_1) = 0$ if and only if $\phi(x_1) = (\pi/2) + k\pi$, as we established in lemma V.3. Then

$$p(x_1)y'(x_1) = z(x_1) = r(x_1)\sin\left(\frac{\pi}{2} + k\pi\right) = (-1)^k r(x_1)$$

(see (V.19)), and therefore

$$y'(x_1) = (-1)^k \frac{r(x_1)}{p(x_1)} \neq 0, \quad \text{q.e.d.}$$

We will now enter the discussion concerning the relative location of the zeros of two solutions of (V.15). As was suggested by the example we discussed in subsection 2 of this section, we will have to assume in addition to $p(x) > 0$ that $q(x) > 0$ in $a \leq x \leq b$. It follows then from (V.20) that

$$\frac{d\phi}{dx} < 0 \quad \text{for all } x \text{ in } a \leq x \leq b.$$

That means that the ϕ-curve (in the x, ϕ–coordinate system) which corresponds to the solution $y(x)$ of (V.15), the ϕ-curve of y (for short), will decrease monotonically (see Fig. 16). Suppose it intersects a line $\phi = (\pi/2) + k\pi$ at x_1 in $a \leq x \leq b$ ($k = 0$ in Fig. 16). If, in addition, it intersects the line $\phi = (\pi/2) + (k - 1)\pi$, then it does so at a point $x_2 > x_1$. This means that the solution $y(x)$ of (V.15) will have two consecutive zeros at x_1, x_2. With a second solution $\bar{y}(x)$ of (V.15) there will correspond a $\bar{\phi}(x) + k\pi$. The curve $\bar{\phi} + k\pi$ cannot possibly intersect the curve $\phi + k\pi$, since it appears from (V.20) that there exists at every point (x, ϕ) one and only one slope ϕ'. Thus, ϕ, $\bar{\phi}$ do not intersect at all or are identical. We can always pick a $\bar{k}$ so that

$$\phi < \bar{\phi} + \bar{k}\pi < \phi + \pi.$$

From this it follows that $\bar{\phi} + \bar{k}\pi$ has to intersect the line $\phi = (\pi/2) + k\pi$ at a point $\bar{x}$ between x_1 and x_2 (see Fig. 16).

Let us state this result as a theorem:

Theorem V.2. *If $y(x)$ and $\bar{y}(x)$ are any two solutions of (V.15) and sign $p(x) =$ sign $q(x)$ and if $y(x)$ vanishes at two consecutive points x_1, x_2 where $x_1 < x_2$, then $\bar{y}(x)$ has to vanish at a point $\bar{x}$, where $x_1 < \bar{x} < x_2$.*

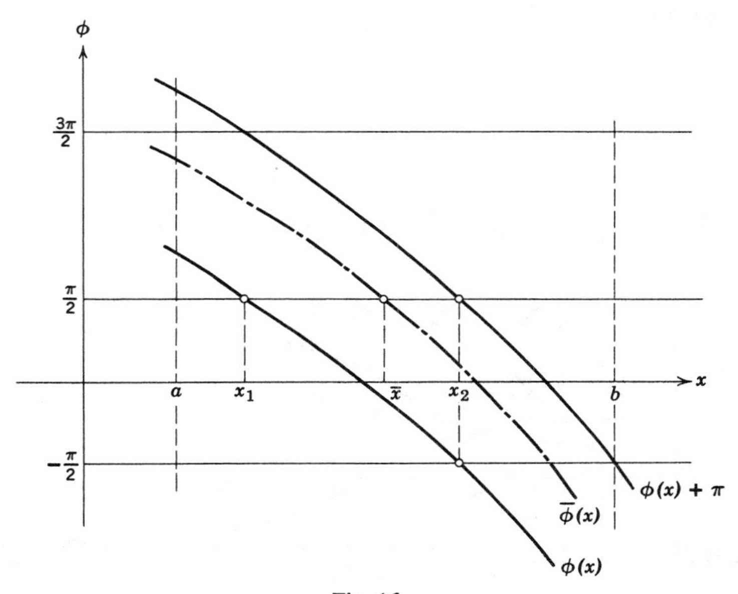

Fig. 16

To supplement our treatment, we have to discuss now the case where sign $p \neq$ sign q, i.e., $q(x) < 0$, if we assume $p(x) > 0$.

In this case

$$\left.\frac{d\phi}{dx}\right|_{\phi = k\pi} = -q > 0;$$

i.e., the ϕ-curves have to pass through the lines $\phi = k\pi$ with a positive slope, while

$$\left.\frac{d\phi}{dx}\right|_{\phi = (\pi/2) + k\pi} = -\frac{1}{p} < 0;$$

i.e., the ϕ-curves pass through the lines $\phi = (\pi/2) + k\pi$ with a negative slope. The direction field is illustrated in Fig. 17. It looks like a trap and that's what it is. We can see that a ϕ-curve which is in a strip $(k - 1)\pi$, $(k - 1)\pi + (\pi/2)$, can never get out of it since a doubling over of the ϕ-curve would mean that $\phi(x)$ is a multiple-valued function with the consequence that the corresponding $y(x)$ would be multiple-valued also, a case which we cannot admit. If the ϕ-curve is in the strip $(k - 1)\pi + (\pi/2)$ $k\pi$, it either passes $\phi = k\pi$ and cannot leave the strip $k\pi$, $k\pi + (\pi/2)$, or it passes $(k - 1)\pi + (\pi/2)$ and cannot leave $(k - 1)\pi$, $(k - 1)\pi + (\pi/2)$, or it does not cross at all. In the first case it cannot possibly have a $\pi/2$-intercept and in the second case it can have one $\pi/2$-intercept at most. We can therefore state

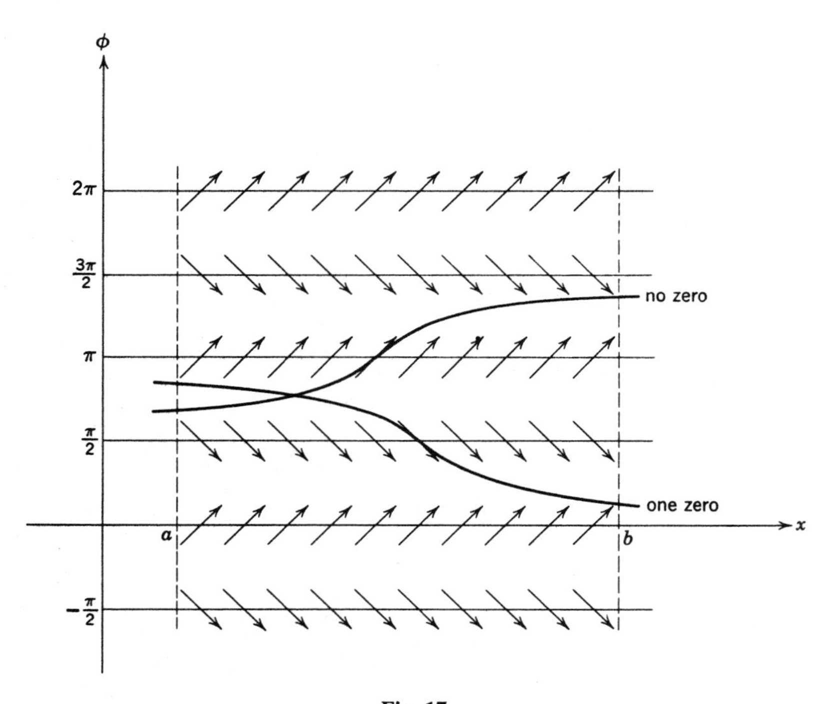

Fig. 17

Theorem V.3. *If $y(x)$ is a solution of (V.15) and sign $p(x) \neq$ sign $q(x)$, then it has either only one zero or no zero at all.*

Theorem V.4. *The zeros of the solutions of (V.15) are discrete; i.e.: in any finite interval there are only finitely many zeros, if any at all.*

Proof. (a) sign $p(x) \neq$ sign $q(x)$. This case is trivial, since there exists either no zero at all or only one zero according to theorem V.3.

(b) sign $p(x) =$ sign $q(x)$. In this instance, according to (V.20), $d\phi/dx < 0$ and on the other hand

$$\frac{d\phi}{dx} > -\left\{\frac{1}{m} + M\right\} \tag{V.23}$$

where

$$m = \min_{[a,b]} p(x), \tag{V.24}$$

$$M = \max_{[a,b]} q(x). \tag{V.25}$$

Let x_1 be the first zero of $y(x)$ in $a \leq x \leq b$. Then its ϕ-curve $\phi(x)$ will have a $\pi/2$-intercept at this point. Let us now consider the line

$$\bar{\phi} = -\left(\frac{1}{m} + M\right)x + C$$

where we choose C such that it passes through the $\pi/2$-intercept $(x_1, (\pi/2) + k\pi)$ of $\phi(x)$. This line will intersect the lines $\phi = (\pi/2) + (k - i)\pi$,

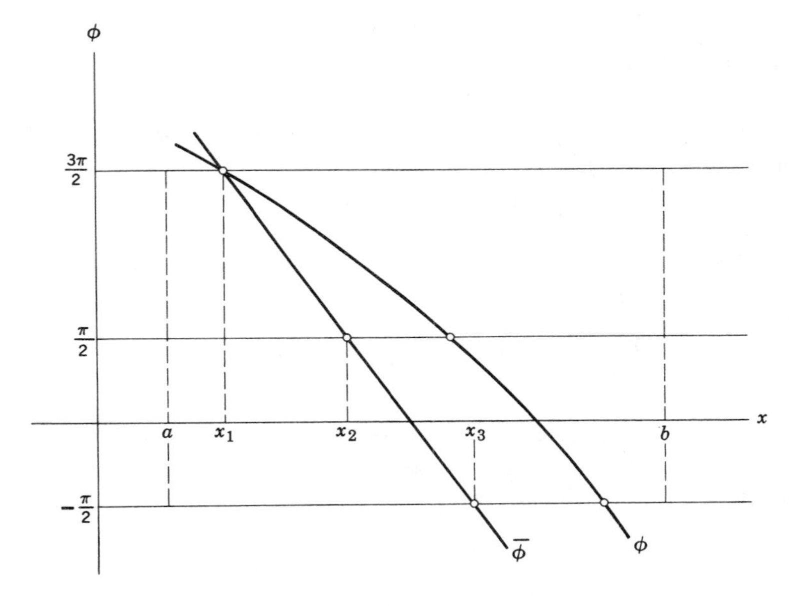

Fig. 18

$(i = 1, 2, 3, \cdots)$ at the points $x_i = x_1 + \dfrac{\pi i}{(1/m) + M}$ (see Fig. 18). If $a \leq x \leq b$ is an interval of finite length, there can be only a finite number of such intersection points in $a \leq x \leq b$. Since the slope of $\phi(x)$ is less steep than the slope of $\bar{\phi}(x)$ according to (V.23), $\phi(x)$ cannot intersect the lines $\phi = (\pi/2) + (k - i)\pi$ in more points than $\bar{\phi}$, q.e.d.

It is to be expected that a comparison of the solutions of (V.15) with the known solutions of (V.11) will furnish valuable information and this is indeed the case. The following "comparison lemma" will constitute a key to a number of rather rich results:

Lemma V.4 (comparison lemma). *If $y_1(x)$ is a solution of*

$$\frac{d}{dx}[p_1(x)y'] + q_1(x)y = 0$$

and $y_2(x)$ is a solution of

$$\frac{d}{dx}[p_2(x)y'] + q_2(x)y = 0,$$

where $p_1(x) \geq p_2(x)$, $q_1(x) \leq q_2(x)$, ($p(x)$ and $q(x)$ may have different or equal signs) and if $\phi_1(x)$ is the ϕ-curve of $y_1(x)$ and $\phi_2(x)$ the ϕ-curve of $y_2(x)$ and for a certain x_0 in $a \leq x \leq b$

$$\phi_1(x_0) = \phi_2(x_0),$$

then for all $x > x_0$

$$\phi_2(x) \leq \phi_1(x).$$

(In case $p_1 \geq p_2$, $q_1 < q_2$ hold, $\phi_2 < \phi_1$, as will become apparent in the following proof. See also problem V.9.)

Proof. According to (V.20)

$$\begin{aligned}\frac{d(\phi_2 - \phi_1)}{dx} &= -\left(\frac{1}{p_2}\sin^2\phi_2 - \frac{1}{p_1}\sin^2\phi_1 + q_2\cos^2\phi_2 - q_1\cos^2\phi_1\right)\\ &= -\left(\frac{1}{p_2} - \frac{1}{p_1}\right)\sin^2\phi_2 - (q_2 - q_1)\cos^2\phi_2 \qquad \text{(V.26)}\\ &\quad + \frac{1}{p_1}(\sin^2\phi_1 - \sin^2\phi_2) + q_1(\cos^2\phi_1 - \cos^2\phi_2).\end{aligned}$$

Let us now introduce the following abridged notation:

$$h(x) = -\left(\frac{1}{p_2} - \frac{1}{p_1}\right)\sin^2\phi_2 - (q_2 - q_1)\cos^2\phi_2$$

where

$$h(x) \leq 0 \qquad \text{(V.27)}$$

in view of

$$\frac{1}{p_2} - \frac{1}{p_1} \geq 0, \quad q_2 - q_1 \geq 0,$$

(note that we can assume without loss of generality that $p_2(x) > 0$) and let

$$k(x) = \frac{1}{p_1}\frac{\sin^2\phi_1 - \sin^2\phi_2}{\phi_2 - \phi_1} + q_1\frac{\cos^2\phi_1 - \cos^2\phi_2}{\phi_2 - \phi_1}.$$

$h(x)$ and $k(x)$ are continuous functions. This is evident in the case of $h(x)$. $k(x)$, however, to make it continuous, has to be defined for $\phi_1 = \phi_2$ by its limit. We have

$$\lim_{\phi_2 \to \phi_1} \frac{\sin^2 \phi_1 - \sin^2 \phi_2}{\phi_2 - \phi_1} = -\sin 2\phi_1$$

and

$$\lim_{\phi_2 \to \phi_1} \frac{\cos^2 \phi_1 - \cos^2 \phi_2}{\phi_2 - \phi_1} = \sin 2\phi_1$$

as the reader can easily verify (see problem V.10).

Hence, in order to make $k(x)$ everywhere continuous, we define

$$k(x)\Big|_{\phi_2 = \phi_1} = -\frac{1}{p_1} \sin 2\phi_1 + q_1 \sin 2\phi_1.$$

We can write (V.26) in terms of these functions $h(x)$ and $k(x)$ in the form

$$\frac{d(\phi_2 - \phi_1)}{dx} = h(x) + (\phi_2 - \phi_1)k(x). \tag{V.28}$$

If we multiply this equation by its integrating factor $e^{-\int k(x)\,dx}$ we obtain

$$\frac{d}{dx}\{e^{-\int k(x)\,dx} \cdot (\phi_2 - \phi_1)\} = h(x)e^{-\int k(x)\,dx}.$$

(see problem V.6). In view of (V.27)

$$\frac{d}{dx}\{e^{-\int k(x)\,dx} \cdot (\phi_2 - \phi_1)\} \leq 0.$$

Hence, the function

$$s(x) = e^{-\int k(x)\,dx} \cdot (\phi_2 - \phi_1)$$

is monotonically decreasing. $s(x_0) = 0$ since $\phi_1(x_0) = \phi_2(x_0)$ by hypothesis, and therefore

$$s(x) \leq 0 \quad \text{for all } x > x_0.$$

Since $e^{-\int k(x)\,dx} > 0$, it has to be true that $\phi_2(x) \leq \phi_1(x)$ for all $x > x_0$, q.e.d.

This result will enable us to obtain some important information about the relative position of the zeros of solutions of differential equations of the type (V.15).

Theorem V.5. *Let $y_1(x)$ be a solution of*

$$\frac{d}{dx}[p_1(x)y'] + q_1(x)y = 0$$

and let us assume that x_1, x_2 *are two consecutive zeros of* $y_1(x)$. *Then, any solution* $y_2(x)$ *of*

$$\frac{d}{dx}\left[p_2(x)y'\right] + q_2(x)y = 0$$

with $p_1(x) \geq p_2(x)$, $q_1(x) \leq q_2(x)$ *where* $p_i(x) > 0$, $q_i(x) > 0$ *has to vanish at least once in* $x_1 < x \leq x_2$; *i.e.: there has to exist at least one* $\bar{x}$, $x_1 < \bar{x} \leq x_2$, *such that* $y_2(\bar{x}) = 0$.

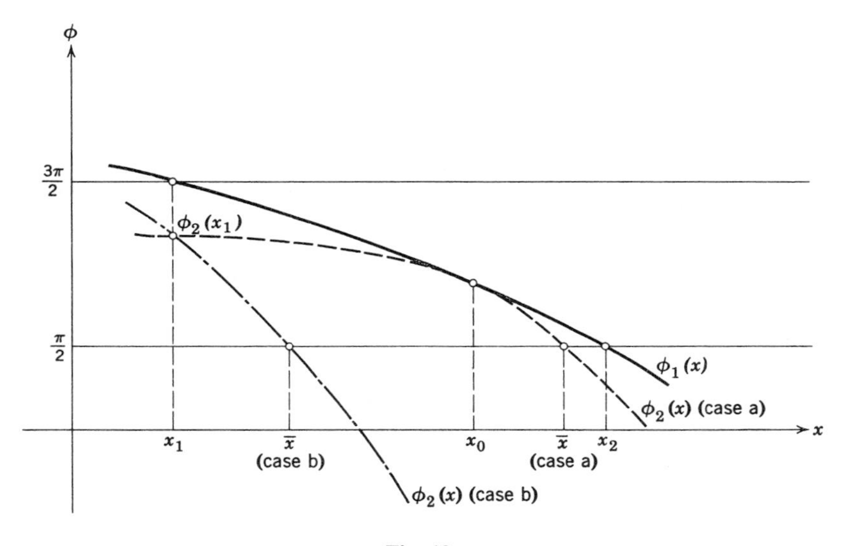

Fig. 19

Proof. If x_1, x_2 are two consecutive zeros of $y_1(x)$, then the ϕ-curve $\phi_1(x)$ of $y_1(x)$ has to intersect some line $\phi = (\pi/2) + k\pi$ at $x = x_1$ and the line $\phi = (\pi/2) + (k - 1)\pi$ at $x = x_2$. (Note that $\dfrac{d\phi_1}{dx} < 0$ in view of $p_1 > 0$, $q_1 > 0$). We can assume without loss of generality that $k = 1$, i.e.: $\phi_1(x_1) = 3\pi/2$, $\phi_1(x_2) = \pi/2$. If $\phi_2(x)$ is the ϕ-curve of $y_2(x)$, we can assume without loss of generality that $\pi/2 < \phi_2(x_1) \leq 3\pi/2$ by a proper choice of n in $\phi_2(x) + n\pi$.

Now, we have to consider the following two possibilities:

(a) There is some point x_0 in $x_1 \leq x < x_2$ such that $\phi_1(x_0) = \phi_2(x_0)$ (see Fig. 19). Then, we have according to lemma V.4

$$\phi_2(x) \leq \phi_1(x) \quad \text{for all } x > x_0.$$

Therefore, $\phi_2(x)$ has to intersect the line $\phi = \pi/2$ at some point $\bar{x}$ with $x_0 < \bar{x} \leq x_2$; i.e.: $y_2(\bar{x}) = 0$ where $x_1 < \bar{x} \leq x_2$.

(b) $\phi_2(x)$ does not intersect $\phi_1(x)$ in $x_1 \leq x < x_2$ (see Fig. 19). Then $\phi_2(x) < \phi_1(x)$ for all x in $x_1 \leq x < x_2$ since $\phi_2(x_1) < \phi_1(x_1)$. Hence, $\phi_2(x)$ has to intersect the line $\phi = \pi/2$ at some point $\bar{x}$ in $x_1 < \bar{x} \leq x_2$; i.e., $y_2(\bar{x}) = 0$ where $x_1 < \bar{x} \leq x_2$.

We have seen in the example we discussed in subsection 2 of this section that the distance between zeros of the solutions of (V.11) for $\lambda > 0$ is given by $k\pi/\sqrt{\lambda}$. This is of course due to the fact that p, q, in (V.11) are both constants. We cannot expect an equally simple result in case p, q are functions of x, as is the case in all the other equations we listed in subsection 1 of this section. We can, however, estimate the distance between zeros of solutions of (V.15) by comparing the general equation (V.15) with an equation of the type (V.11) and applying theorem V.5. Let us consider the two equations

$$\frac{d}{dx}(M_1 y') + m_2 y = 0, \tag{V.29}$$

$$\frac{d}{dx}(m_1 y') + M_2 y = 0 \tag{V.30}$$

where the constants m_1, m_2, M_1, M_2 are chosen so that equation (V.15) with $\operatorname{sign} p(x) = \operatorname{sign} q(x)$ appears as squeezed in between these two equations, or mathematically speaking

$$m_1 \leq p(x) \leq M_1,$$
$$m_2 \leq q(x) \leq M_2.$$

Let us first apply theorem V.5 to the equations (V.15) and (V.29). A solution of (V.29) is

$$y_1^{(1)} = \sin \sqrt{m_2/M_1}(x - \delta) \quad \text{for any } \delta.$$

The zeros of this solution follow each other at distances $\pi\sqrt{M_1/m_2}$. Thus x_1, $x_1 + \pi\sqrt{M_1/m_2}$ would be two consecutive zeros if x_1 is a zero. The solution $y(x)$ of (V.15) has to have a zero $\bar{x}_1$ such that

$$x_1 < \bar{x}_1 \leq x_1 + \pi\sqrt{M_1/m_2}$$

according to theorem V.5.

Let us now consider by choosing δ appropriately, a solution $y_2^{(1)}$ of (V.29) which has a zero at $\bar{x}_1$. The next zero of $y_2^{(1)}$ is then at $\bar{x}_1 +$

$\pi\sqrt{M_1/m_2}$ and the solution $y(x)$ of (V.15) has to again have a zero $\bar{x}_2$ such that

$$\bar{x}_1 < \bar{x}_2 \leq \bar{x}_1 + \pi\sqrt{M_1/m_2},$$

i.e.,

$$0 < \bar{x}_2 - \bar{x}_1 \leq \pi\sqrt{M_1/m_2}.$$

If we apply the same argument to (V.30) and (V.15) in the reverse order (see problem V.13), we will obtain $\bar{x}_2 - \bar{x}_1 \geq \pi\sqrt{m_1/M_2}$. Let us state this extremely interesting result in form of a theorem:

Theorem V.6. *If m_1, m_2, M_1, M_2 are fixed numbers such that*

$$m_1 \leq p(x) \leq M_1$$
$$m_2 \leq q(x) \leq M_2$$

then

$$\pi\sqrt{m_1/M_2} \leq |x_2 - x_1| \leq \pi\sqrt{M_1/m_2}$$

where x_1, x_2 are two consecutive zeros of a solution of (V.15) with sign p = sign q.

Let us apply this theorem to Bessel's equation (V.12):

We will consider the interval $A \leq x \leq A + d$ where A, d may be any positive numbers. Within this interval

$$m_1 = A \leq x = p(x) \leq A + d = M_1,$$

and consequently

$$m_2 = \lambda A - \frac{\mu^2}{A} \leq \lambda x - \frac{\mu^2}{x} = q(x) \leq \lambda(A + d) = M_2.$$

Thus, for two consecutive zeros of a solution of the Bessel equation within an interval $A \leq x \leq A + d$,

$$\frac{\pi}{\sqrt{\lambda}}\sqrt{\frac{A}{A + d}} \leq |x_2 - x_1| \leq \frac{\pi}{\sqrt{\lambda}}\sqrt{\frac{A + d}{A - (\mu^2/\lambda A)}}. \tag{V.31}$$

If A tends to infinity, the upper and lower bound of this inequality both approach $\pi/\sqrt{\lambda}$; i.e.: the distance between two consecutive zeros of a Bessel function approaches $\pi/\sqrt{\lambda}$ as x becomes large.

Problems V.9–15

***9.** Prove lemma V.4 in the following modified form: $\phi_2(x) < \phi_1(x)$ for all $x > x_0$ if $p_1(x) \geq p_2(x)$ and $q_1(x) < q_2(x)$.

10. Find

$$\lim_{\phi_2 \to \phi_1} \frac{\sin^2 \phi_1 - \sin^2 \phi_2}{\phi_2 - \phi_1}$$

and

$$\lim_{\phi_2 \to \phi_1} \frac{\cos^2 \phi_1 - \cos^2 \phi_2}{\phi_2 - \phi_1}.$$

***11.** Let y_1 be a solution of

$$\frac{d}{dx}[p_1(x)y'] + q_1(x)y = 0$$

and y_2 a solution of

$$\frac{d}{dx}[p_2(x)y'] + q_2(x)y = 0$$

and let $p_1 \geq p_2$, $q_1 < q_2$, $p_2 > 0$, $q_1 > 0$.
Let x_1 be a zero of y_1. Prove that if $x_2 > x_1$ is another zero of y_1, then there has to exist an $\bar{x}$ in $x_1 < \bar{x} < x_2$ such that $y_2(\bar{x}) = 0$. Compare this theorem with theorem V.5 and point out the differences.

12. $y^{(2)}(x) = \sin \sqrt{\frac{M_2}{m_1}}(x - \delta)$ is a solution of (V.30). Choose δ such that $y(\bar{x}_1) = 0$ where $\bar{x}_1$ is a zero of the solution of (V.15).

13. Find the lower bound for the difference of two consecutive zeros of the solution of (V.15). (See proof of theorem V.6 and problem 12.)

14. Find an upper and a lower bound for two consecutive zeros of a solution of (V.13), where there are any.

15. Discuss equation (V.13) from the standpoint of theorems V.2 and V.3.

§2. THE SELF-ADJOINT BOUNDARY VALUE PROBLEM OF STURM AND LIOUVILLE

1. Eigenvalues and Eigenfunctions

For the following investigation we assume that the function $q(x)$ in (V.15) has the form

$$q(x) = Q(x) + \lambda R(x)$$

where $R(x) > 0$ in $a \leq x \leq b$.

That this specification is still general enough for our purposes is clear from what was said in subsection 1 of the preceding section about the sources of this equation. (See equations (V.11, 12, 13, and 14).)

Instead of (V.15) we will henceforth consider the equation

$$\frac{d}{dx}[P(x)y'] + [Q(x) + \lambda R(x)]y = 0. \qquad \text{(V.32)}$$

In most instances, the boundary conditions which we encountered in Chapter II have been linear and homogeneous; i.e., no term in the conditions was free from either the unknown function or its derivative and both appear to the first power only. In general, we can write such linear homogeneous boundary conditions in the form

$$a_{11}y(x_1) + a_{12}y'(x_1) = 0, \tag{V.33a}$$

$$a_{21}y(x_2) + a_{22}y'(x_2) = 0. \tag{V.33b}$$

The problem of solving equation (V.32) under the boundary conditions (V.33) is called a *self-adjoint boundary value problem*, or a *Sturm-Liouville problem*, in honor of those two mathematicians, who investigated this problem with great success.

Before we go into any detail, we wish to state an immediate consequence of the corollary to lemma V.1 for functions which satisfy boundary conditions of the type (V.33):

Lemma V.5 (Green's lemma). *If $y(x)$ and $z(x)$ are any two twice differentiable functions and satisfy the boundary conditions (V.33), then we have for any self-adjoint operator L*

$$\int_{x_1}^{x_2} (zL[y] - yL[z])\, dx = 0.$$

Proof. In view of (V.16) we have for a self-adjoint operator L

$$zL[y] - yL[z] = \frac{d}{dx}[P(x)(y'z - yz')].$$

We integrate from x_1 to x_2

$$\int_{x_1}^{x_2} (zL[y] - yL[z])\, dx = [P(x)(y'z - yz')]_{x_1}^{x_2} = \left[Pyz\left(\frac{y'}{y} - \frac{z'}{z}\right)\right]_{x_1}^{x_2}$$

and note that it follows from (V.33a) that $\left(\frac{y'}{y}\right)_{x=x_1} = -\frac{a_{11}}{a_{12}}$ and $\left(\frac{z'}{z}\right)_{x=x_1} = -\frac{a_{11}}{a_{12}}$. Hence the term $\left(\frac{y'}{y} - \frac{z'}{z}\right)$ vanishes at the lower limit. In view of the second condition in (V.33) it vanishes at the upper limit also, q.e.d. (For the case where a_{12} and/or a_{22} vanishes see problem V.16.) If we impose the more general linear homogeneous boundary conditions

$$\begin{aligned} a_{11}y(x_1) + a_{12}y'(x_1) + b_{11}y(x_2) + b_{12}y'(x_2) &= 0, \\ a_{21}y(x_1) + a_{22}y'(x_1) + b_{21}y(x_2) + b_{22}y'(x_2) &= 0, \end{aligned} \tag{V.34}$$

then Green's lemma does not necessarily apply, as we can see from the following example:

$$y'' + \mu^2 y = 0,$$

$$y(0) - ay(1) = 0, \qquad y'(0) - y'(1) = 0.$$

We obtain

$$[P(x)(y'z - yz')]_0^1 = (1 - a)(y'(1)z(1) - y(1)z'(1))$$

and we see that we cannot claim in general that this expression vanishes unless $a = 1$.

One calls the boundary value problem (V.32), (V.34) *self-adjoint* if

$$\int_{x_1}^{x_2} (zL[y] - yL[z])\, dx = 0$$

for any two functions y and z which satisfy the boundary conditions (V.34). If the boundary conditions (V.34) reduce to (V.33), then the problem is self-adjoint in view of lemma V.5. Otherwise, it may or may not be.

To obtain an idea of what we have to expect in the general case, let us first consider the simple boundary value problem

$$y'' + \mu^2 y = 0,$$

$$y(0) = 0, \qquad y(\pi) = 0.$$

The general solution

$$y = c_1 \cos \mu x + c_2 \sin \mu x$$

will satisfy the boundary conditions if $c_1 = 0$ and if $\mu = k$, where $k = 0, \pm 1, \pm 2, \pm 3, \cdots$.

If μ should assume any other value, we would not obtain any solution at all. We call the values $\mu^2 = 1, 4, 9, 16, 25, \cdots$ for which the problem has a nontrivial solution the *eigenvalues*[5] of the given boundary value problem, and we call the corresponding solutions $\sin x$, $\sin 2x$, $\sin 3x$, $\cdots$ the *eigenfunctions*. We will give the following general definition:

A value λ for which the problem (V.32), (V.33) has a nontrivial solution is called an *eigenvalue* and the corresponding solution an *eigensolution* or *eigenfunction* of the Sturm-Liouville problem. The problem of finding such a value or such values λ is called an *eigenvalue problem*.

While it was rather easy to find the eigenvalues in the problem we just discussed, it is in general rather complicated to do this and we will devote Chapter VII exclusively to one particular method which is concerned with their numerical determination. What we are interested in at present are questions as to the existence of eigenvalues, their number, and their distribution. These questions could also be dealt with in Chapter VII from the standpoint of the calculus of variations. However, the discussion

[5] German: "Die dem Problem *eigen*tuemlichen Werte" = "The values belonging to this problem." *Eigenwert* is the accepted German word for eigenvalue.

Eigenvalues are also called by some authors *characteristic values*. Most recently it was proposed (see *Notices of the AMS*, Vol. 6, No. 4, p. 361) to call them *autovalues*, the reason for this being that the word eigenvalue is a linguistic monstrosity while the term characteristic value applies already to too many other concepts and besides is not as easily pronounced.

which we will subsequently present is more elementary and more suitable for an introduction to this delicate subject.

We have seen that the problem just discussed possesses infinitely many eigenvalues and consequently infinitely many eigenfunctions. The zeros of these eigenfunctions have the following property: Between any two consecutive zeros of one eigenfunction there lies one zero of the next higher eigenfunction, i.e., the eigenfunction belonging to the next higher eigenvalue, a phenomenon which is not too surprising after theorem V.5.

On the other hand, we do not want to convey the impression that there always exist many eigenvalues. If we consider the problem

$$y'' - \mu^2 y = 0$$

$$y(0) = 0, \quad b_1 y(x_2) + b_2 y'(x_2) = 0$$

we see immediately that there are difficulties in solving this problem. The differential equation has the general solution

$$y = c_1 e^{\mu x} + c_2 e^{-\mu x}$$

and it follows from the first boundary condition that $c_1 = -c_2$. This gives for the second boundary condition after division by c_1

$$b_1(e^{\mu x} - e^{-\mu x}) + \mu b_2(e^{\mu x} + e^{-\mu x}) = 0$$

or

$$\frac{b_2}{b_1}\mu = -\tanh \mu x_2.$$

In order to solve this equation geometrically, we have to intersect the line $y = (b_2/b_1)\mu$ with the curve $y = -\tanh \mu x_2$ in an μ, y–coordinate system (see Fig. 20). The slope of $y = (b_2/b_1)\mu$ is b_2/b_1 and the slope of $-\tanh \mu x_2$ at $\mu = 0$ is $-x_2$. As is apparent, $\mu = 0$ is a solution under all circumstances but the corresponding solution is the trivial solution $y = 0$. The only possibility of obtaining a nontrivial solution is for $-x_2 < b_2/b_1 < 0$. In this case the two values for μ differ only in sign; i.e., we obtain only one real eigenvalue μ^2.

Problems V.16–18

16. Prove that

$$\frac{d}{dx}(Py') + (Q + \lambda R)y = 0,$$

$$a_{11}y(x_1) + a_{12}y'(x_1) = 0,$$

$$a_{21}y(x_2) + a_{22}y'(x_2) = 0,$$

is a self-adjoint boundary value problem if $a_{12} = 0$ and/or $a_{22} = 0$; i.e.: show that

$$\int_{x_1}^{x_2} (zL[y] - yL[z])\, dx = 0$$

for any two functions y, z which satisfy the boundary conditions. (Note that if $a_{12} = 0$, then $a_{11} \neq 0$ and if $a_{22} = 0$ then $a_{21} \neq 0$.)

17. Prove that

$$\frac{d}{dx}(xy') + \lambda xy = 0,$$

$$|y(0)| < \infty, \quad y(1) = 0$$

is a self-adjoint boundary value problem.

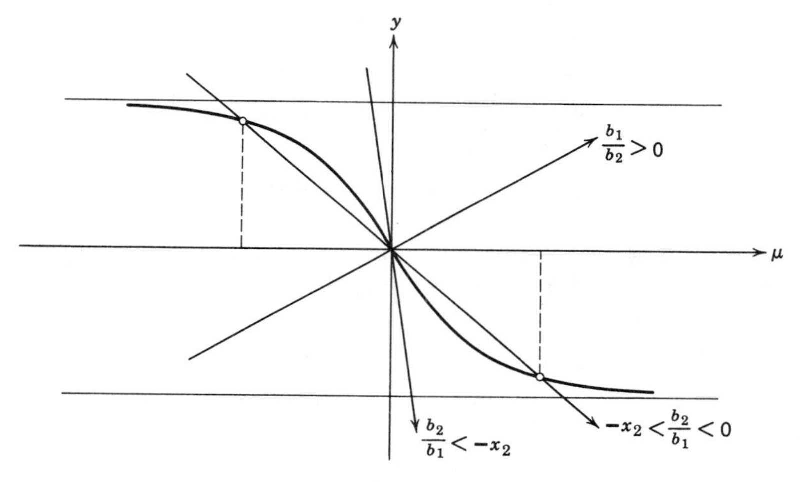

Fig. 20

18. Prove that

$$\frac{d}{dx}[(1 - x^2)y'] + \lambda y = 0,$$

$$|y(-1)| < \infty, \quad |y(1)| < \infty$$

is a self-adjoint boundary value problem.

2. The Sturm-Liouville Problem in the x, ϕ–Diagram

It is apparent that there exists a rather close connection between the zeros of the solutions of a self-adjoint differential equation and the eigenvalues of a self-adjoint boundary value problem. Since the x, ϕ–diagram proved to be of considerable help in our investigations of the first problem, we shall employ it again in our investigations into the latter problem. Introducing the function $z(x)$ according to (V.17) transforms the boundary conditions (V.33) to the form

$$a_{11}y(x_1) + \frac{a_{12}}{P(x_1)}z(x_1) = 0,$$

$$a_{21}y(x_2) + \frac{a_{22}}{P(x_2)}z(x_2) = 0.$$

Let us now introduce four constants α, β, A, B according to

$$a_{11} = A \sin \alpha, \qquad a_{21} = B \sin \beta,$$

$$\frac{a_{12}}{P(x_1)} = -A \cos \alpha, \quad \frac{a_{22}}{P(x_2)} = -B \cos \beta;$$

i.e.,

$$A = \sqrt{a_{11}^2 + \frac{a_{12}^2}{P^2(x_1)}}, \qquad B = \sqrt{a_{21}^2 + \frac{a_{22}^2}{P^2(x_2)}},$$

$$\alpha = -\text{arc tan}\,\frac{a_{11}P(x_1)}{a_{12}}, \quad \beta = -\text{arc tan}\,\frac{a_{21}P(x_2)}{a_{22}}$$

(The signs of A and B depend on whether we use the principal values α and β or $\alpha + \pi$ and $\beta + \pi$.) In terms of these quantities the boundary conditions become

$$\sin \alpha \cos \phi(x_1) - \cos \alpha \sin \phi(x_1) = 0,$$
$$\sin \beta \cos \phi(x_2) - \cos \beta \sin \phi(x_2) = 0,$$

or

$$\sin [\phi(x_1) - \alpha] = 0,$$
$$\sin [\phi(x_2) - \beta] = 0,$$

that is

$$\left.\begin{aligned} \phi(x_1) &= \alpha + \nu\pi, \\ \phi(x_2) &= \beta + \mu\pi, \end{aligned}\right\} \quad \nu, \mu = 0, 1, 2, 3, \cdots \qquad \begin{aligned} &(\text{V.35a}) \\ &(\text{V.35b}) \end{aligned}$$

It is clear that the boundary conditions can only be of consequence for the function $\phi(x)$, because once we have $\phi(x)$, $r(x)$ is determined but for a multiplicative constant (see (V.22)).

3. Number and Distribution of the Eigenvalues of the Sturm-Liouville Boundary Value Problem

We have seen in the example which we discussed in subsection 1 of this section that there are infinitely many eigenvalues in the case where

$\lambda = \mu^2 > 0$ and only one or no eigenvalues at all in the case where $\lambda = -\mu^2 < 0$.

The situation which prevails in general and which is quite clearly reflected in this example is described in the following two theorems:

Theorem V.7. *Let $P(x)$, $Q(x)$, $R(x)$ be continuous, $P(x)$ differentiable and $P(x) > 0$, $R(x) > 0$ in $a \leq x \leq b$. Then, there exists a value λ_0 such that for all $\lambda < \lambda_0$, the Sturm-Liouville problem (V.32), (V.33) with x_1, x_2 in $a \leq x \leq b$ has no solution.*

Proof. Let $M > 0$ be any number, the specific choice of which is left to a later occasion and let us choose a λ_0 (which depends on M) such that

$$Q(x) + \lambda_0 R(x) < -M \quad \text{in } a \leq x \leq b.$$

Since $P(x)$ is continuous in $a \leq x \leq b$, there exists an m such that

$$P(x) \geq m \quad \text{in } a \leq x \leq b.$$

Now, we consider the differential equation

$$\frac{d}{dx}[my'] - My = 0. \tag{V.36}$$

We denote the solution of (V.36) by $y_2(x)$ and its ϕ-curve by $\phi_2(x)$. Clearly

$$y_2(x) = c_1 e^{\sqrt{(M/m)}x} + c_2 e^{-\sqrt{(M/m)}x}$$

and its corresponding ϕ-curve $\phi_2(x)$ is, according to $z(x) = P(x)y'(x)$ (see (V.17) and (V.19)), obtained from

$$\tan \phi_2(x) = \frac{z(x)}{y(x)} = \frac{P(x)y'(x)}{y(x)} = \sqrt{mM}\,\frac{c_1 e^{\sqrt{(M/m)}x} - c_2 e^{-\sqrt{(M/m)}x}}{c_1 e^{\sqrt{(M/m)}x} + c_2 e^{-\sqrt{(M/m)}x}}.$$

Let us determine the c_1, c_2 such that $y_2(x)$ satisfies (V.33a); i.e., $\phi_2(x)$ has to satisfy (V.35a), namely $\phi_2(x_1) = \alpha$. Then

$$\tan \alpha = \sqrt{mM}\,\frac{c_1 e^{\sqrt{(M/m)}x_1} - c_2 e^{-\sqrt{(M/m)}x_1}}{c_1 e^{\sqrt{(M/m)}x_1} + c_2 e^{-\sqrt{(M/m)}x_1}} \quad \text{has to hold.}$$

By this condition, the ratio c_1/c_2 is uniquely determined; i.e., the ϕ-curve

$$\phi_2(x) = \arctan \sqrt{mM}\,\frac{c_1 e^{\sqrt{(M/m)}x} - c_2 e^{-\sqrt{(M/m)}x}}{c_1 e^{\sqrt{(M/m)}x} + c_2 e^{-\sqrt{(M/m)}x}} \tag{V.37}$$

is completely determined by the first boundary condition at x_1 and the second condition at x_2 could only be satisfied accidentally.

Let us now discuss (V.37). Since

$$\lim_{x\to\infty} \frac{c_1 \exp(\sqrt{M/m}\, x) - c_2 \exp(-\sqrt{M/m}\, x)}{c_1 \exp(\sqrt{M/m}\, x) + c_2 \exp(-\sqrt{M/m}\, x)} = 1,$$

we have

$$\lim_{x\to\infty} \phi_2(x) = \text{arc tan} \sqrt{mM};$$

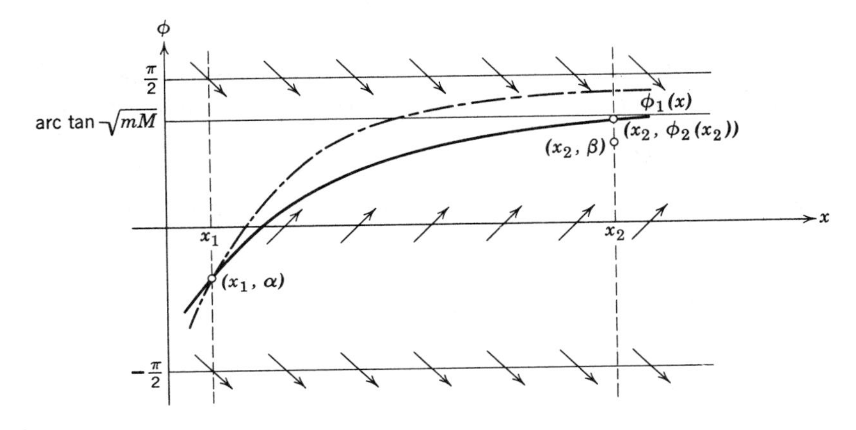

Fig. 21

i.e.: $\phi_2(x)$ approaches the line $\phi = \text{arc tan} \sqrt{mM}$ asymptotically. Further, we note that if ϕ_0 is any given value such that $-\pi/2 < \phi_0 < \pi/2$, we can find an M such that

$$\phi_2(x_2) \geq \phi_0, \quad -\pi/2 < \phi_0 < \pi/2 \qquad \text{(see problem V.19).} \qquad \text{(V.38)}$$

Now, we consider the solution $y_1(x)$ of (V.32) and determine the integration constants such that the boundary condition (V.33a) is satisfied. Then, the ϕ-curve $\phi_1(x)$ of $y_1(x)$ satisfies the boundary condition (V.35a), namely $\phi_1(x_1) = \alpha$. We can now invoke lemma V.4 (problem V.9) according to which

$$\phi_2(x) < \phi_1(x) \quad \text{for all } x > x_1$$

since with $p_2 = m$, $q_2 = -M$ the conditions of lemma V.4 in the modification of problem V.9 are all met. This means that the curve $\phi_1(x)$ has to stay above $\phi_2(x)$ for all $x > x_1$ (see Fig. 21). It is our aim to show that there exists a λ_0 such that for all $\lambda < \lambda_0$, the Sturm-Liouville problem has no solution. We will accomplish our goal by showing that there is an M such that whenever $Q(x) + \lambda R(x) < -M$ in $a \leq x \leq b$, $y_1(x)$ cannot possibly satisfy the boundary condition at x_2, (V.33b).

We can assume without loss of generality that $-\pi/2 < \alpha \leq \pi/2$ and $-\pi/2 < \beta \leq \pi/2$ (see (V.35)). Clearly, the case $\beta = \pi/2$ is impossible, since $\phi_1(x)$ cannot cross the line $\phi = \pi/2$ from below because of the peculiarity of the direction field. Hence, we can restrict our considerations to the case

$$-\frac{\pi}{2} < \beta < \frac{\pi}{2}.$$

Now, we choose a ϕ_0 such that $\beta < \phi_0 < \pi/2$ and choose M such that $\phi_2(x_2) \geq \phi_0$. (see (V.38)). Since $\phi_1(x) > \phi_2(x)$ for all $x > x_1$ and since $\phi_1(x)$ cannot cross $\phi = \pi/2$ from below or, in case $\alpha = \pi/2$, has to proceed into the strip $-\pi/2 < \phi < \pi/2$ because of the nature of the direction field along $\phi = \pi/2$, it is impossible to satisfy $\phi_1(x_2) = \beta$. Now, all we have to do is determine a λ_0 such that

$$\lambda_0 < \frac{-M - \max\limits_{[a,b]} |Q(x)|}{\min\limits_{[a,b]} R(x)}$$

and our theorem is proved.

Theorem V.8. *Let $P(x)$, $Q(x)$, $R(x)$ be continuous, $P(x)$ differentiable, and $P(x) > 0$, $R(x) > 0$ in $a \leq x \leq b$. Then, there are infinitely many values $\lambda_1 < \lambda_2 < \lambda_3 < \cdots$ (eigenvalues) for which the Sturm-Liouville problem (V.32), (V.33) has a nontrivial solution. If $y(x, \lambda_k)$ is the solution corresponding to the* k*th eigenvalue λ_k, then $y(x, \lambda_k)$ has one more zero in $x_1 \leq x \leq x_2$ than $y(x, \lambda_{k-1})$, which is the solution corresponding to the $(k-1)$th eigenvalue λ_{k-1} and $\lim\limits_{k\to\infty} \lambda_k = \infty$.*

To prove this theorem, we need the following two lemmas:

Lemma V.6. *Let $\phi(x, \lambda)$ be the ϕ-curve corresponding to the solution $y(x, \lambda)$ of (V.32) which satisfies the boundary condition (V.33a) at x_1 for any λ. Under the conditions stated in theorem V.8, $\phi(x_2, \lambda)$ is strictly monotonically decreasing in λ, i.e.:*

$$\phi(x_2, \lambda_1) > \phi(x_2, \lambda_2) \quad \text{if } \lambda_1 < \lambda_2.$$

Proof. Let $y(x, \lambda)$ be a solution of (V.32) for any λ which satisfies the boundary condition (V.33a) at x_1. Then, the corresponding ϕ-curve $\phi(x, \lambda)$ satisfies the boundary condition (V.35a) at x_1.

Let $\lambda_1 < \lambda_2$. Then $Q(x) + \lambda_1 R(x) < Q(x) + \lambda_2 R(x)$ in $a \leq x \leq b$ and in view of lemma V.4 in the modification of problem V.9

$$\phi(x, \lambda_1) > \phi(x, \lambda_2) \quad \text{for all } x > x_1,$$

since $\phi(x_1, \lambda_1) = \phi(x_1, \lambda_2)$. Since $x_2 > x_1$, we have

$$\phi(x_2, \lambda_1) > \phi(x_2, \lambda_2), \quad \text{q.e.d.}$$

Lemma V.7. *Let $\phi(x, \lambda)$ be the ϕ-curve corresponding to the solution $y(x, \lambda)$ of (V.32) which satisfies the boundary condition (V.33a) at x_1 for any λ.*

Then

$$\lim_{\lambda \to \infty} \phi(x_2, \lambda) = -\infty.$$

Proof (*by contradiction*). According to lemma V.6, $\phi(x_2, \lambda)$ is strictly monotonically decreasing in λ. Hence, either there exists a limit of $\phi(x_2, \lambda)$ or $\phi(x_2, \lambda)$ approaches $-\infty$. Let us assume that

$$\lim_{\lambda \to \infty} \phi(x_2, \lambda) = L.$$

Since $y(x, \lambda)$ satisfies the boundary condition (V.33a), $\phi(x, \lambda)$ satisfies (V.35a); i.e.: $\phi(x_1, \lambda) = \alpha$ and we can assume without loss of generality that $-(\pi/2) \leq \alpha < (\pi/2)$. $d\phi/dx < 0$ in view of $R(x) > 0$, $P(x) > 0$ for all λ which are sufficiently large (see (V.20)). Hence, we can assume that

$$\frac{\pi}{2} - k\pi \leq L < \frac{\pi}{2} - (k-1)\pi$$

for some positive integer k. Then, no ϕ-curve which originates at (x_1, α) can have more than k $\pi/2$-intercepts; i.e., no solution $y(x, \lambda)$ which satisfies the first boundary condition (V.33a) at x_1 can have more than k zeros in $x_1 \leq x \leq x_2$. We will now show that this leads to a contradiction inasmuch as we will demonstrate a solution $y(x, \lambda)$ which has at least $k + 1$ zeros in $x_1 \leq x \leq x_2$, or, as a matter of fact, any number of zeros. For this purpose we proceed as follows:

Since $P(x)$, $Q(x)$, $R(x)$ are continuous in $a \leq x \leq b$ there are constants M, m_q, m_r such that

$$P(x) \leq M,$$

$$Q(x) + \lambda R(x) \geq m_q + \lambda m_r > 0 \quad \text{for sufficiently large } \lambda.$$

The differential equation

$$\frac{d}{dx}[My'] + (m_q + \lambda m_r)y = 0$$

has the solution

$$\bar{y} = A \sin \sqrt{\frac{m_q + \lambda m_r}{M}}(x - \delta).$$

We determine δ such that $\bar{y}$ satisfies the boundary condition (V.33a) at x_1;

i.e., the ϕ-curve of y is such that $\bar{\phi}(x_1) = \alpha$. Next, we invoke lemma V.4 again, according to which

$$\bar{\phi}(x) \geq \phi(x) \quad \text{for all } x > x_1.$$

Clearly, we can choose λ such that $\bar{y}$ has at least $k + 1$ zeros in $x_1 \leq x \leq x_2$ and its ϕ-curve $\bar{\phi}$ accordingly has at least $k + 1$ $\pi/2$-intercepts; i.e., $y(x, \lambda)$ has to have at least $k + 1$ zeros in $x_1 \leq x \leq x_2$, q.e.d.

Proof of Theorem V.8. First we note, that $y(x, \lambda)$ which is a solution of (V.32) and satisfies the boundary condition (V.33a) at x_1, is continuous in λ (see AIII.B5) and consequently, its ϕ-curve $\phi(x, \lambda)$ with $\phi(x_1, \lambda) = \alpha$ is continuous in λ. Now, we choose a λ_0 such that $Q(x) + \lambda_0 R(x) > 0$. The ϕ-curve $\phi(x, \lambda_0)$ of the solution $y(x, \lambda_0)$ which satisfies (V.33a) will intersect the line $x = x_2$ at some point ϕ_0. If $\phi_0 = \beta$, then $\phi(x_2, \lambda_0) = \beta$; i.e., $y(x, \lambda_0)$ satisfies also (V.33b); hence, $\lambda = \lambda_0$ is the first eigenvalue.

If $\phi_0 \neq \beta$, we can assume without loss of generality that $\beta - \pi \leq \phi_0 < \beta$. Now, we let λ increase and consequently, $\phi(x_2, \lambda)$ will decrease monotonically in view of lemma V.6 and continuously until for a certain $\lambda = \lambda_1$, $\phi(x_2, \lambda_1) = \beta - \pi$. Then, λ_1 is the first eigenvalue. We let λ continue to increase until for some $\lambda = \lambda_2$, $\phi(x_2, \lambda_2) = \beta - 2\pi$; i.e., λ_2 is the second eigenvalue, and so on. Clearly, there is no end to this process because $\lim_{\lambda \to \infty} \phi(x_2, \lambda) = -\infty$ in view of lemma V.7 and $\phi(x_2, \lambda)$ is continuous in λ. Thus, we obtain successively values λ_k such that

$$\phi(x_2, \lambda_k) = \beta - k\pi \qquad \text{(see Fig. 22).}$$

Furthermore, since $\phi(x_2, \lambda_k) = \beta - k\pi$ while $\phi(x_2, \lambda_{k-1}) = \beta - (k - 1)\pi$, it follows that $\phi(x, \lambda_k)$ has to have one more $\pi/2$-intercept than $\phi(x, \lambda_{k-1})$ in $x_1 \leq x \leq x_2$. Hence, $y(x, \lambda_k)$ has to have one more zero in $x_1 \leq x \leq x_2$ than $y(x, \lambda_{k-1})$.

We still have to show that $\lim_{k \to \infty} \lambda_k = \infty$. Suppose that $\lim_{k \to \infty} \lambda_k = \Lambda < \infty$. Then, there exists for any arbitrary small $\epsilon > 0$ an N_ϵ such that

$$|\lambda_n - \lambda_m| < \epsilon \quad \text{for all } n, m > N_\epsilon.$$

Since $\phi(x_2, \lambda)$ is continuous in λ, we have for any arbitrary small $\epsilon_1 > 0$ a δ_{ϵ_1} such that

$$|\phi(x_2, \lambda_n) - \phi(x_2, \lambda_m)| < \epsilon_1 \quad \text{for } |\lambda_n - \lambda_m| < \delta_{\epsilon_1}.$$

Let $\epsilon_1 < \pi$ and choose $\epsilon = \delta_{\epsilon_1}$. Then

$$|\phi(x_2, \lambda_n) - \phi(x_2, \lambda_m)| < \pi.$$

However,

$$|\phi(x_2, \lambda_n) - \phi(x_2, \lambda_m)| \geq \pi$$

since $\phi(x_2, \lambda_n) = \beta - n\pi$ and $\phi(x_2, \lambda_m) = \beta - m\pi$, q.e.d.

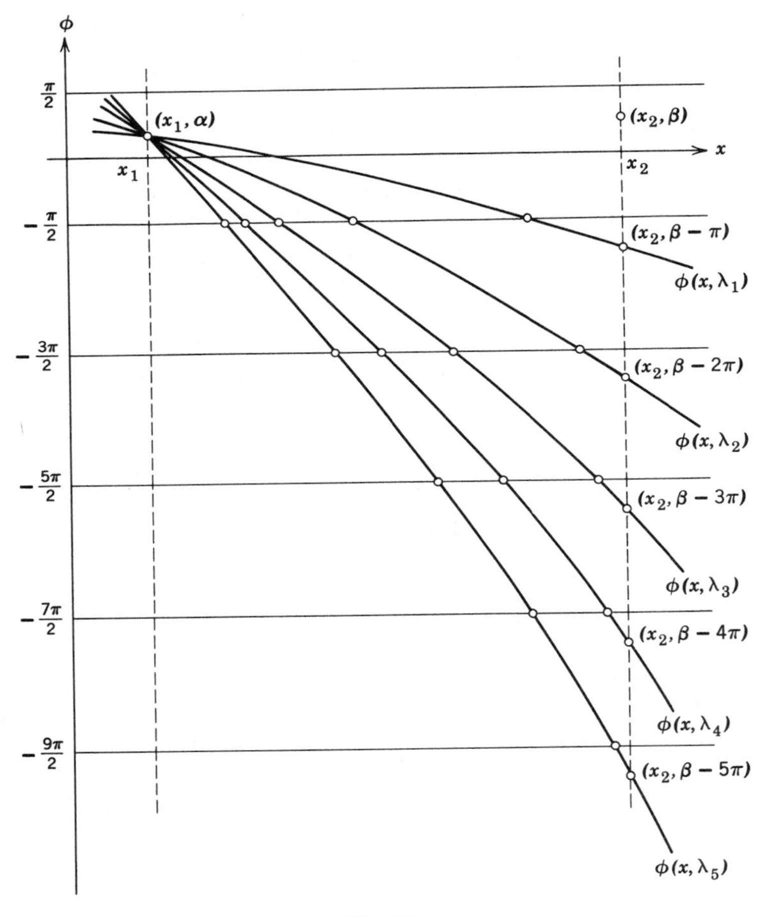

Fig. 22

Summarizing, we can state: The Sturm-Liouville problem has, in general, no solution, but there exist infinitely many values $\lambda_1, \lambda_2, \lambda_3, \cdots$ for which the problem has a nontrivial solution. These solutions, called eigenfunctions, have the property that every one has one more zero in the interval given by the boundary conditions than the next lower one, i.e., the one corresponding to the next lower eigenvalue.

Problems V.19–23

***19.** Given a value ϕ_0 where $-\pi/2 < \phi_0 < \pi/2$. Show that there exists an M such that $\phi_2(x_2) \geq \phi_0$ where $\phi_2(x)$ is given by (V.37).

20. Find all solutions of the problem

$$y'' + \lambda y = 0,$$
$$y(0) + y'(0) = 0, \quad y(2\pi) + y'(2\pi) = 0.$$

21. Find an approximation for the eigenvalue of the problem

$$y'' - \mu^2 y = 0,$$
$$y(0) = 0, \quad 2y(1) - y'(1) = 0.$$

22. Write the boundary conditions in problems 20 and 21 in polar coordinates.

23. Find the eigenvalues of the problem

$$x^2 y'' + 2xy' + \lambda y = 0, \qquad \lambda > \tfrac{1}{4},$$
$$y(1) = 0, \quad y(e^\pi) = 0.$$

4. The Orthogonality of Eigenfunctions

The simple Sturm-Liouville problem

$$y'' + \lambda y = 0,$$
$$y(0) = 0, \quad y(\pi) = 0$$

has the solutions (eigenfunctions) $y_k = \sin kx$, where $k = 1, 2, 3, \cdots$. These functions, as one can easily verify, constitute an orthogonal system on the interval $0 \leq x \leq \pi$. As we have seen in Chapter IV, the system $\{\cos nx, \sin mx\}$ is orthogonal on any interval of length 2π. These functions are also solutions of a Sturm-Liouville problem which, however, is of a more general nature than the one we discussed in the preceding subsections, namely, it is of the type (V.32), (V.34) (see problem V.24).

We will show in this subsection that the orthogonality is a characteristic property of eigenfunctions of the Sturm-Liouville problem.

Using the abbreviation

$$L[y] = \frac{d}{dx}[P(x)y'] + Q(x)y,$$

we can write (V.32) in the form

$$L[y] + \lambda R(x)y = 0$$

where $L[y]$ is a self-adjoint differential expression (see lemma V.2). (As a matter of fact, the entire left side of this equation is a self-adjoint differential expression but this fact is immaterial for the following discussion.)

Let λ_i, λ_k be two different eigenvalues of (V.32), (V.33) and y_i, y_k the corresponding eigenfunctions. Then, the equations

$$L[y_i] + \lambda_i R(x) y_i = 0, \tag{V.39}$$

$$L[y_k] + \lambda_k R(x) y_k = 0 \tag{V.40}$$

and the boundary conditions

$$\left.\begin{aligned} a_{11} y_\nu(x_1) + a_{12} y'_\nu(x_1) = 0 \\ a_{21} y_\nu(x_2) + a_{22} y'_\nu(x_2) = 0 \end{aligned}\right\}, \qquad \nu = i, k \tag{V.41}$$

are identically satisfied. If we multiply (V.39) by y_k and (V.40) by y_i and subtract (V.40) from (V.39), we obtain

$$y_k L[y_i] - y_i L[y_k] + (\lambda_i - \lambda_k) R(x) y_i y_k = 0. \tag{V.42}$$

In view of (V.16) the first two terms in this identity can be written in the form

$$y_k L[y_i] - y_i L[y_k] = \frac{d}{dx} [P(x)(y'_i y_k - y_i y'_k)]$$

and after integrating (V.42) between x_1 and x_2 and invoking Green's lemma (lemma V.5)

$$(\lambda_i - \lambda_k) \int_{x_1}^{x_2} R(x) y_i y_k \, dx = 0.$$

Thus we have proved

$$\int_{x_1}^{x_2} R(x) y_i y_k \, dx = 0, \qquad i \neq k,$$

since we assumed that $\lambda_i \neq \lambda_k$.

Theorem V.9. *If y_i, y_k are two different eigenfunctions of the Sturm-Liouville problem (V.32), (V.33), then*

$$\{\sqrt{R(x)}\, y_i(x), \sqrt{R(x)}\, y_k(x)\}$$

are orthogonal on the interval $x_1 \leq x \leq x_2$.

Remark: As we can see from the preceding proof, orthogonal functions can be obtained without imposing boundary conditions at all, if there exist two values x_1, x_2 such that

$$P(x_1) = P(x_2) = 0.$$

Then, $[P(x)(y'_i y_k - y_i y'_k)]_{x_1}^{x_2}$ vanishes whatever boundary conditions we may impose, provided y_i, y_k are quadratically integrable.

In equation (V.14) (Legendre's equation)

$$P(x) = 1 - x^2$$

and therefore

$$P(-1) = P(+1) = 0;$$

i.e., the solutions of (V.14) will constitute an orthogonal system in the interval $-1 \leq x \leq 1$ without having to satisfy any specific boundary conditions at all. (See also problems V.17, 18, and 27.)

Problems V.24–27

24. Find all solutions of

$$y'' + \lambda y = 0,$$
$$y(0) - y(2\pi) = 0,$$
$$y'(0) - y'(2\pi) = 0$$

and show that this is a self-adjoint boundary value problem.

25. Find all solutions of

$$y'' + \lambda y = 0,$$
$$y'(0) = 0,$$
$$y'(\pi) = 0.$$

What are the orthogonality properties of the solutions?

26. Show that the boundary conditions in problems V.20, 21, 23, 24, and 25 are special cases of the general boundary conditions (V.34).

27. Prove that any two solutions y_i, y_k of

$$\frac{d}{dx}(xy') + \lambda xy = 0,$$

$$|y(0)| < \infty, \quad y(1) = 0,$$

are such that

$$\int_0^1 xy_iy_k\,dx = 0.$$

5. Fourier Expansions in Terms of Eigenfunctions

One reason that we can easily expand a function into an infinite series with terms $\sin kx$, $\cos kx$ is that these functions constitute an orthogonal system. We meet the same situation—as far as the orthogonality is concerned—in dealing with any system of eigenfunctions arising out of the Sturm-Liouville problem in view of its self-adjointness (see §2.1). This makes us expect that we can establish similar expansions in terms of eigenfunctions in general, which is indeed the case.

Without going into any details as far as the existence and convergence of such an expansion is concerned (see also the following subsection—6), we will nevertheless establish the formal theory of such a general Fourier expansion. Let $y_1(x), y_2(x), y_3(x), \cdots$ be the denumerable number of eigenfunctions of a Sturm-Liouville problem of the type (V.32), (V.33). Then, we have according to theorem V.9

$$(\sqrt{R(x)}y_i \cdot \sqrt{R(x)}y_k) = 0 \quad \text{for } i \neq k \tag{V.43}$$

where the inner product $(f_i \cdot f_k)$ is defined as

$$(f_i \cdot f_k) = \int_{x_1}^{x_2} f_i f_k \, dx \qquad \text{(see Chapter IV, §1.3).}$$

Since the y_k are solutions of a linear homogeneous boundary value problem, $c_k y_k$ are also solutions where the c_k are any arbitrary constants.

Let us choose

$$c_k = (\sqrt{R(x)}y_k \cdot \sqrt{R(x)}y_k)^{-1/2}$$

Then, the system $\{c_k\sqrt{R(x)}y_k\} = \{\sqrt{R(x)}\bar{y}_k\}$ is orthonormal; i.e.,

$$(\sqrt{R(x)}\bar{y}_k \cdot \sqrt{R(x)}\bar{y}_k) = 1 \tag{V.44}$$

(see Chapter IV, §1.3).

Let $f(x)$ be a function which is "well behaved" within the interval $x_1 \leq x \leq x_2$. If we assume that an expansion of this function of the type

$$f(x) = a_1\bar{y}_1(x) + a_2\bar{y}_2(x) + a_3\bar{y}_3(x) + \cdots \tag{V.45}$$

is possible, and if we assume that this series converges uniformly, then we can obtain the a_k's by the same method we employed in Chapter IV, §1.5; i.e., we multiply (V.45) by $R(x)\bar{y}_k$ and integrate the result termwise over the interval $x_1 \leq x \leq x_2$.

This gives

$$a_k = \int_{x_1}^{x_2} f(x)R(x)\bar{y}_k(x) \, dx$$

in view of (V.43) and (V.44).

Hence,

$$f(x) = \sum_{k=1}^{\infty} a_k\bar{y}_k(x) \tag{V.46}$$

with

$$a_k = (f(x) \cdot R(x)\bar{y}_k(x))$$

would appear to represent the function $f(x)$.

6. Completeness and the Expansion Problem

In accordance with the definition which we gave in Chapter IV, §2.6, we say that the orthogonal system $\{\phi_k\}$ is complete on $x_1 \leq x \leq x_2$ if one can find for every $\epsilon > 0$ an N_ϵ such that

$$\int_{x_1}^{x_2} \left[f(x) - \sum_{k=1}^{n} C_k\phi_k(x) \right]^2 dx < \epsilon \tag{V.47}$$

for all $n > N_\epsilon$, where $f(x)$ is any sectionally continuous function.[6]

[6] The more advanced reader may replace the Riemann Integral by the Lebesgue integral and replace the condition of sectional continuity on $f(x)$ by the Lebesgue integrability of the square of $f(x)$. ($f(x) \in L^2[x_1, x_2]$.)

For reasons of convenience we will assume for the following discussion that the system $\{\phi_k\}$ is orthonormal; i.e.,

$$|\phi_k|^2 = (\phi_k \cdot \phi_k) = \int_{x_1}^{x_2} \phi_k^2\, dx = 1 \quad \text{and} \quad (\phi_i \cdot \phi_k) = 0, \qquad i \neq k.$$

(For notation see Chapter IV, §1.3.) It follows then from (V.47) with

$$C_k = (f \cdot \phi_k) = \int_{x_1}^{x_2} f(x)\phi_k(x)\, dx$$

that

$$\int_{x_1}^{x_2} f^2(x)\, dx = \sum_{k=1}^{\infty} C_k^2 \tag{V.48}$$

or in the more sophisticated notation of Chapter IV, §1.3,

$$|f|^2 = \sum_{k=1}^{\infty} (f \cdot \phi_k)^2,$$

which is the completeness relation (see Chapter IV, §2.6), also referred to as the *Parseval equality*. (See also problem IV.27.)

If we start out with the completeness relation and retrace our steps, we see that (V.47) is satisfied. (See also problem V.29.)

Let us now give the following definition: The orthogonal (orthonormal) system $\{\phi_k\}$ is *closed* on $x_1 \leq x \leq x_2$ if for any sectionally continuous function (square integrable function) $\psi(x)$ it follows from

$$(\psi \cdot \phi_k) = \int_{x_1}^{x_2} \psi(x)\phi_k(x)\, dx = 0 \quad \text{for all } k \tag{V.49}$$

that $\psi(x) = 0$ on $x_1 \leq x \leq x_2$ with the exception of finitely many points (except on a set of measure zero).

As can be shown, the concepts of completeness and closedness are equivalent. We will only indicate a proof that the closedness follows from the completeness and refer the reader who is interested in pursuing this discussion further to the literature.[7]

Let us assume that $\{\phi_k\}$ is complete, i.e., (V.48) is satisfied, but not closed. Then, there exists a function $\psi(x) \not\equiv 0$ which is sectionally continuous (square integrable) such that

$$(\psi \cdot \phi_k) = 0 \tag{V.50}$$

[7] See Fr. Riesz and B. Sz-Nagy: *Functional Analysis*, F. Ungar Publishing Company, 1955; and A. Taylor: *Introduction to Functional Analysis*, John Wiley and Sons, New York, 1958, p. 115, Theorem 3.2.K.

for all k. Since $\{\phi_k\}$ is complete, we have, in view of (V.48),

$$|\psi|^2 = \sum_{k=1}^{\infty} (\psi \cdot \phi_k)^2$$

and because of (V.50)

$$|\psi|^2 = 0,$$

which contradicts our assumption that $\{\phi_k\}$ is not closed.

We are now in a position to relate the completeness of a system of orthogonal functions to the possibility of expanding a given function in terms of the members of the orthogonal system.

Clearly, if the series $\sum_{k=1}^{\infty} (f \cdot \phi_k)\phi_k$ converges uniformly (see Chapter IX, §3.1), then the processes of passage to the limit and integration in (V.47) can be interchanged (see **AII.C4**) and it follows that

$$\lim_{n \to \infty} \sum_{k=1}^{n} (f \cdot \phi_k)\phi_k = f(x),$$

i.e., the series represents $f(x)$.

Since completeness is equivalent to closedness, the validity of (V.47) can be established by establishing the given orthogonal system as closed and thus divorcing the problem of whether the uniformly convergent series $\sum_{k=1}^{\infty} (f \cdot \phi_k)\phi_k$ represents $f(x)$ or not entirely from convergence considerations. One can show that, if $\{\phi_k\}$ is a complete (closed) orthogonal system, then it is possible to expand any continuous function $f(x)$ which is sectionally smooth and vanishes in x_1 and/or x_2 if ϕ_k vanishes in x_1 and/or x_2, into a uniformly convergent series

$$f(x) = \sum_{k=1}^{\infty} c_k \phi_k$$

with coefficients

$$c_k = \frac{(f \cdot \phi_k)}{|\phi_k|^2} = \frac{\displaystyle\int_{x_1}^{x_2} f(x)\phi_k(x)\, dx}{\displaystyle\int_{x_1}^{x_2} \phi_k^2(x)\, dx}.$$

Problems V.28–29

***28.** Show that a sufficient condition for the completeness of $\{\phi_k\}$ (with respect to the class of all sectionally continuous functions) is that the completeness relation is satisfied for all continuous functions (see also Chapter IV, §2.6, proof of theorem IV.5).

29. Show that, if $\{\phi_k\}$ is a complete orthogonal system on $x_1 \le x \le x_2$, then the completeness relation is satisfied and vice versa. (See also Chapter IV, §2.6 and problem IV.27.)

RECOMMENDED SUPPLEMENTARY READING

E. Kamke: *Differentialgleichungen reeller Funktionen*, Akademische Verlags Gesellschaft, Leipzig, 1930.

E. A. Coddington and N. Levinson: *Theory of Ordinary Differential Equations*, McGraw-Hill Book Co., New York, 1955.

R. V. Churchill: *Fourier Series and Boundary Value Problems*, McGraw-Hill, New York, 1941.

R. Courant and D. Hilbert: *Methods of Mathematical Physics*, Interscience Publishers, New York, 1953.

E. C. Titchmarsh: *Eigenfunction Expansions*, The Clarendon Press, Oxford University, 1946.

VI

LEGENDRE POLYNOMIALS AND BESSEL FUNCTIONS

§1. THE LEGENDRE EQUATION AND LEGENDRE POLYNOMIALS

1. Series Solution of a Linear Differential Equation of the Second Order with Analytic Coefficients

In Chapter II, §3.4 we arrived at the equation

$$(1 - x^2)y'' - 2xy' + \lambda y = 0 \tag{VI.1}$$

in our attempt to solve the problem of the stationary temperature distribution which is generated by a spherical stove.

Equation (VI.1) is called the *Legendre equation* in honor of the French mathematician *A. M. Legendre* who investigated in his papers "Recherches sur la figure des planètes"[1] and "Recherches sur l'attraction des sphéroides homogènes"[2] a specific type of solution of this equation which will also be of special interest for our investigations.

If we divide (VI.1) by $(1 - x^2)$, we obtain

$$y'' - \frac{2x}{1 - x^2}\, y' + \frac{\lambda}{1 - x^2}\, y = 0. \tag{VI.2}$$

This equation is a special case of the more general type of equation

$$y'' - a(x)y' - b(x)y = 0 \tag{VI.3}$$

where the coefficients $a(x)$ and $b(x)$ are analytic, i.e., have a convergent Taylor expansion in a neighborhood of $x = 0$.

[1] "Investigations of the Shape of Planets," *Mem. Math. Phys., tires des rg. de l'Acad. sc.*, pp. 370–384 (1783).

[2] "Investigations of the Attraction of Homogeneous Spheres," *Mem. Math. Phys., pres. a l'Acad. sc. par divers sav.*, Vol. 10, pp. 411–434 (1785).

Clearly, the series

$$a(x) = \frac{2x}{1 - x^2} = 2(x + x^3 + x^5 + \cdots)$$

and

$$b(x) = -\frac{\lambda}{1 - x^2} = -\lambda(1 + x^2 + x^4 + \cdots)$$

both converge in $|x| < 1$ (see **AII.C1**).

In general, one can state concerning the solutions of (VI.3) the following:

Theorem VI.1. *If the coefficients $a(x)$ and $b(x)$ in*

$$y'' - a(x)y' - b(x)y = 0$$

have a convergent Taylor expansion

$$a(x) = \sum_{k=0}^{\infty} a_k x^k, \quad b(x) = \sum_{k=0}^{\infty} b_k x^k$$

in $|x| < r$, then the solution $y(x)$ can be represented by a Taylor expansion $y(x) = \sum_{k=0}^{\infty} c_k x^k$ which converges also in $|x| < r$.

(*Remark:* If the coefficients $a(x)$, $b(x)$ are analytic in some neighborhood of x_0, then the solution will be analytic in the same neighborhood of x_0. In this case, the terms x^n have to be replaced by $(x - x_0)^n$ and $|x| < r$ has to be replaced by $|x - x_0| < r$ in the subsequent proof.)

Proof. Let us first assume that $y = \sum_{k=0}^{\infty} c_k x^k$ is a solution of (VI.3) and converges in $|x| < r$. Then, we can differentiate termwise as often as we please in the interval of convergence (see **AII.D2**) and obtain

$$y' = \sum_{k=1}^{\infty} c_k \cdot k \cdot x^{k-1},$$

$$y'' = \sum_{k=2}^{\infty} c_k \cdot k(k-1)x^{k-2}.$$

If we substitute the series representations for $y, y', y'', a(x), b(x)$ into (VI.3), we obtain

$$\sum_{k=2}^{\infty} c_k k(k-1)x^{k-2} - \sum_{k=0}^{\infty} a_k x^k \sum_{k=1}^{\infty} c_k k x^{k-1} - \sum_{k=0}^{\infty} b_k x^k \sum_{k=0}^{\infty} c_k x^k = 0$$

and after changing the summation subscript in the first term and in the second factor of the second term

$$\sum_{n=0}^{\infty} c_{n+2}(n+2)(n+1)x^n - \sum_{n=0}^{\infty} a_n x^n \sum_{n=0}^{\infty} c_{n+1}(n+1)x^n$$
$$- \sum_{n=0}^{\infty} b_n x^n \sum_{n=0}^{\infty} c_n x^n = 0. \quad \text{(VI.4)}$$

Since the series $\sum_{n=0}^{\infty} a_n x^n$ and $\sum_{n=0}^{\infty} b_n x^n$ converge in $|x| < r$ and since we assumed that also $\sum_{n=0}^{\infty} c_n x^n$ and consequently the series for its first and second derivative converge in $|x| < r$, we can form the products in the second and third term of (VI.4) as follows (see **AII.D3**):

$$\sum_{n=0}^{\infty} a_n x^n \sum_{n=0}^{\infty} c_{n+1}(n+1)x^n = \sum_{n=0}^{\infty} [a_0 c_{n+1}(n+1) + \cdots + a_n c_1]x^n,$$

$$\sum_{n=0}^{\infty} b_n x^n \sum_{n=0}^{\infty} c_n x^n = \sum_{n=0}^{\infty} (b_0 c_n + \cdots + b_n c_0)x^n$$

and we obtain for (VI.4)

$$\sum_{n=0}^{\infty} [c_{n+2}(n+1)(n+2) - a_0 c_{n+1}(n+1) - \cdots$$

$$- a_n c_1 - b_0 c_n - \cdots - b_n c_0]x^n$$

$$= \sum_{n=0}^{\infty} \Big[c_{n+2}(n+1)(n+2) - c_{n+1}[(n+1)a_0]$$

$$- c_n[na_1 + b_0] - \cdots - c_1[a_n + b_{n-1}] - c_0 b_n \Big] x^n = 0.$$

This power series has to vanish identically; hence, all the coefficients have to vanish

$$c_{n+2}(n+1)(n+2) - c_{n+1}(n+1)a_0 - c_n(na_1 + b_0) - \cdots$$
$$- c_1(a_n + b_{n-1}) - c_0 b_n = 0.$$

This relation yields for $n = 0$

$$c_2 = \tfrac{1}{2}(c_1 a_0 + c_0 b_0).$$

Thus, we see that c_0 and c_1 can be chosen arbitrarily and all the subsequent coefficients are determined by

$$c_{n+2} = \frac{1}{(n+1)(n+2)} [c_{n+1}(n+1)a_0 + c_n(na_1 + b_0) + \cdots$$
$$+ c_1(a_n + b_{n-1}) + c_0 b_n]. \quad \text{(VI.5)}$$

Therefore, if $\sum_{n=0}^{\infty} c_n x^n$ is a solution of (VI.3) which has the property that it converges in $|x| < r$, then its coefficients are recursively determined by (VI.5) in terms of the arbitrarily chosen constants c_0 and c_1.

Now, we can proceed to the second part of our proof, namely, to establish that $\sum_{n=0}^{\infty} c_n x^n$ with coefficients c_n which are determined by (VI.5) is a solution of (VI.3) and converges for $|x| < r$.

We proceed for this purpose as follows:

Since $\sum_{n=0}^{\infty} a_n x^n$ converges in $|x| < r$, the series $\sum_{n=0}^{\infty} a_n \rho^n$ where $0 < \rho < r$ converges and it necessarily follows that $\lim_{n\to\infty} a_n \rho^n = 0$ (see **AII.B1**).

Hence, there exists an M_a such that

$$|a_n \rho^n| \leq M_a \quad \text{for all } n. \tag{VI.6}$$

$\sum_{n=0}^{\infty} b_n x^n$ also converges in $|x| < r$. Hence, we can integrate termwise in $|x| < r$ (see **AII.D2**) and obtain $\sum_{n=0}^{\infty} \frac{b_n x^{n+1}}{n+1}$, which also converges in $|x| < r$. Hence, $\sum_{n=0}^{\infty} \frac{b_n \rho^{n+1}}{n+1}$, where $0 < \rho < r$, converges and consequently $\lim_{n\to\infty} \frac{b_n \rho^{n+1}}{n+1} = 0$. It follows that there exists an M_b such that

$$\left| \frac{b_n \rho^{n+1}}{n+1} \right| \leq M_b \quad \text{for all } n. \tag{VI.7}$$

Let $M = \max(M_a, M_b)$. Then, in view of (VI.6) and (VI.7)

$$|a_n \rho^n| \leq M \quad \text{and} \quad \left| \frac{b_n \rho^{n+1}}{n+1} \right| \leq M \quad \text{for all } n \text{ and } 0 < \rho < r.$$

Thus, we obtain

$$|a_n| \leq \frac{M}{\rho^n} = A_n \quad \text{and} \quad |b_n| \leq \frac{(n+1)M}{\rho^{n+1}} = B_n \quad \text{for all } n. \tag{VI.8}$$

We will now majorize the solution $\sum_{n=0}^{\infty} c_n x^n$ by a series $\sum_{n=0}^{\infty} C_n x^n$, i.e., show that $|c_n| \leq C_n$ for all n, and show that the latter converges in $|x| < r$.

It follows from (VI.5) that

$$|c_{n+2}| \leq \frac{1}{(n+1)(n+2)} [(n+1)|a_0||c_{n+1}| + |c_n|(n|a_1| + |b_0|) + \cdots + |c_0||b_n|]. \tag{VI.9}$$

Let us choose $C_0 = |c_0|$, $C_1 = |c_1|$ and all the following C_k according to

$$C_{n+2} = \frac{1}{(n+1)(n+2)} [(n+1)A_0 C_{n+1} + C_n(nA_1 + B_0) + \cdots + C_0 B_n]. \tag{VI.10}$$

Clearly, in view of (VI.8) and (VI.9)

$$|c_n| \leq C_n$$

and the series $\sum_{n=0}^{\infty} c_n x^n$ appears majorized ($\ll$) as follows

$$\sum_{n=0}^{\infty} c_n x^n \ll \sum_{n=0}^{\infty} C_n x^n. \tag{VI.11}$$

The solution of the differential equation (*dominating equation*)

$$Y'' - A(x)Y' - B(x)Y = 0 \tag{VI.12}$$

with

$$A(x) = \sum_{n=0}^{\infty} A_n x^n = M \sum_{n=0}^{\infty} \left(\frac{x}{\rho}\right)^n \qquad \text{(see (VI.8))}$$

and

$$B(x) = \sum_{n=0}^{\infty} B_n x^n = M \sum_{n=0}^{\infty} \frac{n+1}{\rho^{n+1}} x^n \qquad \text{(see (VI.8))}$$

—if one exists—is clearly $Y(x) = \sum_{n=0}^{\infty} C_n x^n$, where the C_n are determined by (VI.10) since (VI.10) bears the same relation to (VI.12) that (VI.5) does to (VI.3).

Since

$$A(x) = M \sum_{n=0}^{\infty} \left(\frac{x}{\rho}\right)^n = \frac{M\rho}{\rho - x} \quad \text{for } |x| < \rho$$

and

$$B(x) = M \sum_{n=0}^{\infty} \frac{n+1}{\rho^{n+1}} x^n = M \sum_{n=1}^{\infty} \frac{nx^{n-1}}{\rho^n} = A'(x) = \frac{M\rho}{(\rho - x)^2} \quad \text{for } |x| < \rho,$$

equation (VI.12) is identical with

$$Y'' - \frac{M\rho}{\rho - x} Y' - \frac{M\rho}{(\rho - x)^2} Y = 0, \tag{VI.13}$$

which is an equation of the Euler-Cauchy type (see Chapter II, §3.4) and can be reduced to a linear differential equation with constant coefficients via the transformation

$$t = \log(\rho - x), \quad |x| < \rho.$$

Then

$$Y' = -\frac{dY}{dt} \cdot \frac{1}{\rho - x}$$

and

$$Y'' = \frac{d^2Y}{dt^2} \cdot \frac{1}{(\rho - x)^2} - \frac{dY}{dt} \cdot \frac{1}{(\rho - x)^2}.$$

Hence, we obtain for (VI.13) after multiplication by $(\rho - x)^2$

$$\frac{d^2Y}{dt^2} + (M\rho - 1)\frac{dY}{dt} - M\rho Y = 0.$$

The roots of the characteristic equation

$$\lambda^2 + (M\rho - 1)\lambda - M\rho = 0$$

are $\lambda_1 = -M\rho$, $\lambda_2 = 1$. Therefore,

$$e^{-M\rho t} = (\rho - x)^{-M\rho}, \quad e^t = (\rho - x)$$

are two linearly independent solutions of (VI.12) and we obtain for the general solution of (VI.12) with two arbitrary constants D_1, D_2

$$Y = D_1(\rho - x) + D_2(\rho - x)^{-M\rho} = E_1\left(1 - \frac{x}{\rho}\right) + E_2\left(1 - \frac{x}{\rho}\right)^{-M\rho}$$

where $E_1 = D_1\rho$ and $E_2 = \dfrac{D_2}{\rho^{M\rho}}$.

We can expand $\left(1 - \dfrac{x}{\rho}\right)^{-M\rho}$ for $|x| < \rho$ into a binomial series and obtain

$$Y = E_1\left(1 - \frac{x}{\rho}\right) + E_2(1 + Mx + \cdots) = (E_1 + E_2) + \left(E_2M - \frac{E_1}{\rho}\right)x + \cdots.$$

Let us choose E_1 and E_2 such that

$$E_1 + E_2 = C_0,$$
$$-\frac{E_1}{\rho} + ME_2 = C_1,$$

which is certainly possible since

$$\begin{vmatrix} 1 & 1 \\ \dfrac{-1}{\rho} & M \end{vmatrix} = M + \frac{1}{\rho} \neq 0.$$

Then, the remaining coefficients of the series are determined by (VI.10) and we see that the series $\sum_{n=0}^{\infty} C_n x^n$ converges in $|x| < \rho$. Since ρ may assume any value in $0 < \rho < r$, it converges in $|x| < r$ and in view of (VI.11), the series solution $\sum_{n=0}^{\infty} c_n x^n$ of (VI.3) also converges in $|x| < r$, q.e.d.

If we choose $c_0 = 1$, $c_1 = 0$, we obtain from (VI.5)

$$y_1(x) = 1 + c_2x^2 + \cdots \tag{VI.14}$$

and if we choose $c_0 = 0$, $c_1 = 1$, we obtain from (VI.5)

$$y_2(x) = x + c_3x^3 + \cdots. \tag{VI.15}$$

As is shown in the theory of ordinary differential equations, a necessary and sufficient condition for two solutions y_1, y_2 of a linear differential equation of the second order to be linearly independent in an interval $a \leq x \leq b$ is that the *Wronskian*

$$W[y_1, y_2] = \begin{vmatrix} y_1 & y_2 \\ y_1' & y_2' \end{vmatrix}$$

does not vanish for some x in $a \leq x \leq b$.[3] (See problems VI.1, 2.)

Clearly,

$$W[y_1, y_2]_{x=0} = \begin{vmatrix} 1 & 0 \\ 0 & 1 \end{vmatrix} = 1$$

where y_1 and y_2 are the functions determined in (VI.14) and (VI.15) and we see that these solutions are linearly independent in any interval including the point $x = 0$.

Problems VI.1–4

1. Let y_1, y_2 be two solutions of a linear homogeneous differential equation of second order such that $W[y_1, y_2] \neq 0$ for $x = x_0$. Show that it is possible to find two constants c_1, c_2 such that $y = c_1y_1 + c_2y_2$ satisfies the initial conditions $y(x_0) = y_0$, $y'(x_0) = y_0'$, where y_0, y_0' are given values.

2. Show that if it is possible to find two constants c_1, c_2 such that $y(x_0) = y_0$, $y'(x_0) = y_0'$, where $y = c_1y_1 + c_2y_2$ and y_0, y_0' are any given values, then $W[y_1, y_2] \neq 0$ at $x = x_0$.

3. Reduce

$$4y'' + \frac{1}{x}y' - \frac{1}{x^2}y = 0$$

to a differential equation with constant coefficients and solve.

4. What is the general solution of

$$a_0y'' + \frac{a_1}{x}y' + \frac{a_2}{x^2}y = 0$$

where the a_k are real constants, if the roots of the equation

$$a_0\lambda^2 + (a_1 - a_0)\lambda + a_2 = 0$$

are complex?

2. Series Solutions of the Legendre Equation

As pointed out in the preceding section, the Legendre equation (VI.1) is of the type (VI.3) and the coefficients $a(x)$ and $b(x)$ can be expanded into Taylor series at $x = 0$ which converge in $|x| < 1$. Hence, we expect, in view of theorem VI.1 that series solutions of (VI.1) can be obtained and will converge in $|x| < 1$.

[3] See L. R. Ford: *Differential Equations*, McGraw-Hill, 1955, p. 64; or E. A. Coddington and N. Levinson: *Theory of Ordinary Differential Equations*, McGraw-Hill, 1955, p. 83.

In order to obtain a series solution, we do not have to follow the general pattern as outlined in the preceding section. We can arrive at a recursion formula for the c_n taking a shortcut: by substituting $y = \sum_{n=0}^{\infty} c_n x^n$ and the series representations of its derivatives directly into (VI.1) without expanding $a(x)$ and $b(x)$.

We obtain after collection of terms of the same order

$$\sum_{\nu=2}^{\infty} c_\nu \nu(\nu - 1)x^{\nu-2} - \sum_{\nu=0}^{\infty} [c_\nu \nu(\nu - 1) + 2\nu c_\nu - \lambda c_\nu]x^\nu = 0. \quad \text{(VI.16)}$$

If, in the first sum, we change the summation subscript from $\nu - 2$ to μ and write afterwards ν instead of μ, we obtain

$$\sum_{\nu=0}^{\infty} [c_{\nu+2}(\nu + 2)(\nu + 1) - c_\nu \nu(\nu - 1) - 2\nu c_\nu + \lambda c_\nu]x^\nu = 0.$$

This equation has to be identically satisfied; hence the coefficients of all powers of x have to vanish and we see that

$$c_{\nu+2}(\nu + 2)(\nu + 1) - c_\nu \nu(\nu - 1) - 2\nu c_\nu + \lambda c_\nu = 0. \quad \text{(VI.17)}$$

Before we go any further, let us write the arbitrary parameter λ in the form $n(n + 1)$ as was suggested by the developments in Chapter II, §3.4. Then, from (VI.17) the following recursion formula for the c_ν in terms of n appears:

$$c_{\nu+2} = c_\nu \frac{\nu(\nu + 1) - n(n + 1)}{(\nu + 1)(\nu + 2)} = -c_\nu \frac{(n - \nu)(n + \nu + 1)}{(\nu + 1)(\nu + 2)}. \quad \text{(VI.18)}$$

We can see immediately from this formula that:

in case n is an even integer, $n = 2\nu$,

$$c_{2\nu+2} = c_{2\nu+4} = \cdots = 0$$

and in case n is an odd integer, $n = 2\nu + 1$,

$$c_{2\nu+3} = c_{2\nu+5} = \cdots = 0;$$

i.e., if n is an integer, there occur only finitely many terms of either even or odd order in the series expansion of $y(x)$.

In general, whether n is an integer or not, for $c_0 = 1$, $c_1 = 0$, we arrive at a solution of the form

$$y_1(x) = 1 + c_2 x^2 + c_4 x^4 + \cdots, \quad y_1(-x) = y_1(x)$$

and for $c_0 = 0$, $c_1 = 1$, the solution is

$$y_2(x) = x + c_3 x^3 + c_5 x^5 + \cdots, \quad y_2(-x) = -y_2(x),$$

where $W[y_1, y_2] = 1$ at $x = 0$; i.e., y_1, y_2 are linearly independent in any interval containing the point $x = 0$.

In case n is an even integer, $y_1(x)$ degenerates into a polynomial and in case n is an odd integer, $y_2(x)$ degenerates into a polynomial.

The series for $y_1(x)$ and $y_2(x)$ converge according to theorem VI.1 for all $|x| < 1$. However, the polynomial solutions of the Legendre equation which are obtained for integral values of n are the only even and odd solutions of the Legendre equation which are continuous in the closed interval $|x| \leq 1$, as we will deduce from the following lemma:

Lemma VI.1. *Let* $y(x) = \sum_{\nu=0}^{\infty} c_\nu x^\nu$ *be a solution of the Legendre equation where either* $c_0 = 0$ *or* $c_1 = 0$ *if* n *in* $\lambda = n(n+1)$ *is not an integer, or* $c_0 = 0, c_1 \neq 0$ *if* n *is an even integer, or* $c_1 = 0, c_0 \neq 0$ *if* n *is an odd integer; i.e.,* $y(x)$ *is an odd or an even function and infinitely many* c_ν*'s in its series expansion are different from zero.*

In this case

$$\lim_{x \to \pm 1} |y(x)| = \infty.$$

Proof. It follows from the recursion formula (VI.18) that

$$c_\nu = \frac{(\nu-1)(\nu-2)-\lambda}{\nu(\nu-1)} c_{\nu-2} = \frac{\nu-2}{\nu}\left[1 - \frac{\lambda}{(\nu-1)(\nu-2)}\right] c_{\nu-2}$$

where we write again λ instead of $n(n+1)$.

By the same token

$$c_{\nu-2} = \frac{(\nu-3)(\nu-4)-\lambda}{(\nu-2)(\nu-3)} c_{\nu-4} = \frac{\nu-4}{\nu-2}\left[1 - \frac{\lambda}{(\nu-3)(\nu-4)}\right] c_{\nu-4},$$

$$\cdot$$
$$\cdot$$
$$\cdot$$

$$c_{\nu-2\mu+2} = \frac{(\nu-2\mu+1)(\nu-2\mu)-\lambda}{(\nu-2\mu+2)(\nu-2\mu+1)}$$
$$= \frac{\nu-2\mu}{(\nu-2\mu+2)}\left[1 - \frac{\lambda}{(\nu-2\mu+1)(\nu-2\mu)}\right] c_{\nu-2\mu},$$

where $\nu - 2\mu \geq 0$.

By successive substitution,

$$c_\nu = \frac{\nu-2}{\nu}\left[1 - \frac{\lambda}{(\nu-1)(\nu-2)}\right]\frac{\nu-4}{\nu-2}\left[1 - \frac{\lambda}{(\nu-3)(\nu-4)}\right]\cdots$$
$$\frac{(\nu-2\mu)}{(\nu-2\mu+2)}\left[1 - \frac{\lambda}{(\nu-2\mu+1)(\nu-2\mu)}\right] c_{\nu-2\mu}$$

and after cancellations

$$c_\nu = \frac{\nu - 2\mu}{\nu} c_{\nu-2\mu} \prod_{\kappa=1}^{\mu} \left[1 - \frac{\lambda}{(\nu - 2k + 1)(\nu - 2k)}\right]. \quad \text{(VI.19)}$$

λ has a fixed value. Therefore, we certainly can choose an m (even or odd if n is not an integer depending on whether c_1 or c_0 is zero, and even if n is odd, and odd if n is even) large enough so that

$$1 - \frac{\lambda}{(m + 1)m} > 0.$$

Then, $1 - \dfrac{\lambda}{(m + l + 1)(m + l)} > 0$ for all $l = 0, 2, 4, 6, \cdots$ and all the terms in

$$\prod_{l=0}^{2\mu-2}\left[1 - \frac{\lambda}{(m + l + 1)(m + l)}\right]$$

are positive. We can assume without loss of generality that $c_m > 0$ for this particular m and see that

$$c_{m+2\mu} = \frac{m}{m + 2\mu} c_m \prod_{l=0}^{2\mu-2}\left[1 - \frac{\lambda}{(m + l + 1)(m + l)}\right] > 0, \qquad l \text{ even},$$

for all μ. (In case $c_m < 0$, then $c_{m+2\mu} < 0$ for all μ and the subsequent proof can be carried out just as well with reversed signs.)

Since a product $\prod_{\nu=0}^{\infty} (1 + a_\nu)$ converges if $\sum_{\nu=0}^{\infty} |a_\nu|$ converges (see AII.C7) and in view of the convergence of

$$\sum_{l=0}^{\infty} \frac{\lambda}{(m + l + 1)(m + l)}$$

(see AII.C2), we have

$$\lim_{\mu\to\infty} \prod_{l=0}^{2\mu-2}\left[1 - \frac{\lambda}{(m + l + 1)(m + l)}\right] = c > 0,$$

where c is some constant; i.e., for any arbitrary small $\epsilon > 0$ there exists an N_ϵ such that

$$\left|\prod_{l=0}^{2\mu-2}\left[1 - \frac{\lambda}{(m + l + 1)(m + l)}\right] - c\right| < \epsilon$$

for all $\mu > N_\epsilon$.

Let $\gamma = c - \epsilon$. Then

$$\prod_{l=0}^{2\mu-2}\left[1 - \frac{\lambda}{(m + l + 1)(m + l)}\right] > \gamma \quad \text{for all } \mu > N_\epsilon$$

and

$$c_{m+2\mu} = \frac{m}{m+2\mu} c_m \prod_{l=0}^{2\mu-2} \left[1 - \frac{\lambda}{(m+l+1)(m+l)} \right] > c_m \gamma \frac{m}{m+2\mu}$$

where c_m and γ have fixed values.

Hence,

$$c_k > c_m \gamma m \frac{1}{k} \quad \text{for all } k > m + 2N_\epsilon$$

and consequently

$$\sum_{k=m+2N_\epsilon+1}^{\infty} c_k x^k > c_m \gamma m \sum_{k=m+2N_\epsilon+1}^{\infty} \frac{x^k}{k} \quad \text{for } x > 0.$$

Hence

$$\lim_{x\to 1} \sum_{k=m+2N_\epsilon+1}^{\infty} c_k x^k \geq c_m \gamma m \sum_{k=m+2N_\epsilon+1}^{\infty} \frac{1}{k} = \infty$$

since $\sum_{k=1}^{\infty} 1/k$ diverges (see **AII.C2**).

Hence,

$$\lim_{x\to 1} y(x) = \infty. \tag{VI.20}$$

Now we have to discuss the case where $x \to -1$, i.e., $x < 0$.

By hypothesis, $y(x)$ is either even or odd. If $y(x)$ is even, then $y(-x) = y(x)$ and it follows from (VI.20) that

$$\lim_{x\to -1} y(x) = \lim_{x\to -1} y(-x) = \lim_{\xi\to 1} y(\xi) = \infty.$$

If $y(x)$ is odd, then $y(-x) = -y(x)$ and it follows from(VI. 20) that

$$\lim_{x\to -1} y(x) = -\lim_{x\to -1} y(-x) = -\lim_{\xi\to 1} y(\xi) = -\infty.$$

Thus, $\lim_{x\to \pm 1} |y(x)| = \infty$, q.e.d.

From this lemma there follows immediately

Theorem VI.2. *Solutions of the Legendre equation (VI.1) which are continuous in $|x| \leq 1$ can be obtained if and only if n in $\lambda = n(n + 1)$ is an integer.*[4]

Proof. If n is an integer, then an even or an odd polynomial of nth order is a solution which is clearly everywhere continuous.

If n is not an integer, then, both the even solution $y_1(x)$ and the odd

[4] This theorem can also be proved independently of lemma VI.1 by showing that the polynomial solutions of the Legendre equation constitute a closed orthogonal system on $-1 \leq x \leq 1$. See R. V. Churchill: *Fourier Series and Boundary Value Problems*, McGraw-Hill Book Co., New York, 1941, p. 185.

solution $y_2(x)$ of (VI.1) are not continuous at $x = \pm 1$, as was proved in lemma VI.1, and we obtain for the general solution $y = \alpha y_1 + \beta y_2$ for $(\alpha, \beta) \neq (0, 0)$ that $\lim |y(x)| = \infty$, as $x \to 1$ or $x \to -1$ or both.

Problems VI.5–8

5. Let n be an integer and let $c_0 \neq 0$ and $c_1 \neq 0$. Prove that $\lim_{x \to -1} |y(x)| = \infty$ where $y(x)$ is the general solution of the Legendre equation (VI.1). (*Hint:* Use lemma VI.1.)

6. Show that if $y(x)$ is a solution of Legendre's equation, then also $y(-x)$, $y(x) + y(-x)$, and $y(x) - y(-x)$ are solutions of the Legendre equation. Show that $y(x) + y(-x)$ is even and $y(x) - y(-x)$ is odd.

7. Let $c_1 = 0$ and find the solution of (VI.1) for $n = 0, 2, 4$.

8. Let $c_0 = 0$ and find the solution of (VI.1) for $n = 1, 3, 5$.

3. Polynomial Solutions of the Legendre Equation

We now consider the case where n is an integer. We can assume without loss of generality, that $n \geq 0$, because if $n < 0$, then $n(n + 1) = |n|(|n| - 1) \geq 0$ for all integers n and the status quo is restored by substituting $|n| \to n + 1$.

We have seen in the preceding subsection that in case n is even and $c_1 = 0$, the solution is a polynomial of nth order. The condition $c_1 = 0$ is clearly equivalent to the boundary conditions $y(0) = c_0$, $y'(0) = 0$. In case n is odd and $c_0 = 0$, a polynomial of nth order again furnishes a solution and the condition $c_0 = 0$ is clearly equivalent to the boundary conditions $y(0) = 0$, $y'(0) = c_1$.

Thus, for integral values of n the following polynomials are solutions of (VI.1):

$$\pi_0 = c_0, \qquad \pi_1 = c_1 x,$$

$$\pi_2 = c_0(1 - 3x^2), \qquad \pi_3 = c_1\left(x - \frac{5x^3}{3}\right),$$

$$\pi_4 = c_0\left(1 - 10x^2 + \frac{35}{3}x^4\right), \qquad \pi_5 = c_1\left(x - \frac{14x^3}{3} + \frac{21x^5}{5}\right)$$

$$\vdots$$

(see problems VI.7, 8).

These polynomials constitute, according to theorem V.9 and in view of the remark on p. 177, an orthogonal system on the interval $-1 \leq x \leq 1$, a fact which we will re-establish by an independent method in subsection

5 of this section. In order to normalize these polynomials in the sense of Chapter V, §2.5 we have to evaluate the integral

$$\int_{-1}^{1} \pi_n^2(x)\, dx,$$

which will also be done in subsection 5 of this section.

Before going into all these questions, we have to establish a general expression for the nth polynomial π_n.

The last nonvanishing coefficient of π_n is c_n. Instead of writing all the coefficients in terms of c_0 or c_1 as we have done in the preceding development, we will express them all in terms of the last nonvanishing coefficient c_n which has the definite advantage that it is then no longer necessary to distinguish between even and odd order.

It follows from (VI.18) that

$$c_{n-2} = -c_n \frac{n(n-1)}{2(2n-1)},$$

$$c_{n-4} = c_n \frac{n(n-1)(n-2)(n-3)}{2 \cdot 4(2n-1)(2n-3)},$$

$$\vdots$$

In general we will have

$$c_{n-2k} = (-1)^k \frac{n(n-1)\cdots(n-2k+1)\cdot c_n}{(2n-1)(2n-3)\cdots(2n-2k+1)\cdot 2\cdot 4\cdots(2k)}.$$

Thus, we can write the nth polynomial π_n in the form

$$\pi_n(x) = c_n \sum_{k=0}^{[n/2]} (-1)^k \frac{n(n-1)\cdots(n-2k+1)}{(2n-1)(2n-3)\cdots(2n-2k+1)\cdot 2\cdot 4\cdots(2k)} x^{n-2k} \tag{VI.21}$$

where $[n/2]$ stands for the largest integer $\leq n/2$.

4. Generating Function—Legendre Polynomials

We consider the following simple problem of potential theory: Let us assume two unit masses one of which is considered to be located at a fixed reference point and the other of which rotates along a circle of radius r about a point C which is at unit distance from the reference point (see Fig. 23).

The potential of m_1 with respect to m_2 is but for a multiplicative constant

$$\Phi(\theta) = \frac{1}{z(\theta)}$$

where $z(\theta)$ denotes the distance between m_1 and m_2 and θ denotes the angle between the line joining C and m_1 and the line joining C and m_2.

According to the law of cosines

$$z(\theta) = (1 - 2r\cos\theta + r^2)^{1/2};$$

hence,

$$\Phi(\theta) = [1 - (2r\cos\theta - r^2)]^{-1/2}. \tag{VI.22}$$

Let us consider Φ as a function of r. The denominator vanishes for

$$r_{1,2} = \cos\theta \pm i\sin\theta, \qquad i = \sqrt{-1},$$

as one can immediately see. Hence, the only real values of r for which the denominator vanishes are $r = \pm 1$ (for $\theta = 0, \pi$) and the same holds true

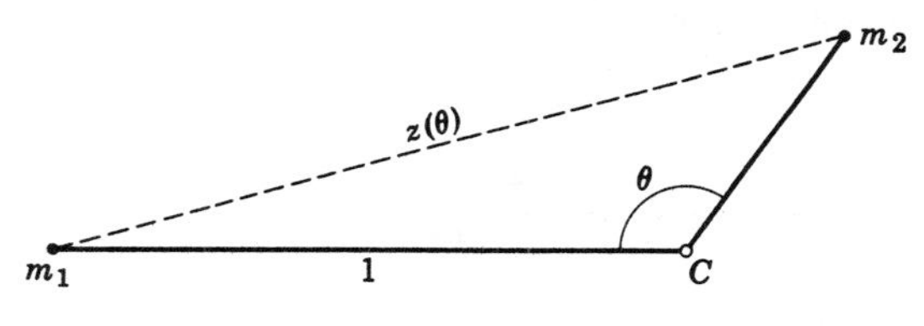

Fig. 23

for all the derivatives of Φ with respect to r. In $-1 < -R \le r \le R < 1$, $\left|\frac{\partial^{(n)}\Phi}{\partial r^n}\right| \le M$ for all n; hence Φ has a uniformly convergent Taylor expansion in $-R \le r \le R$:

$$\Phi = \sum_{n=0}^{\infty}\left(\frac{\partial^n\Phi}{\partial r^n}\right)_{r=0} r^n.$$

To find the derivatives $\partial^n\Phi/\partial r^n$ and evaluate them at $r = 0$ is a very tedious process. However, we can arrive at the series expansion of Φ in terms of r via a shortcut in the true spirit of 18th-century mathematics by expanding Φ in (VI.22) into a binomial series

$$[1 - (2r\cos\theta - r^2)]^{-1/2} = \sum_{j=0}^{\infty}(-1)^j\binom{-\frac{1}{2}}{j}(2r\cos\theta - r^2)^j$$

where $\binom{n}{m} = \dfrac{n(n-1)\cdots(n-m+1)}{1\cdot 2\cdots m}$, (note that this series does not converge for $|2r\cos\theta - r^2| > 1$, e.g., $r = \frac{3}{4}$, $\theta = \pi$) and expand again according to the binomial theorem

$$(2r\cos\theta - r^2)^j = \sum_{k=0}^{j}(-1)^k\binom{j}{k}2^{j-k}\cos^{j-k}\theta\, r^{j+k}.$$

Thus, we obtain the right expansion by not quite legitimate means, namely

$$[1 - (2r\cos\theta - r^2)]^{-1/2} = \sum_{j=0}^{\infty}\sum_{k=0}^{j}\binom{-\frac{1}{2}}{j}\binom{j}{k}2^{j-k}\cos^{j-k}\theta(-1)^{j+k}r^{j+k} \tag{VI.23}$$

The coefficient of r^n, which we designate by P_n, is a function of $\cos\theta$. We can easily find this coefficient because all the contributions to it come from the terms in (VI.23) for which $j + k = n$, where $j - k \geq 0$. It follows that

$$P_n(\cos\theta) = \sum_{k=0}^{[n/2]}(-1)^n\binom{-\frac{1}{2}}{n-k}\binom{n-k}{k}2^{n-2k}\cos^{n-2k}\theta. \tag{VI.24}$$

Let us now examine the coefficient $c_k^{(n)}$ of $\cos^{n-2k}\theta$ for any fixed k:

$$c_k^{(n)} = (-1)^n 2^{n-2k} \cdot \frac{(-\frac{1}{2})(-\frac{1}{2}-1)\cdots(-\frac{1}{2}-n+k+1)(n-k)(n-k-1)\cdots(n-2k+1)}{(n-k)!k!}. \tag{VI.25}$$

After some manipulations (see problems VI.9, 10) we find

$$c_k^{(n)} = (-1)^k\frac{(2n-2k)!}{2^n(n-k)!(n-2k)!k!}. \tag{VI.26}$$

Thus, we can write (VI.24) in the form

$$P_n(\cos\theta) = \sum_{k=0}^{[n/2]}(-1)^k\frac{(2n-2k)!}{2^n(n-k)!(n-2k)!k!}\cos^{n-2k}\theta. \tag{VI.27}$$

On comparing the coefficient of x^{n-2k} in (VI.21) with the coefficient of $\cos^{n-2k}\theta$ in (VI.27), we see that P_n and π_n are for $x = \cos\theta$ identical except for a multiplicative factor. In order to determine this factor, we have to solve the equation

$$c_n\frac{n(n-1)\cdots(n-2k+1)}{(2n-1)(2n-3)\cdots(2n-2k+1)\cdot 2\cdot 4\cdots(2k)} = \frac{(2n-2k)!}{2^n(n-k)!(n-2k)!k!} \tag{VI.28}$$

for c_n (we recall that we left the choice of c_n open for such a later decision), obtaining

$$c_n = \frac{(2n-1)(2n-3)\cdots 3\cdot 1}{n!}$$

(see problem VI.11).

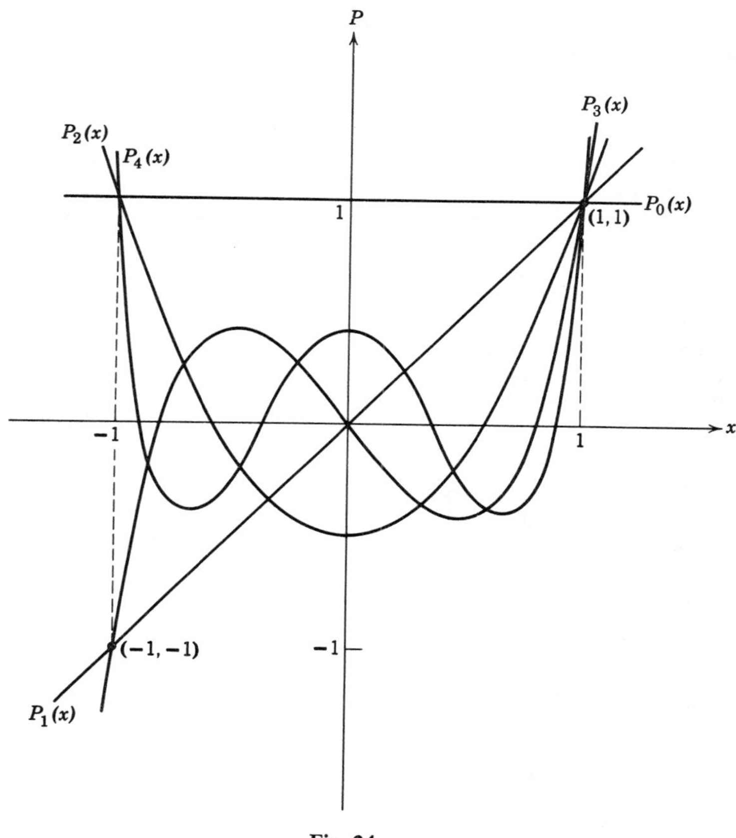

Fig. 24

The polynomials P_n are called the *Legendre Polynomials*. We see from (VI.27) that for $\cos\theta = x$, the first five Legendre polynomials have the form

$$P_0(x) = 1, \qquad P_1(x) = x,$$

$$P_2(x) = \tfrac{3}{2}x^2 - \tfrac{1}{2}, \qquad P_3(x) = \tfrac{5}{2}x^3 - \tfrac{3}{2}x,$$

$$P_4(x) = \tfrac{35}{8}x^4 - \tfrac{15}{4}x^2 + \tfrac{3}{8}, \qquad P_5(x) = \tfrac{63}{8}x^5 - \tfrac{35}{4}x^3 + \tfrac{15}{8}x$$

$$\vdots$$

(see problem VI.12). The graphs of the first four of these polynomials are represented in Fig. 24 (see problem VI.21).

Problems VI.9–14

9. Show that (VI.25) can be written in the form

$$c_k^{(n)} = (-1)^k \frac{1 \cdot 3 \cdot 5 \cdots (2n-2k-1)}{2^k(n-2k)!k!}.$$

(*Hint*: Take out a factor $\dfrac{(-1)^{n-k}}{2^{n-k}}$ and cancel by $(n-k)(n-k-1)\cdots(n-2k+1)$.)

10. Show that the result of problem 9 can be written in the form (VI.26). (*Hint*: Multiply numerator and denominator by $2 \cdot 4 \cdot 6 \cdots (2n-2k)$.)

11. Solve (VI.28) for c_n. (*Hint*: Observe that $2 \cdot 4 \cdot 6 \cdots (2k) = 2^k k!$; hence $(2n-2k)! = (2n-2k-1)\cdots 3 \cdot 1 \cdot 2^{n-k}(n-k)!$ and $n(n-1)\cdots(n-2k+1)\cdot(n-2k)! = n!$.)

12. Find the Legendre polynomials for $n = 0, 1, 2, 3, 4, 5$.

13. Assuming that $y(x)$ is a solution of Legendre's equation (VI.1), find a differential equation of second order for $u(x)$ which is satisfied by

$$u(x) = \frac{d^m}{dx^m} y(x).$$

(*Hint*: Differentiate (VI.1) m times.)

***14.** Write the equation which is obtained in problem 13 in self-adjoint form. Find a function $v(x)$ such that for any integer n in $\lambda = n(n+1)$, the functions $w_n(x) = v(x)u_n(x)$ are orthogonal on $-1 \le x \le 1$, where $u_n(x)$ are the solutions of the equation which is obtained in problem 13. (These functions $w_n(x)$ are called the *associated Legendre functions*, see also Chapter VIII, §1.4.)

5. Orthogonality and Normalization

We will now give a proof for the orthogonality of the Legendre polynomials which is independent of the one presented in Chapter V, §2.4. This proof will yield at the same time the value of $\int_{-1}^{1} P_n^2(x)\,dx$.

In the preceding subsection we saw that

$$[1-(2rx-r^2)]^{-1/2} = \sum_{n=0}^{\infty} P_n(x)r^n, \quad |x| \le 1, \tag{VI.29}$$

where we have written again x instead of $\cos\theta$. Let us think of this equation written again in terms of ρ instead of r, let us multiply corresponding sides of the equations, and integrate the product between -1 and 1:

$$\int_{-1}^{1} \frac{dx}{\sqrt{1-2xr+r^2}\sqrt{1-2x\rho+\rho^2}} = \sum_{n=0}^{\infty}\sum_{m=0}^{\infty} \int_{-1}^{1} P_n(x)P_m(x)\,dx \cdot r^n \rho^m. \tag{VI.30}$$

We will now evaluate the integral on the left side, expand the result into a series in terms of (ρr), and finally compare the coefficients of the thus-obtained series with the coefficients of the series on the right side of (VI.30). This will furnish finally the desired results.

To evaluate the integral on the left side, we substitute as follows

$$u^2 = \frac{1 + r^2}{2r} - x,$$

$$v^2 = \frac{1 + \rho^2}{2\rho} - x;$$

It follows then that $dx = -u\,du - v\,dv$ and $u\,du = v\,dv$. By adding to both sides of the latter equation the term $u\,dv$ we find

$$\frac{du}{v} = \frac{dv}{u} = \frac{d(u + v)}{u + v}$$

and thus we have for the integral that is under consideration

$$-\frac{1}{2\sqrt{r\rho}}\int \frac{u\,du + v\,dv}{uv} = -\frac{1}{2\sqrt{r\rho}}\left(\int \frac{du}{v} + \int \frac{dv}{u}\right)$$

$$= -\frac{1}{2\sqrt{r\rho}} \cdot 2\int \frac{d(u + v)}{u + v} = -\frac{1}{\sqrt{r\rho}} \log |u + v|.$$

We have at the upper limit $x = 1$

$$\left.\begin{aligned} u\big|_{x=1} &= \sqrt{\frac{1 + r^2}{2r} - 1} = \frac{1}{\sqrt{2r}}(1 - r) \\ v\big|_{x=1} &= \sqrt{\frac{1 + \rho^2}{2\rho} - 1} = \frac{1}{\sqrt{2\rho}}(1 - \rho) \end{aligned}\right\}, \quad r, \rho < 1,$$

and at the lower limit $x = -1$

$$\left.\begin{aligned} u\big|_{x=-1} &= \sqrt{\frac{1 + r^2}{2r} + 1} = \frac{1}{\sqrt{2r}}(1 + r) \\ v\big|_{x=-1} &= \sqrt{\frac{1 + \rho^2}{2\rho} + 1} = \frac{1}{\sqrt{2\rho}}(1 + \rho) \end{aligned}\right\}, \quad r, \rho < 1.$$

After some cumbersome algebraic manipulations we will arrive at

$$\int_{-1}^{1} \frac{dx}{\sqrt{1 - 2rx + r^2}\sqrt{1 - 2\rho x + \rho^2}} = \frac{1}{\sqrt{\rho r}} \log \frac{1 + \sqrt{r\rho}}{1 - \sqrt{r\rho}}.$$

Since $\rho < 1$, $r < 1$, we can expand this result into a convergent series as follows:

$$\frac{1}{\sqrt{r\rho}} \log \frac{1 + \sqrt{r\rho}}{1 - \sqrt{r\rho}} = 2\left(1 + \frac{r\rho}{3} + \cdots + \frac{(r\rho)^v}{2v + 1} + \cdots\right)$$

$\left(\text{because } \log \dfrac{1+\xi}{1-\xi} = \log(1+\xi) - \log(1-\xi) \text{ and } \log(1 \pm \xi) = \pm\xi - \dfrac{\xi^2}{2} \pm \dfrac{\xi^3}{3} - \cdots\right)$. Hence, we have for (VI.30)

$$2\sum_{\nu=0}^{\infty} \frac{(r\rho)^\nu}{2\nu+1} = \sum_{n=0}^{\infty}\sum_{m=0}^{\infty} \int_{-1}^{1} P_n(x)P_m(x)\,dx\, r^n\rho^m$$

and can see by comparing the coefficients that

$$\int_{-1}^{1} P_n(x)P_m(x)\,dx = 0, \quad m \neq n,$$

and

$$\int_{-1}^{1} P_n^2(x)\,dx = \frac{2}{2n+1}. \tag{VI.31}$$

It follows that the polynomials $[\sqrt{(2n+1)/2}]P_n(x)$ where $P_n(x)$ is defined by (VI.27) constitute an orthonormal system on the interval $-1 \leq x \leq 1$.

Problems VI.15–22

***15.** Find $P_n(1)$ for any integer n. (*Hint:* Use the expansion (VI.29), expand the left side into a power series, and compare the coefficients.)

16. Show that $P_n(x)$ cannot have more than one zero in $|x| > 1$. (*Hint:* Use theorem V.3.)

17. Show that $P_n(-1) = (-1)^n$. (*Hint:* Use the result in problem 15.)

18. Show that

$$\sum_{k=0}^{[n/2]} (-1)^k \frac{(2n-2k)!}{2^n(n-k)!(n-2k)!k!} x^{n-2k}$$
$$= \frac{1}{2^n n!}\frac{d^n}{dx^n}\sum_{k=0}^{n}(-1)^k \frac{n(n-1)\cdots(n-k+1)}{k!} x^{2n-2k}.$$

19. Use the result in problem 18 to show that

$$P_n(x) = \frac{1}{2^n n!}\frac{d^n}{dx^n}(x^2-1)^n.$$

(*Hint:* Expand $(x^2-1)^n$ into a binomial series.) This formula is called *Rodrigues' formula.*

***20.** Show that the zeros of $P_n(x)$ are all real and lie in $-1 \leq x \leq 1$. (*Hint:* Use Rodrigues' formula.)

21. Sketch the $P_n(x)$ for $n = 0, 1, 2, 3, 4, 5$ in $-1 \leq x \leq 1$. (*Hint:* Use all the results obtained in problems 15, 17, and 20 and theorem V.5.)

***22.** According to theorem V.1 all the roots of $P_n(x)$ are simple. Give an independent proof of this fact, using the equation obtained in V.13. (*Hint:* Assume that there is a multiple root and note that $P_n(\pm 1) \neq 0$.)

6. Legendre Functions of the Second Kind

For every integer n only one polynomial solution of the Legendre equation is obtained. The other solution which is linearly independent of $P_n(x)$ is an infinite series, namely

$$c_1x + c_3x^3 + c_5x^5 + \cdots$$

in the case where n is even, and

$$c_0 + c_2x^2 + c_4x^4 + \cdots$$

in the case where n is odd.

The c_n are computed in terms of the c_0 and c_1 according to (VI.18) and we have for $n = 2\nu$

$$Q_{2\nu}(x) = c_1\left[x - \frac{(2\nu - 1)(2\nu + 2)}{3!}x^3 + \frac{(2\nu - 1)(2\nu - 3)(2\nu + 2)(2\nu + 4)}{5!}x^5 - \cdots\right]$$

and for $n = 2\nu + 1$

$$Q_{2\nu+1}(x) = c_0\left[1 - \frac{(2\nu + 1)(2\nu + 2)}{2!}x^2 + \frac{(2\nu + 1)(2\nu - 1)(2\nu + 2)(2\nu + 4)}{4!}x^4 - \cdots\right].$$

These series converge for $|x| < 1$, according to theorem VI.1, and grow beyond all bounds as $x \to \pm 1$ according to lemma VI.1.

In order to obtain a solution which is linearly independent of $P_n(x)$ and converges for $|x| > 1$ one can proceed as follows:

A transformation of the independent variable x in (VI.1) according to

$$x = \frac{1}{\xi} \tag{VI.32}$$

yields

$$y'' - \frac{2\xi}{1 - \xi^2}y' - \frac{n(n + 1)}{(1 - \xi^2)\xi^2}y = 0 \tag{VI.33}$$

where $'$ and $''$ now indicate derivatives with respect to ξ. In §2.2 of this chapter we will prove a theorem (theorem VI.3) which guarantees the convergence of a series solution of (VI.33) for $|\xi| < 1$. From this we can obtain a series solution of (VI.1) which converges in $|x| > 1$ in view of (VI.32), namely

$$Q_n(x) = cx^{-n}\left[\frac{1}{x} + \frac{(n+1)(n+2)}{2(2n+3)} \cdot \frac{1}{x^3} + \frac{(n+1)(n+2)(n+3)(n+4)}{2 \cdot 4(2n+3)(2n+5)} \cdot \frac{1}{x^5} + \cdots\right] \quad \text{(VI.34)}$$

(see problems VI.37 and VI.38).

The functions $Q_{2\nu}(x)$, $Q_{2\nu+1}(x)$ for $|x| < 1$ and $Q_n(x)$ for $|x| > 1$ are called *Legendre functions of the second kind.*

Problems VI.23–25

23. Show that $Q_n(x)$ is even if n is odd, and odd if n is even.

24. Transform (VI.1) according to (VI.32) into (VI.33).

25. Prove that $P_{2\nu}(x)$, $Q_{2\nu}(x)$ and $P_{2\nu+1}(x)$, $Q_{2\nu+1}(x)$ in $|x| < 1$ and $P_n(x)$, $Q_n(x)$ in $|x| > 1$ are linearly independent.

7. Legendre Polynomials as Potentials of Multipoles with Respect to an Action Point at Unit Distance.

We will shortly discuss in this section still another process of generating the Legendre polynomials. The basic idea of this process is due to J. C. Maxwell.[5] We consider an electric *dipole*, i.e., a positive charge $1/h$ and a negative charge $-(1/h)$ at a distance h apart and let h approach zero. We choose our coordinate system such that both charges are located on the x-axis with the negative charge in the origin. The potential Φ_2 with respect to an action point P on the unit sphere $x^2 + y^2 + z^2 = 1$ (see Fig. 25) is

$$\Phi_2 = \lim_{h \to 0} \frac{1}{h}\left\{\frac{1}{\sqrt{(x-h)^2 + y^2 + z^2}} - \frac{1}{\sqrt{x^2 + y^2 + z^2}}\right\}$$

$$= \frac{\partial}{\partial x} \cdot \frac{1}{R}\bigg|_{R=1} = \frac{-x}{R^3}\bigg|_{R=1} = -x = -P_1(x) \cdot 1!.$$

Next, we find the potential Φ_4 of a *quadrupole* at the origin with respect to an action point on the unit sphere: A negative double charge $-(2/h^2)$

[5] J. C. Maxwell: *A Treatise on Electricity and Magnetism*, Oxford at the Clarendon Press, 1881, Chapter IX.

is placed at the origin and two positive charges $1/h^2$ are placed at the points $(h, 0, 0)$ and $(-h, 0, 0)$ respectively. Then

$$\Phi_4 = \lim_{h\to 0} \frac{1}{h^2}\left\{\frac{1}{\sqrt{(x+h)^2+y^2+z^2}} - \frac{2}{\sqrt{x^2+y^2+z^2}} + \frac{1}{\sqrt{(x-h)^2+y^2+z^2}}\right\}_{R=1}$$

$$= \frac{\partial^2}{\partial x^2}\cdot\frac{1}{R}\bigg|_{R=1} = -\frac{R^2-3x^2}{R^5}\bigg|_{R=1} = -1+3x^2 = P_2(x)\cdot 2!.$$

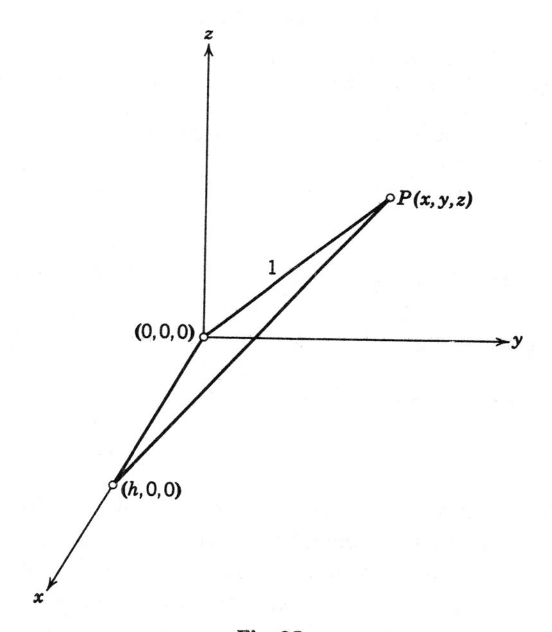

Fig. 25

As the reader can see for himself, one obtains for the potential Φ_8 of an *octopole*

$$\Phi_8 = \frac{\partial^3}{\partial x^3}\cdot\frac{1}{R}\bigg|_{R=1} = -15x^3+9x = -P_3(x)\cdot 3!$$

(see problem VI.26) and in general for the potential Φ_{2^n} of a *2^n-pole* with respect to an action point on the unit sphere

$$\Phi_{2^n} = \frac{\partial^n}{\partial x^n}\cdot\frac{1}{R}\bigg|_{R=1} = (-1)^n P_n(x)n!$$

The reader can see that these potentials are essentially the Legendre polynomials except for the factor $(-1)^n n!$

$$P_n(x) = \frac{(-1)^n}{n!} \frac{\partial^n}{\partial x^n} \cdot \frac{1}{R}\bigg|_{R=1}.$$

We do not wish to go into further details along these lines. We merely want to point out that a more general approach of this sort—more general insofar as one considers the charges arranged in star patterns around the origin—will lead to the so-called *spherical surface harmonics* (see Chapter VIII, §1).[6]

Problems VI.26–28

26. Find Φ_8 by the procedure outlined in subsection 7.

***27.** Find the stationary temperature distribution in a homogeneous and isotropic hemisphere of radius 1, where the temperature on the spherical part is a given function of the elevation ϕ only and the bottom is insulated.

28. Same as in problem 27 if the temperature on the spherical part is constant (c_0) and the bottom is insulated.

§2. THE BESSEL EQUATION AND BESSEL FUNCTIONS

1. Introductory Remarks

In this section we deal with the equation

$$x^2y'' + xy' + (\lambda^2x^2 - \mu^2)y = 0 \tag{VI.35}$$

where we write λ^2 instead of λ as we have done in equation (II.61). We can do this without loss of generality in view of theorems V.7 and V.8, according to which infinitely many eigenvalues of the Sturm-Liouville problem involving equation (VI.35) can be expected only if $\lambda x^2 - \mu^2 > 0$.

As was mentioned already in Chapter II, one calls the equation which is derived from (V.35) after elimination of λ, the *Bessel equation* in honor of the German mathematician and astronomer *F. W. Bessel*, who investigated systematically the solutions of this equation for integral values of μ in his paper "Untersuchungen des Theils der planetarischen Stoerungen, welcher aus der Bewegung der Sonne entsteht."[7]

Bessel was not the first one to deal with equations of this type. As early

[6] We suggest that the reader study these matters with A. Sommerfeld: *Partial Differential Equations in Physics*, Academic Press, New York, 1949.

[7] "Investigations of That Part of Planetary Disturbances Which Is Due to the Motion of the Sun," *Berliner Abhandlungen* 1824, pp. 1–52.

as 1703, *Jakob Bernoulli* had obtained a series solution of a special type of differential equation now bearing *Riccati*'s name, namely

$$y' = x^2 + y^2, \tag{VI.36}$$

which can be reduced (or complicated, depending on the point of view) into a Bessel equation with $\mu = \frac{1}{4}$.[8]

We will outline how one can transform (VI.36) into a Bessel equation:

One introduces a new variable ξ and a new function η according to

$$x = \sqrt{2\xi}$$

$$y = -\left(\frac{1}{2\sqrt{2\xi}} + \frac{\eta'\sqrt{2\xi}}{\eta}\right);$$

(VI.36) then becomes

$$\xi^2\eta'' + \xi\eta' + (\xi^2 - \tfrac{1}{16})\eta = 0,$$

as the reader can easily verify (see problems VI.29–33). This is a Bessel equation for $\mu = \frac{1}{4}$ in ξ, η.

We refer the student who is interested in the history of Bessel functions to the monograph by *G. N. Watson*.[9] This volume is also recommended for references to practically everything that has some connection with the Bessel functions. To obtain some insight into the nature of the solutions of (VI.35), we will first consider an equation of a more general type which embraces the Bessel equation as a special case, namely

$$y'' + \frac{a(x)}{x}y' + \frac{b(x)}{x^2}y = 0 \tag{VI.37}$$

where $a(x)$, $b(x)$ are analytic functions at $x = 0$, i.e., $a(x)$ and $b(x)$ have a Taylor expansion at $x = 0$ which converges in a neighborhood of this point:

$$\begin{aligned} a(x) &= a_0 + a_1x + a_2x^2 + \cdots, \\ b(x) &= b_0 + b_1x + b_2x^2 + \cdots. \end{aligned} \tag{VI.38}$$

The coefficients of (VI.37) are singular at $x = 0$ and these singularities are such that x times the coefficient of y' and x^2 times the coefficent of y are analytic according to (VI.38) at $x = 0$. One calls a singular point with

[8] It is actually difficult to find a branch of classical mathematics in which at least one of the Bernoullis was not involved in a rather prominent way.

[9] G. N. Watson: *Bessel Functions*, 2nd ed., The Macmillan Company, New York, 1945.

these specific properties a *regular singular point*. If we substitute the expansions for $a(x)$ and $b(x)$ into (VI.37), we obtain

$$y'' + \left(\frac{a_0}{x} + a_1 + a_2 x + \cdots\right) y' + \left(\frac{b_0}{x^2} + \frac{b_1}{x} + b_2 + b_3 x + \cdots\right) y = 0. \tag{VI.39}$$

In a neighborhood of $x = 0$

$$\frac{a_0}{x} >> a_1 + a_2 x + \cdots$$

$$\frac{b_0}{x^2} >> \frac{b_1}{x} + b_2 + b_3 x + \cdots$$

where $>>$ stands for "much greater than."

This suggests that the solution of (VI.37) will behave in a neighborhood of $x = 0$ like the solution of

$$y'' + \frac{a_0}{x} y' + \frac{b_0}{x^2} y = 0 \tag{VI.40}$$

because the terms $a_1 + a_2 x + \cdots$ and $\dfrac{b_1}{x} + b_2 + b_3 x + \cdots$ respectively will be of only minor influence. Equation (VI.40) is an equation of the Euler-Cauchy type and can be reduced to a linear differential equation with constant coefficients, as we pointed out in Chapter II, §3.4 and again in the proof of theorem VI.1. We introduce a new variable t according to

$$t = \log x$$

which transforms (VI.40) into

$$\frac{d^2 y}{dt^2} + (a_0 - 1)\frac{dy}{dt} + b_0 y = 0$$

with solutions of the form $e^{\lambda t}$. Consequently, the solutions of (VI.40) in terms of x will have the form $y = x^\lambda$ where λ is a solution of the so-called *indicial equation*

$$\lambda(\lambda - 1) + a_0 \lambda + b_0 = 0 \tag{VI.41}$$

as follows after substitution of x^λ into (VI.40).

Being inclined to think that the solutions of (VI.37) will behave like x^λ in a neighborhood of the origin, we conjecture that we will obtain the solution of (VI.37) itself by adding to x^λ infinitely many correction terms of higher order than λ

$$y = c_0 x^\lambda + c_1 x^{\lambda+1} + c_2 x^{\lambda+2} + \cdots = \sum_{\nu=0}^{\infty} c_\nu x^{\lambda+\nu}. \tag{VI.42}$$

That this series is indeed a solution of (VI.37) can be seen by substitution and consequent determination of coefficients and convergence investigations. Assuming that termwise differentiation is permissible, we obtain

$$y' = \sum_{\nu=0}^{\infty} c_\nu(\lambda + \nu)x^{\lambda+\nu-1},$$

$$y'' = \sum_{\nu=0}^{\infty} c_\nu(\lambda + \nu)(\lambda + \nu - 1)x^{\lambda+\nu-2}.$$

Substitution of (VI.42) and the series representations for y' and y'' into (VI.37) yields

$$\sum_{\nu=0}^{\infty} [c_\nu(\lambda + \nu)(\lambda + \nu - 1) + c_\nu(\lambda + \nu)a(x) + c_\nu b(x)]x^{\lambda+\nu} = 0$$

or in view of (VI.38)

$$\sum_{\nu=0}^{\infty} [c_\nu(\lambda + \nu)(\lambda + \nu - 1) + a_0(\lambda + \nu)c_\nu + b_0 c_\nu + a_1(\lambda + \nu - 1)c_{\nu-1} + b_1 c_{\nu-1} + \cdots + a_\nu \lambda c_0 + b_\nu c_0]x^{\lambda+\nu} = 0. \quad \text{(VI.43)}$$

This has to be an identity in x; i.e., the coefficient of every power of x has to vanish. If we arrange the terms according to the c_ν's, we find for the coefficient of $x^{\lambda+\nu}$

$$c_\nu[(\lambda + \nu)(\lambda + \nu - 1) + a_0(\lambda + \nu) + b_0] + c_{\nu-1}[a_1(\lambda + \nu - 1) + b_1] + \cdots + c_0[a_\nu\lambda + b_\nu] = 0.$$

Introducing the abbreviated notation

$$\begin{aligned} f_0(m) &= m(m - 1) + a_0 m + b_0, \\ f_k(m) &= a_k m + b_k \end{aligned} \quad \text{(VI.44)}$$

we may write this equation in the simpler form

$$c_\nu f_0(\lambda + \nu) + c_{\nu-1} f_1(\lambda + \nu - 1) + \cdots + c_0 f_\nu(\lambda) = 0 \quad \text{(VI.45)}$$

with the solution for c_ν

$$c_\nu = -\frac{c_{\nu-1}f_1(\lambda + \nu - 1) + \cdots + c_0 f_\nu(\lambda)}{f_0(\lambda + \nu)}. \quad \text{(VI.46)}$$

Since (VI.45) degenerates for $\nu = 0$ to

$$c_0 f_0(\lambda) = 0$$

and we wish to leave the choice of c_0 to the time when we impose certain boundary conditions, we have to choose λ such that $f_0(\lambda) = 0$, which is, according to (VI.44), nothing but the indicial equation (VI.41).

Since this is a quadratic equation for λ which in general has two different solutions, we accordingly expect that we will obtain two linearly independent solutions of (VI.37). (That this is not always the case will become apparent in subsection 4 of this section.)

After we have made a choice of λ, say λ_1, we will find the rest of the coefficients $c_1^{(1)}, c_2^{(1)}, c_3^{(1)}, \cdots$ from (VI.46) in terms of a $c_0^{(1)}$. For the second solution λ_2 of (VI.41), we obtain another set of constants $c_1^{(2)}, c_2^{(2)}, c_3^{(2)}, \cdots$ in terms of $c_0^{(2)}$ and the general solution will therefore contain two arbitrary constants $c_0^{(1)}$ and $c_0^{(2)}$.

In the successive computation of the coefficients according to (VI.46) one thing we have to guard against is the vanishing of the denominator:

$$f_0(\lambda + \nu) = 0. \tag{VI.47}$$

The solutions of $f_0(\lambda) = 0$, i.e., $\lambda^2 + (a_0 - 1)\lambda + b_0 = 0$ are

$$\lambda_1 = \frac{-a_0 + 1 + \sqrt{(a_0 - 1)^2 - 4b_0}}{2}$$

and

$$\lambda_2 = \frac{-a_0 + 1 - \sqrt{(a_0 - 1)^2 - 4b_0}}{2}.$$

Thus, $\lambda + \nu$ could be a solution of (VI.47) if and only if

$$\nu = \sqrt{(a_0 - 1)^2 - 4b_0};$$

i.e.: since ν is an integer, it follows that the square root of the discriminant has to be an integer. In this case, the second solution λ_2 has to be disregarded in order to avoid the vanishing of the denominator in (VI.46) and only one solution (corresponding to λ_1) will be obtained by this process.

Now that we know how to find a solution of the form $\sum_{\nu=0}^{\infty} c_\nu x^{\lambda+\nu}$ provided such a solution exists, we have to turn to the problem of its existence, to which we devote the next subsection.

Problems VI.29–33

29. Show that

$$x = \sqrt{2\xi},$$

$$y = -\left(\frac{1}{2\sqrt{2\xi}} + \frac{\eta'\sqrt{2\xi}}{\eta}\right)$$

transforms the Riccati equation $y' = x^2 + y^2$ into a Bessel equation with $\mu = \frac{1}{4}$.

30. Apply the transformation

$$x = a\xi^{\alpha}$$
$$y = \xi^{\beta}\eta$$

to the Bessel equation

$$x^2y'' + xy' + (x^2 - \mu^2)y = 0.$$

31. Given

$$x^2y'' + Axy' + (B + Cx^{\nu})y = 0.$$

Show that this equation can be transformed into a Bessel equation via the transformation

$$x = \left(\frac{\nu\xi}{2\sqrt{C}}\right)^{2/\nu},$$

$$y = \eta\left(\frac{\nu\xi}{2\sqrt{C}}\right)^{(1-A)/\nu}$$

32. Let $A = B = 0$ in problem 31 and let $\nu = 4$. Transform the equation in problem 31 under these conditions into a Riccati equation by means of the transformation

$$\eta = \frac{1}{y}\cdot\frac{dy}{dx},$$

$$\xi = -x.$$

***33.** Derive the transformation in problem 29 from the results obtained in problems 30, 31, and 32.

2. Solution of a Linear Differential Equation of the Second Order at a Regular Singular Point

We have seen in the preceding subsection that if

$$y = \sum_{\nu=0}^{\infty} c_{\nu}x^{\lambda+\nu} = x^{\lambda}\sum_{\nu=0}^{\infty} c_{\nu}x^{\nu} \tag{VI.48}$$

is a solution of (VI.37), then λ is a solution of the indicial equation (VI.41) and the coefficients c_ν are to be determined in terms of c_0 from the recursion formula (VI.46).

We will devote this section to an investigation of the question of whether (VI.48) is really a solution of (VI.37). The proof is really very simple, once we have established the following lemma:

Lemma VI.2. *Let λ be a solution of the indicial equation (VI.41) for which $f_0(\lambda + \nu) \neq 0$ for all positive integers ν and let c_ν represent the coefficients which are determined according to the recursion formula (VI.46) in terms of c_0. If $a(x) = \sum_{\nu=0}^{\infty} a_\nu x^\nu$ and $b(x) = \sum_{\nu=0}^{\infty} b_\nu x^\nu$ converge for all $|x| < r$, then the series $\sum_{\nu=0}^{\infty} c_\nu x^\nu$ converges for all $|x| < r$ also.*

Proof. We will establish the proof for this lemma by majorizing the series $\sum_{\nu=0}^{\infty} c_\nu x^\nu$ by another series $\sum_{\nu=0}^{\infty} C_\nu x^\nu$ whose convergence in $|x| < r$ will be demonstrated. We proceed for this purpose as follows:

From (VI.46)

$$|c_\nu| \leq \frac{1}{|f_0(\lambda + \nu)|} (|c_{\nu-1}||f_1(\lambda + \nu - 1)| + \cdots + |c_0||f_\nu(\lambda)|).$$

Since $f_0(\lambda + \nu) \neq 0$ for all $\nu = 1, 2, 3, \cdots$ we can find for any arbitrary small $\epsilon > 0$ an N_ϵ such that

$$0 < \frac{|\lambda| + \nu}{|f_0(\lambda + \nu)|} = \frac{|\lambda| + \nu}{|\nu^2 + (2\lambda - 1 + a_0)\nu + \lambda^2 + a_0\lambda - \lambda + b_0|} < \epsilon$$

for all $\nu > N_\epsilon$.

Hence

$$|c_\nu| \leq \frac{\epsilon}{|\lambda| + \nu} [|c_{\nu-1}||a_1(\lambda + \nu - 1) + b_1| + \cdots + |c_0||a_\nu\lambda + b_\nu|],$$
$$\text{for } \nu > N_\epsilon.$$

Since

$$\frac{|a_k(\lambda + \nu - k) + b_k|}{|\lambda| + \nu} \leq \frac{|a_k|(|\lambda| + \nu - k) + |b_k|}{|\lambda| + \nu} \leq |a_k| + |b_k|$$

for all $k \leq \nu$, we obtain

$$|c_\nu| \leq \epsilon[|c_{\nu-1}|(|a_1| + |b_1|) + \cdots + |c_0|(|a_\nu| + |b_\nu|)]. \quad \text{(VI.49)}$$

The Taylor expansions for $a(x)$ and $b(x)$ converge in $|x| < r$ by hypothesis. Hence if $0 < \rho < r$, the series

$$\sum_{\nu=0}^{\infty} a_\nu \rho^\nu \quad \text{and} \quad \sum_{\nu=0}^{\infty} b_\nu \rho^\nu$$

converge and consequently (see AII.B1)

$$\lim_{n\to\infty} a_n\rho^n = 0 \quad \text{and} \quad \lim_{n\to\infty} b_n\rho^n = 0;$$

it follows that there exists an M such that

$$|a_n\rho^n| \leq M \quad \text{and} \quad |b_n\rho^n| \leq M \quad \text{for all } n.$$

Therefore

$$|a_n| \leq \frac{M}{\rho^n}, \qquad |b_n| \leq \frac{M}{\rho^n} \quad \text{for all } n. \quad \text{(VI.50)}$$

Thus, we obtain for (VI.49) in view of (VI.50)

$$|c_\nu| \leq 2\epsilon M\left[\frac{|c_{\nu-1}|}{\rho} + \cdots + \frac{|c_0|}{\rho^\nu}\right]$$

for all $\nu > N_\epsilon$.

Let us now consider a series $\sum_{n=0}^{\infty} C_n x^n$ where the C_n are determined as follows:

$$C_0 = |c_0|,\ C_1 = |c_1|, \cdots C_{N_\epsilon} = |c_{N_\epsilon}| \quad \text{and}$$

$$C_k = 2\epsilon M\left(\frac{C_{k-1}}{\rho} + \cdots + \frac{C_0}{\rho^k}\right) \quad \text{for all } k \geq N_\epsilon + 1. \tag{VI.51}$$

Clearly,

$$|c_k| \leq |C_k| \quad \text{and} \quad \sum_{n=0}^{\infty} c_n x^n \ll \sum_{n=0}^{\infty} C_n x^n \tag{VI.52}$$

($\ll$ stands for "majorized by").

We will now show that $\sum_{n=0}^{\infty} C_n x^n$ converges for all $|x| < r$. We assume for the moment that $\sum_{n=0}^{\infty} C_n x^n$ does converge in $|x| < r$ and that $\sum_{n=0}^{\infty} C_n x^n = \Phi(x)$. Now we determine $\Phi(x)$ directly as follows:

Since

$$\sum_{n=0}^{\infty}\left(\frac{x}{\rho}\right)^n = \frac{\rho}{\rho - x} \quad \text{for } |x| < \rho,$$

we have

$$\sum_{k=1}^{\infty}\left(\frac{x}{\rho}\right)^k = \frac{x}{\rho}\sum_{k=0}^{\infty}\left(\frac{x}{\rho}\right)^k = \frac{x}{\rho - x}$$

and therefore

$$\sum_{n=0}^{\infty} C_n x^n\left(1 - 2M\epsilon\sum_{k=1}^{\infty}\left(\frac{x}{\rho}\right)^k\right) = \Phi(x)\left(1 - \frac{2M\epsilon x}{\rho - x}\right).$$

Assuming that $\sum_{n=0}^{\infty} C_n x^n$ converges in $|x| < r$ and knowing that $\sum_{k=1}^{\infty}\left(\frac{x}{\rho}\right)^k$ converges in $|x| < \rho$, we can form the product of these two series as follows (see AII.D3):

$$\sum_{n=0}^{\infty} C_n x^n \sum_{k=1}^{\infty}\left(\frac{x}{\rho}\right)^k = \sum_{n=0}^{\infty} x^n\left(\frac{C_{n-1}}{\rho} + \cdots + \frac{C_0}{\rho^n}\right).$$

Hence

$$\sum_{n=0}^{\infty} C_n x^n\left(1 - 2M\epsilon\sum_{k=1}^{\infty}\left(\frac{x}{\rho}\right)^k\right) = \sum_{n=0}^{\infty} x^n\left[C_n - 2M\epsilon\left(\frac{C_{n-1}}{\rho} + \cdots + \frac{C_0}{\rho^n}\right)\right].$$

Since (VI.51) holds for all $k \geq N_\epsilon + 1$, we have

$$C_n - 2M\epsilon\left(\frac{C_{n-1}}{\rho} + \cdots + \frac{C_0}{\rho^n}\right) = 0 \quad \text{for all } n \geq N_\epsilon + 1$$

and

$$\sum_{n=0}^{\infty} C_n x^n \left(1 - 2M\epsilon \sum_{k=1}^{\infty} \left(\frac{x}{\rho}\right)^k\right) = \text{pol}\,(x)$$

where pol (x) stands for a polynomial of the N_ϵth order in x.

Thus

$$\Phi(x) = \frac{\text{pol}\,(x)}{1 - \dfrac{2M\epsilon x}{\rho - x}} = \frac{(\rho - x)\,\text{pol}\,(x)}{\rho - x - 2M\epsilon x} = \frac{(\rho - x)\,\text{pol}\,(x)}{\rho\left[1 - \dfrac{1 + 2M\epsilon}{\rho}\,x\right]}$$

$$= \frac{1}{\rho}(\rho - x)\,\text{pol}\,(x) \sum_{k=0}^{\infty} \left(\frac{1 + 2M\epsilon}{\rho}\,x\right)^k.$$

The latter series converges for all

$$|x| \left|\frac{1 + 2M\epsilon}{\rho}\right| < 1$$

or

$$|x| < \frac{\rho}{1 + 2M\epsilon}.$$

Since we can choose ϵ arbitrarily small, we obtain for $\Phi(x)$ a convergent series in $|x| < \rho$ and since ρ is any number in $0 < \rho < r$, it follows that the series for $\Phi(x)$ converges in $|x| < r$. Retracing our steps we see that $\sum_{n=0}^{\infty} C_n x^n$ converges in $|x| < r$ and it follows in view of (VI.52) that $\sum_{n=0}^{\infty} c_n x^n$ converges in $|x| < r$, q.e.d.

Note that in case $r = \infty$, i.e., if the Taylor expansions of $a(x)$ and $b(x)$ converge everywhere, then $\sum_{n=0}^{\infty} c_n x^n$ is also convergent everywhere.

Theorem VI.3. $y = \sum_{n=0}^{\infty} c_n x^{n+\lambda}$ *is a solution of (VI.37) in* $|x| < r$ *if* λ *is a solution of the indicial equation (VI.41) such that* $f_0(\lambda + \nu) \neq 0$ *for all* $\nu = 1, 2, 3, \cdots$, *the coefficients are determined in terms of* c_0 *by (VI.46), and the coefficients* $a(x)$ *and* $b(x)$ *have convergent Taylor expansions in* $|x| < r$.

Proof. $\sum_{n=0}^{\infty} c_n x^n$ converges in $|x| < r$ according to lemma VI.2. Since a convergent power series can be differentiated termwise as often as one pleases in the interval of convergence (see AII.D2), we obtain for the derivatives of

$$y(x) = \sum_{n=0}^{\infty} c_n x^{n+\lambda} = x^\lambda \sum_{n=0}^{\infty} c_n x^n$$

according to the product rule

$$y'(x) = \lambda x^{\lambda-1} \sum_{n=0}^{\infty} c_n x^n + x^\lambda \sum_{n=0}^{\infty} c_n n x^{n-1} = \sum_{n=0}^{\infty} (\lambda + n) c_n x^{\lambda+n-1},$$

$$y''(x) = \lambda(\lambda - 1) x^{\lambda-2} \sum_{n=0}^{\infty} c_n x^n + 2\lambda x^{\lambda-1} \sum_{n=0}^{\infty} c_n n x^{n-1} + x^\lambda \sum_{n=0}^{\infty} c_n n(n-1) x^{n-2}.$$

Substituting these series representations of y, y', y'' into (VI.37) and carrying out some simple manipulations, we arrive at (VI.43), which is identically satisfied in view of (VI.46), q.e.d.

Problems VI.34–39

34. Find a solution of

$$4y'' + \frac{2}{x} y' + \frac{1}{x} y = 0$$

by the method of subsection 1.

35. Find the solutions of

$$y'' + y' - \frac{2}{x^2} y = 0$$

by the method of subsection 1.

36. Find the solutions of

$$x^2 y'' + 2xy' - xy = 0$$

by the method of subsection 1.

37. Find the solutions of

$$y'' - \frac{2\xi}{1 - \xi^2} y' - \frac{n(n+1)}{(1 - \xi^2)\xi^2} y = 0, \quad \left(\frac{d}{d\xi} = {}'\right)$$

and their radius of convergence.

38. Transform in the solution of problem 37 the independent variable according to $\xi = 1/x$, state radius of convergence, and show that the function thus obtained is identical with $Q_n(x)$ in §1.6.

39. Find the solutions of

$$xy'' + (1 - x)y' + \lambda y = 0.$$

For what values of λ does one obtain polynomials as solutions? (This equation is called *Laguerre's equation* ; see Chapter VIII, §1.6.)

3. Bessel Functions of Integral Order

Let us now use the result which we obtained in the preceding subsection to solve Bessel's equation. However, we will restrict our considerations first to the case where μ is an integer, i.e.,

$$\mu^2 = 0, 1, 4, 9, 16, \cdots.$$

A formal simplification of equation (VI.35) can be obtained by elimination of the second parameter λ by means of the transformation

$$\xi = \lambda x.$$

After having carried out this transformation in (VI.35), we will write again x instead of ξ and have

$$x^2 y'' + x y' + (x^2 - \mu^2) y = 0. \tag{VI.53}$$

Division by x^2 shows (VI.53) to be a special case of (VI.37), namely

$$y'' + \frac{1}{x} y' + \frac{-\mu^2 + x^2}{x^2} y = 0$$

with

$$\begin{aligned} a(x) &= 1, \\ b(x) &= -\mu^2 + x^2. \end{aligned} \tag{VI.54}$$

At least one solution of (VI.53) will have—according to theorem VI.3—the form

$$y(x) = \sum_{\nu=0}^{\infty} c_\nu x^{\lambda+\nu}$$

where λ is a root of the indicial equation (VI.41) with $a_0 = 1$ and $b_0 = -\mu^2$:

$$\lambda^2 - \mu^2 = 0.$$

The solutions of this equation are $\lambda = \pm\mu$ where we have to disregard the solution $-\mu$, since μ is an integer (see p. 208).

Since the Taylor expansions of $a(x)$ and $b(x)$ in (VI.54) are polynomials, we have convergence everywhere and we expect in view of lemma VI.2 that the series $\sum_{n=0}^{\infty} c_n x^n$ is everywhere convergent.

According to (VI.44),

$$\begin{aligned} f_0(m) &= m^2 - \mu^2, \\ f_k(m) &= \begin{cases} 1 & \text{for } k = 2 \\ 0 & \text{for } k \neq 2 \end{cases} \end{aligned} \tag{VI.55}$$

and therefore

$$c_\nu = -\frac{c_{\nu-2}}{\nu(\nu + 2\mu)} \tag{VI.56}$$

where μ is a positive integer.

This formula gives the coefficients of even order in terms of c_0 and the coefficients of odd order in terms of c_1. c_0 can be left arbitrary, for $\lambda = \mu$ satisfies (VI.41) and this implies the vanishing of the absolute term in (VI.43). As far as c_1 is concerned, we have to undertake a separate investigation owing to the fact that all the f_k, $k \neq 2$ vanish.

Equation (VI.45) becomes in view of (VI.55) for $\nu = 1$

$$c_1[(\lambda + 1)^2 - \mu^2] + c_0(\lambda^2 - \mu^2) = 0$$

and for $\lambda = \mu$

$$c_1[2\mu + 1] = 0.$$

Since $2\mu + 1$ cannot vanish for a positive integral value of μ, we have to take $c_1 = 0$. Consequently, all coefficients of odd order in the series solution will vanish.

The coefficients of even order can be successively obtained from (VI.56) as

$$c_2 = -\frac{c_0}{4(1+\mu)},$$

$$c_4 = \frac{c_0}{4 \cdot 8(1+\mu)(2+\mu)},$$

$$c_6 = -\frac{c_0}{4 \cdot 8 \cdot 12(1+\mu)(2+\mu)(3+\mu)},$$

$$\vdots$$

$$c_{2k} = (-1)^k \frac{c_0 \mu!}{4^k k!(\mu+k)!}.$$

Therefore, the solution of (VI.53), where μ is an integer, will have the form

$$y_\mu = c_0 \sum_{k=0}^{\infty} \frac{(-1)^k 2^\mu (x/2)^{2k+\mu} \mu!}{k!(\mu+k)!}.$$

It is customary to choose for c_0 the specific value

$$c_0 = \frac{1}{2^\mu \mu!}. \tag{VI.57}$$

With this value for c_0, we obtain what is called the *Bessel function of* μth *order of the first kind*

$$J_\mu(x) = \sum_{k=0}^{\infty} \frac{(-1)^k (x/2)^{2k+\mu}}{k!(\mu + k)!}. \tag{VI.58}$$

The solution of (VI.35) is therefore

$$J_\mu(\lambda x) = \sum_{k=0}^{\infty} \frac{(-1)^k (\lambda x/2)^{2k+\mu}}{k!(\mu + k)!}. \tag{VI.59}$$

This appears to be the only solution which we can obtain, in case μ is an integer, by the procedure outlined in subsection 1 of this section.

Even if the knowledge of only one solution seems to be sufficient for our purposes (cf. Chapter II, §2.4) we will nevertheless discuss a second solution which is linearly independent of $J_\mu(x)$ in subsection 6.

The series

$$\sum_{n=0}^{\infty} c_n x^n = \sum_{k=0}^{\infty} \frac{(-1)^k (x/2)^{2k}}{k!(\mu + k)!}$$

converges everywhere according to lemma VI.2 (see also problem VI.40). Hence, the series representation of $J_\mu(x)$ which is obtained from the above series by multiplication with x^μ, where μ is a positive integer, converges everywhere too.

Problems VI.40–44

40. Prove that

$$\sum_{k=0}^{\infty} \frac{(-1)^k (x/2)^{2k}}{k!(\mu + k)!}$$

converges everywhere without reference to lemma VI.2. (*Hint:* Use the comparison test, AII.B2.)

41. Write down the first four terms of the series expansion for $J_0(x)$ and $J_1(x)$.

42. Prove that

$$\frac{d}{dx} J_0(x) = -J_1(x).$$

(*Hint:* Use termwise differentiation.)

43. Solve the problem of the initially displaced circular membrane of radius 1 if $\bar{\phi} = \bar{\phi}(r)$, i.e., the displacement function does not depend on θ but on r only (see Chapter II, §2.4).

44. Show that

$$J_\mu(-x) = (-1)^\mu J_\mu(x)$$

where μ is an integer.

4. Bessel Functions of Nonintegral Order and of Negative Order

Let us now free ourselves from the restriction that μ has to be an integer and let us indicate this by writing ρ instead. The expression

$(\rho + k)!$, where ρ is not an integer, is not defined. If we want to save the formal appearance of (VI.58), we have to investigate the possibilities of generalizing the concept of factorials.

If we apply successively integration by parts to the integral

$$\int_0^\infty t^n e^{-t}\,dt,$$

where n represents a positive integer, we obtain

$$\int_0^\infty t^n e^{-t}\,dt = n\int_0^\infty t^{n-1}e^{-t}\,dt = \cdots = n(n-1)\cdots 3\cdot 2\cdot\int_0^\infty e^{-t}\,dt = n!$$

This facts suggests trying a generalization of the factorials for nonintegral arguments by means of this integral.

We introduce for this purpose the function

$$\Gamma(\alpha + 1) = \int_0^\infty t^\alpha e^{-t}\,dt, \quad \alpha > -1, \tag{VI.60}$$

called the Γ-*function* (*gamma function*). This is an improper integral, which exists for all $\alpha > -1$ (see problem VI.45).

For $\alpha = -1$, the function $\Gamma(\alpha + 1)$ will have a vertical asymptote, since the integral diverges. This integral will also diverge for any $\alpha < -1$, which makes definition (VI.60) useless within the range $-\infty < \alpha < -1$ (see problem VI.46). It follows, however, from (VI.60) that the function $\Gamma(\alpha + 1)$ has for any $\alpha > 0$ the factorial property

$$\Gamma(\alpha + 1) = \alpha\Gamma(\alpha), \tag{VI.61}$$

as one can see if one applies integration by parts once. This property enables us to extend the definition of the gamma function for any value of α:

$$\Gamma(\alpha) = \begin{cases} \displaystyle\int_0^\infty t^{\alpha-1}e^{-t}\,dt & \text{for } \alpha > 0, \\ \displaystyle\frac{\Gamma(\alpha+1)}{\alpha} & \text{for } \alpha \le 0. \end{cases} \tag{VI.62}$$

It follows from this definition that $\Gamma(\alpha)$ is regular for $\alpha > 0$ and all nonintegral negative values of α and that it has vertical asymptotes for $\alpha = 0, -1, -2, -3, \cdots$. A more sophisticated discussion of (VI.62) will furnish the geometric representation of the gamma function as we have presented it in Fig. 26.

Using the gamma function to replace the factorials, we can now define the Bessel function of ρth order where ρ may be any real number, as follows:

$$J_\rho(x) = \sum_{k=0}^{\infty} \frac{(-1)^k(x/2)^{2k+\rho}}{k!\Gamma(k+\rho+1)}. \tag{VI.63}$$

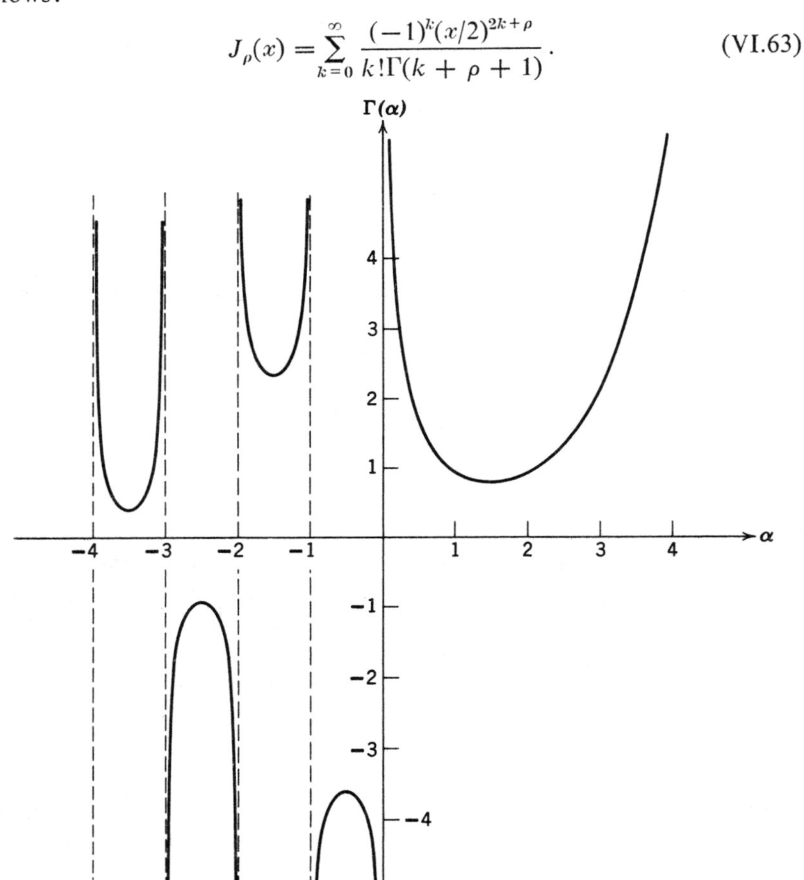

Fig. 26

We can show that this is a solution of (VI.53) for $\mu = \rho$, if we have the understanding that in accordance with the definition of the gamma function $1/\Gamma(-n) = 0$ for all positive integers n.

One obtains by differentiation of

$$J_\rho(x) = \left(\frac{x}{2}\right)^\rho \sum_{k=0}^{\infty} \frac{(-1)^k(x/2)^{2k}}{k!\Gamma(k+\rho+1)}$$

according to the product rule (termwise differentiation of the series is permissible in view of its convergence)

$$\frac{dJ_\rho(x)}{dx} = \sum_{k=0}^{\infty} \frac{(-1)^k(2k+\rho)x^{2k+\rho-1}}{k!2^{2k+\rho}\Gamma(\rho+k+1)},$$

$$\frac{d^2J_\rho(x)}{dx^2} = \sum_{k=0}^{\infty} \frac{(-1)^k(2k+\rho)(2k+\rho-1)x^{2k+\rho-2}}{2^{2k+\rho}k!\Gamma(\rho+k+1)}.$$

If we substitute these expressions into (VI.53), $x^2J''_\rho$, xJ'_ρ, $-\rho^2J_\rho$ may be expressed as power series in terms of $x^{2k+\rho}$. The only term which does not fit in without tampering is x^2J_ρ, but this difficulty can be circumvented in the following way: If we introduce in

$$x^2J_\rho = \sum_{k=0}^{\infty} \frac{(-1)^k x^{2k+\rho+2}}{k!\Gamma(k+\rho+1)2^{2k+\rho}}$$

a new summation subscript l according to $k = l - 1$ and then again write k instead of l, we find

$$x^2J_\rho(x) = -\sum_{k=1}^{\infty} \frac{(-1)^k x^{2k+\rho}}{(k-1)!\Gamma(\rho+k)2^{2k+\rho-2}}$$

$$= -\sum_{k=0}^{\infty} \frac{(-1)^k 4x^{2k+\rho}k(k+\rho)}{k!\Gamma(\rho+k+1)2^{2k+\rho}}.$$

(We can extend the summation at the lower limit to $k = 0$ because the term corresponding to $k = 0$ is zero anyway.)

We thus have

$$x^2J''_\rho + xJ'_\rho + (x^2 - \rho^2)J_\rho$$

$$= \sum_{k=0}^{\infty} \frac{(-1)^k x^{2k+\rho}}{k!\Gamma(k+\rho+1)2^{2k+\rho}} [(2k+\rho)(2k+\rho-1)$$

$$+ 2k + \rho - 4k(k+\rho) - \rho^2] \quad \text{(VI.64)}$$

and the reader can see that the expression in brackets vanishes for any value ρ (see problem VI.47).

Since this is true for any ρ, it is true in particular for $\rho = -\mu$, where μ is a positive integer. We therefore have two solutions J_μ and $J_{-\mu}$ for the Bessel equation of integral order. There is, however, no reason for displaying any smugness since we will show immediately that these two solutions are either identical or differ only in sign:

Since

$$J_{-\mu}(x) = \sum_{k=0}^{\infty} \frac{(-1)^k(x/2)^{2k-\mu}}{k!\Gamma(k-\mu+1)}$$

where

$$\frac{1}{\Gamma(k-\mu+1)} = 0 \quad \text{for } k = 0, 1, 2, 3, \cdots, (\mu - 1),$$

we have

$$J_{-\mu}(x) = \sum_{k=\mu}^{\infty} \frac{(-1)^k (x/2)^{2k-\mu}}{k!\Gamma(k-\mu+1)}.$$

Introduction of a new summation subscript $l = k - \mu$ and changing the notation back again to k yields

$$J_{-\mu}(x) = \sum_{k=0}^{\infty} \frac{(-1)^{k+\mu}(x/2)^{2k+\mu}}{(k+\mu)!\Gamma(k+1)} = (-1)^{\mu} \sum_{k=0}^{\infty} \frac{(-1)^k (x/2)^{2k+\mu}}{k!\Gamma(k+\mu+1)},$$

i.e.:

$$J_{\mu}(x) = (-1)^{\mu} J_{-\mu}(x) \tag{VI.65}$$

where μ is any integer.

We will indicate in subsection 6 how a second linearly independent solution of the Bessel equation of integral order can be found.

In case ρ is not an integer, $J_{\rho}(x)$, $J_{-\rho}(x)$ constitute a system of two linearly independent solutions, as we will show in the following subsection.

Problems VI.45–48

45. Prove that

$$0 < \int_0^{\infty} t^{\alpha} e^{-t}\, dt < \infty$$

for all $\alpha > -1$. $\left(Hint: \int_0^{\infty} = \int_0^1 + \int_1^{\infty}.\right)$

46. Prove that $\int_0^{\infty} t^{\alpha} e^{-t}\, dt$ does not exist for $\alpha \leq -1$. (See hint in problem 45.)

47. Show that (VI.64) vanishes for any ρ.

48. Write down the first four terms of the series expansions for $J_{1/2}(x)$ and $J_{-1/2}(x)$.

5. The Linear Independence of Two Solutions of Bessel's Equation of Nonintegral Order

In order to establish the linear independence of J_{ρ}, $J_{-\rho}$ for nonintegral ρ we have to evaluate the Wronskian $W[J_{\rho}, J_{-\rho}]$. (See also §1.1 and problems VI.1, 2.)

We will write the Bessel equation in the self-adjoint form

$$\frac{d}{dx}(xy') + \left(x - \frac{\rho^2}{x}\right) y = 0$$

(see Chapter V, §1.1).

We denote the left side, which is a self-adjoint differential expression (see Chapter V, §1.1) by $L[y]$:

$$L[y] = \frac{d}{dx}(xy') + \left(x - \frac{\rho^2}{x}\right)y.$$

Since J_ρ, $J_{-\rho}$ are solutions of Bessel's equation,

$$L[J_\rho] = 0, \quad L[J_{-\rho}] = 0. \tag{VI.66}$$

Since $L[y]$ is a self-adjoint differential expression, we have according to (V.16)

$$J_\rho L[J_{-\rho}] - J_{-\rho}L[J_\rho] = \frac{d}{dx}[x(J_\rho J'_{-\rho} - J'_\rho J_{-\rho})]$$

where $p(x) = x$, $y = J_{-\rho}$, $z = J_\rho$, and in view of (VI.66)

$$\frac{d}{dx}[x(J_\rho J'_{-\rho} - J'_\rho J_{-\rho})] = 0. \tag{VI.67}$$

Since

$$W[J_\rho, J_{-\rho}] = \begin{vmatrix} J_\rho & J_{-\rho} \\ J'_\rho & J'_{-\rho} \end{vmatrix} = J_\rho J'_{-\rho} - J_{-\rho}J'_\rho,$$

(VI.67) is equivalent to

$$\frac{d}{dx}[xW[J_\rho, J_{-\rho}]] = 0$$

and consequently

$$W[J_\rho, J_{-\rho}] = \frac{c}{x}. \tag{VI.68}$$

We have to show, in order to establish the linear independence of J_ρ, $J_{-\rho}$, that the constant c in (VI.68) is not zero. We consider for this purpose the series expansions of the Bessel functions involved and their derivatives:

$$J_\rho(x) = \sum_{k=0}^{\infty} \frac{(-1)^k (x/2)^{2k+\rho}}{k!\Gamma(k+\rho+1)}, \tag{VI.69}$$

$$J_{-\rho}(x) = \sum_{k=0}^{\infty} \frac{(-1)^k (x/2)^{2k-\rho}}{k!\Gamma(k-\rho+1)}, \tag{VI.70}$$

$$J'_\rho(x) = \sum_{k=0}^{\infty} \frac{(-1)^k (2k+\rho)(x/2)^{2k+\rho-1}}{2k!\Gamma(k+\rho+1)}, \tag{VI.71}$$

$$J'_{-\rho}(x) = \sum_{k=0}^{\infty} \frac{(-1)^k (2k-\rho)(x/2)^{2k-\rho-1}}{2k!\Gamma(k-\rho+1)}. \tag{VI.72}$$

In evaluating the Wronskian, we multiply (VI.69) by (VI.72), and (VI.70) by (VI.71), compute the difference, and have to obtain according to (VI.68) a constant, divided by x.

Multiplication of the $(k + 1)$th term of (VI.69) with the $(m + 1)$th term of (VI.72) gives for the exponent of $x/2$

$$\rho + 2k - \rho + 2m - 1 = 2k + 2m - 1.$$

It is apparent that only multiplication of the first term of (VI.69) with the first term of (VI.72) yields a term with the -1th power of x, and the same holds true for the product of (VI.70) and (VI.71). All other terms have to cancel and there remains

$$\frac{c}{x} = \frac{-\rho(x/2)^{-1}}{2\Gamma(\rho + 1)\Gamma(1 - \rho)} - \frac{\rho(x/2)^{-1}}{2\Gamma(1 - \rho)\Gamma(\rho + 1)}.$$

Since $\Gamma(\rho + 1) = \rho\Gamma(\rho)$ we have

$$\frac{c}{x} = \frac{-2}{x\Gamma(1 - \rho)\Gamma(\rho)}.$$

This expression does not vanish since the denominator is finite in view of the assumption that ρ is not an integer.[10]

Problem VI.49.

49. Evaluate $W[J_{1/2}, J_{-1/2}]$.

6. Bessel Functions of the Second Kind.[11]

We have seen in subsection 4 that μ being an integer implies

$$J_\mu = (-1)^\mu J_{-\mu};$$

i.e., J_μ, $J_{-\mu}$ are not linearly independent and we are one solution short. We are now going to outline a method by which one can establish a second

[10] That $J_\rho, J_{-\rho}$ are linearly independent can also be seen from the fact that if there were two nonzero constants c_1, c_2 such that

$$c_1 J_\rho + c_2 J_{-\rho} = 0,$$

it would follow that $J_{-\rho}$ is regular at $x = 0$, which is not the case. However, the proof we gave yields at the same time a method by which the Wronskian can be evaluated (see problems VI.49 and 55).

[11] This subsection may be omitted because the results obtained here are only used in a very limited way in the later developments.

solution of Bessel's equation of integral order, which is linearly independent of J_μ.

The leading idea of the following heuristic discussion shall first be demonstrated with a much simpler example, namely, the case where the characteristic equation of

$$y'' + a_1 y' + a_2 y = 0,$$

where a_1, a_2 are constants, has a double root λ. In this case only one solution is obtained, namely $y_1 = e^{\lambda x}$.

If we change the coefficients of the differential equation slightly such that

$$\lambda + \epsilon, \quad \lambda - \epsilon,$$

where $\epsilon > 0$ is small, appear as solutions of the characteristic equation, then, any linear combination of $e^{(\lambda+\epsilon)x}$, $e^{(\lambda-\epsilon)x}$ will be the general solution of the altered equation. (See also problem VI.50.)

We specifically consider the linear combination

$$\frac{1}{2\epsilon}(e^{(\lambda+\epsilon)x} - e^{(\lambda-\epsilon)x}).$$

Since the altered equation degenerates into the original equation, as $\epsilon \to 0$, we conjecture that the linear combination of the two solutions of the altered equation will become a solution of the original equation as $\epsilon \to 0$,

Now,

$$y_2 = \lim_{\epsilon\to 0} \frac{e^{(\lambda+\epsilon)x} - e^{(\lambda-\epsilon)x}}{2\epsilon} = \frac{d}{d\lambda} e^{\lambda x} = xe^{\lambda x}$$

and this is indeed a solution of the original equation, as one can see by substitution. That it is linearly independent of $e^{\lambda x}$ can be seen by evaluation of the Wronskian (see problems VI.51 and VI.52).

A very similar approach will lead us to the second solution of Bessel's equation of integral order.

We will change the integral order μ to a nonintegral order ρ where

$$\rho = \mu + \epsilon, \quad |\epsilon| < 1.$$

The equation of order ρ thus obtained will have the two linearly independent solutions J_ρ, $J_{-\rho}$ and thus, $c_1 J_\rho + c_2 J_{-\rho}$ is the general solution. The specific linear combination

$$\frac{J_\rho(x) - (-1)^\mu J_{-\rho}(x)}{\rho - \mu}$$

is also a solution of Bessel's equation of order ρ, and we conjecture—in analogy with the argument we used in the previous example—that the limit of this expression as $\epsilon \to 0$ will be a second solution of Bessel's equation of order μ, the so-called *Bessel function of the* μth *order and the second kind*:

$$Y_\mu(x) = \lim_{\substack{\rho \to \mu \\ (\epsilon \to 0)}} \frac{J_\rho(x) - (-1)^\mu J_{-\rho}(x)}{\rho - \mu}.$$

This limit is according to (VI.65) an indeterminate form and we have to use l'Hospital's rule for its evaluation.

We have

$$Y_\mu(x) = \left(\frac{\partial J_\rho}{\partial \rho}\right)_{\rho=\mu} - (-1)^\mu \left(\frac{\partial J_{-\rho}}{\partial \rho}\right)_{\rho=\mu} \tag{VI.73}$$

That this is indeed a solution of Bessel's equation of the μth order we will verify by substitution:

Since the mixed derivatives of J_ρ, $J_{-\rho}$ with respect to x and ρ are continuous for all $x > 0$ and all nonintegral ρ, we can interchange these differentiations[12] and we find (see also problem VI.53)

$$\frac{\partial}{\partial \rho} \bigg| \; x^2 \frac{d^2 J_{\pm\rho}}{dx^2} + x \frac{dJ_{\pm\rho}}{dx} + (\rho^2 - x^2) J_{\pm\rho} = 0,$$

$$x^2 \frac{d^2}{dx^2} \frac{\partial J_{\pm\rho}}{\partial \rho} + x \frac{d}{dx} \frac{\partial J_{\pm\rho}}{\partial \rho} + (\rho^2 - x^2) \frac{\partial J_{\pm\rho}}{\partial \rho} + 2\rho J_{\pm\rho} = 0.$$

Combination of the equation for ρ with the one for $-\rho$ with the coefficient $(-1)^\mu$ leads to

$$L_\rho\left[\frac{\partial J_\rho}{\partial \rho} - (-1)^\mu \frac{\partial J_{-\rho}}{\partial \rho}\right] = 2\rho[J_\rho - (-1)^\mu J_{-\rho}]$$

where $L_\rho[y]$ denotes the left side of Bessel's equation of order ρ. If we now take the limit as $\rho \to \mu$ $(\epsilon \to 0)$,

$$L_\mu\left[\frac{\partial J_\rho}{\partial \rho} - (-1)^\mu \frac{\partial J_{-\rho}}{\partial \rho}\right]_{\rho=\mu} = 0;$$

i.e.: Y_μ is a solution of Bessel's equation of order μ.

[12] See R. Courant: *Differential and Integral Calculus*, Vol. II, Interscience Publishers, New York, 1936, p. 55.

Let us now find the series representation of this function Y_μ. We begin with

$$\frac{\partial}{\partial\rho}J_\rho(x)=\sum_{k=0}^{\infty}\frac{(-1)^k}{k!}$$

$$\cdot\frac{\log(x/2)\cdot(x/2)^{2k+\rho}\Gamma(k+\rho+1)-(x/2)^{2k+\rho}\dfrac{\partial\Gamma(k+\rho+1)}{\partial\rho}}{\Gamma^2(k+\rho+1)}$$

$$=\log\frac{x}{2}\sum_{k=0}^{\infty}\frac{(-1)^k(x/2)^{2k+\rho}}{k!\Gamma(k+\rho+1)}-\sum_{k=0}^{\infty}\frac{(-1)^k(x/2)^{2k+\rho}}{k!\Gamma(k+\rho+1)}\cdot\frac{\partial\log\Gamma(k+\rho+1)}{\partial\rho}.$$

This expression has to be considered for $\rho\to\mu$. That the first term does not cause any difficulties is clear enough. The second term or rather the second factor of the second term requires a little more attention. In view of the factorial property of the gamma function, we have

$$\Gamma(\rho+k+1)=(k+\rho)(k+\rho-1)\cdots(1+\epsilon)\Gamma(1+\epsilon)$$

and therefore

$$\frac{\partial\log\Gamma(k+\rho+1)}{\partial\rho}=\frac{1}{k+\rho}+\frac{1}{k+\rho-1}+\cdots$$
$$+\frac{1}{2+\epsilon}+\frac{1}{1+\epsilon}+\frac{\partial\Gamma(1+\epsilon)}{\partial\epsilon}.$$

Hence,

$$\left.\frac{\partial\log\Gamma(k+\rho+1)}{\partial\rho}\right|_{\rho=\mu}=\frac{1}{k+\mu}+\frac{1}{k+\mu-1}+\cdots+\frac{1}{2}+\frac{1}{1}+\Gamma'(1).$$

It follows from (VI.60) that

$$\Gamma'(1)=\int_0^\infty\log t\,.\,e^{-t}\,dt.$$

The number $\gamma=-\int_0^\infty\log t\,.\,e^{-t}\,dt$ is called *Euler's constant* and its value is found to be

$$\gamma=0.57721566\cdots.^{13}\qquad\text{(see problem VI.54).}$$

We therefore conclude that

$$\left(\frac{\partial J_\rho}{\partial\rho}\right)_{\rho=\mu}=(\log(x/2)+\gamma)J_\mu(x)$$
$$-\sum_{k=0}^{\infty}\frac{(-1)^k(x/2)^{2k+\mu}}{k!\Gamma(k+\mu+1)}\left(\frac{1}{k+\mu}+\cdots\frac{1}{2}+\frac{1}{1}\right).\qquad\text{(VI.74)}$$

[13] See E. T. Wittaker and G. N. Watson: *Modern Analysis*, Cambridge University Press, 1920.

(In case $\mu = 0$, the first term in the latter sum ($k = 0$) has to be set equal to zero.)

Up to now, everything has gone rather smoothly. The real trouble starts when we try to differentiate $J_{-\rho}$ with respect to ρ:

Now

$$J_{-\rho}(x) = \sum_{k=0}^{\infty} \frac{(-1)^k (x/2)^{-\rho+2k}}{k!\Gamma(k-\rho+1)}.$$

The denominator $\Gamma(k - \rho + 1)$ has vertical asymptotes for $\rho = \mu$ and $k = 0, 1, 2, \cdots, (\mu - 1)$, thus making a differentiation with respect to ρ at $\rho = \mu$ impossible. We therefore split the summation at $k = \mu - 1$:

$$J_{-\rho}(x) = \sum_{k=0}^{\mu-1} \frac{(-1)^k (x/2)^{-\rho+2k}}{k!\Gamma(k-\rho+1)} + \sum_{k=\mu}^{\infty} \frac{(-1)^k (x/2)^{-\rho+2k}}{k!\Gamma(k-\rho+1)}$$

and transform the terms of the first sum, using the relation

$$\Gamma(s)\Gamma(1-s) = \frac{\pi}{\sin \pi s}$$ [14]

If we subsequently carry out the following steps: Differentiate with respect to ρ, set $\rho = \mu$, and finally make the substitution $-\mu + 2k = \mu + 2\kappa$, and afterwards write k again instead of κ, we see that

$$\begin{aligned}\left(\frac{\partial J_{-\rho}}{\partial \rho}\right)_{\rho=\mu} &= (-1)^\mu \sum_{k=0}^{\mu-1} \frac{(\mu-k-1)!(x/2)^{-\mu+2k}}{k!} \\ &\quad + \sum_{k=0}^{\infty} \frac{(-1)^{\mu+k}(x/2)^{\mu+2k}}{\Gamma(k+1)(k+\mu)!}\left(\log(x/2) - \frac{1}{k} - \cdots - \frac{1}{2} - \frac{1}{1} + \gamma\right) \\ &= (-1)^\mu \sum_{k=0}^{\mu-1} \frac{(\mu-k-1)!(x/2)^{-\mu+2k}}{k!} + (-1)^{\mu-1}(\log(x/2)+\gamma)J_\mu(x) \\ &\quad + (-1)^\mu \sum_{k=0}^{\infty} \frac{(-1)^k (x/2)^{\mu+2k}}{k!(k+\mu)!}\left(\frac{1}{k} + \cdots + \frac{1}{2} + \frac{1}{1}\right). \qquad \text{(VI.75)}\end{aligned}$$

Combining (VI.74) and (VI.75) according to (VI.73) yields

$$\begin{aligned}Y_\mu(x) &= 2(\log(x/2) + \gamma)J_\mu(x) - \sum_{k=0}^{\mu-1} \frac{(\mu-k-1)!(x/2)^{-\mu+2k}}{k!} \\ &\quad - \sum_{k=0}^{\infty} \frac{(-1)^k (x/2)^{\mu+2k}}{k!(\mu+k)!}\left(\frac{1}{k+\mu} + \cdots + \frac{1}{k+1} + \frac{2}{k} + \cdots + \frac{2}{2} + \frac{2}{1}\right), \qquad \text{(VI.76)}\end{aligned}$$

the *Bessel function of the* μth *order of the second kind.*

[14] See E. T. Wittaker and G. N. Watson: *Modern Analysis*, Cambridge University Press, 1920.

One can show by the same method which was employed in subsection 5 that

$$W[J_\mu, Y_\mu] = \frac{2}{x}.$$

(see problem VI.55).

Thus, the two solutions J_μ, Y_μ constitute a system of two linearly independent solutions of the Bessel equation of integral order. We wish to mention that the function Y_μ, as we introduced it in this section, was first defined by *Hankel* (*Hankel function*), while *Weber* and *Schlaefli* defined a slightly modified Bessel function of the second kind, namely

$$\mathbf{Y}_\mu(x) = \lim_{\rho\to\mu} \frac{\cos \rho\pi J_\rho(x) - J_{-\rho}(x)}{\sin \rho\pi} = \frac{1}{\pi} Y_\mu(x). \qquad \text{(VI.77)}$$

This definition has an advantage insofar as difficulties which arise for $\mu = (2\nu + 1)/2$ in generalizing Hankel's function to nonintegral values of μ are avoided by taking the definition (VI.77).[15]

Problems VI.50–56

50. Suppose λ is a double root of $\lambda^2 + a_1\lambda + a_2 = 0$. State the linear differential equation of second order with constant coefficients which has

$$c_1e^{(\lambda+\epsilon)x} + c_2e^{(\lambda-\epsilon)x}$$

as its general solution.

51. Show that $y = xe^{\lambda x}$ is a solution of $y'' + a_1y' + a_2y = 0$ if λ is a double root of its characteristic equation.

52. Show that $W[e^{\lambda x}, xe^{\lambda x}] \neq 0$.

***53.** Show that

$$\frac{\partial^2 J_{\pm\rho}}{\partial x\,\partial\rho} = \frac{\partial^2 J_{\pm\rho}}{\partial\rho\,\partial x}$$

in $x > 0$ and all ρ.

54. Find an approximative value of Euler's constant γ by using Simpson's rule on an interval $\epsilon \leq t \leq A$ where $\epsilon > 0$ has to be chosen suitably small and A has to be chosen large.

***55.** Show that $W[J_\mu, Y_\mu] = 2/x$. (*Hint:* The terms which are multiplied by $\log(x/2)$ cannot possibly make any contribution to the Wronskian because the logarithm has a branch point at $x = 0$ and the Wronskian has a pole of first order at $x = 0$, as was shown in subsection 5.)

56. Write down the first four terms of the expansion of Y_0.

7. Circular Membrane with Axially Symmetric Initial Displacement

Let us consider a circular membrane with radius 1. Suppose the initial displacement to be a function of the distance r from the center, but

[15] See G. N. Watson: *Bessel Functions*, 2nd ed., The Macmillan Company, 1945, pp. 58 and 63.

independent of the polar angle θ (see problem VI.43). We call such a displacement axially symmetric. The membrane equation in cylindrical coordinates has the form

$$\frac{\partial^2 u}{\partial r^2} + \frac{1}{r}\frac{\partial u}{\partial r} + \frac{1}{r^2}\frac{\partial^2 u}{\partial \theta^2} = \frac{\partial^2 u}{\partial t^2},$$

if we assume that the units are chosen such that $\tau/\rho = 1$ (see **AI.B6**), where $u(r, \theta, t)$ stands for the vertical displacement of the point (r, θ) at the time t.

The membrane, being stretched over the unit circle $r = 1$, gives rise to the boundary condition

$$u(1, \theta, t) = 0 \tag{VI.78}$$

and we assume it to be initially displaced according to

$$u(r, \theta, 0) = \phi(r) \tag{VI.79}$$

where $\phi(1) = 0$, with no initial velocity:

$$\left.\frac{\partial u(r, \theta, t)}{\partial t}\right|_{t=0} = 0. \tag{VI.80}$$

We now apply Bernoulli's separation method, assuming the solution to be of the form

$$u(r, \theta, t) = R(r)\Theta(\theta)T(t).$$

Under this circumstance, the boundary condition (VI.78) will appear in the form

$$R(1) = 0. \tag{VI.81}$$

It follows from (VI.79) that

$$\Theta(\theta) = \text{constant};$$

i.e., the solution u has to be independent of θ. (The symmetry of the initial position is preserved throughout the entire process.)

Thus, we have to deal with the simpler equation

$$\frac{\partial^2 u}{\partial r^2} + \frac{1}{r}\frac{\partial u}{\partial r} = \frac{\partial^2 u}{\partial t^2} \tag{VI.82}$$

or in terms of R and T

$$R''T + \frac{1}{r}R'T = RT''.$$

From this equation we find, by using the familiar procedure and argument,

$$T'' + \lambda^2 T = 0 \tag{VI.83}$$

$$r^2R'' + rR' + \lambda^2 r^2 R = 0. \tag{VI.84}$$

We recognize (VI.84) as Bessel's equation of zeroth order. We did not choose a negative constant $-\lambda^2$ for the following reason: We have to obtain infinitely many solutions of (VI.84) in order to satisfy (VI.79) by invoking the superposition principle. However, the equation

$$r^2R'' + rR' - \lambda^2 r^2 R = 0$$

has, in view of theorem V.7, only finitely many eigenvalues, if any at all, under homogeneous boundary conditions.

In our case we can only impose one boundary condition, namely, $R(1) = 0$ if we require regularity of our solution at $r = 0$, as seems indicated by the nature of our problem. This latter condition excludes $R = Y_0(\lambda r)$ since this function tends to infinity of logarithmic order as $r \to 0$ (see (VI.76)).

The function $R(r) = J_0(\lambda r)$ is regular at $r = 0$, and satisfies the boundary condition (VI.81), if λ is such that

$$J_0(\lambda) = 0. \tag{VI.85}$$

The solutions λ_i of (VI.85) are the zeros of the Bessel function of zeroth order, or, as we may put it, the eigenvalues of the Sturm-Liouville problem from which these solutions originate.

Equation (VI.83) under the condition (VI.80) has the solution

$$T = \cos \lambda t$$

and we therefore have infinitely many solutions

$$u_i(r, t) = c_i \cos \lambda_i t J_0(\lambda_i r),$$

where the λ_i are the zeros of $J_0(\lambda)$.

(VI.79) finally yields the condition

$$\phi(r) = \sum_{i=1}^{\infty} c_i J_0(\lambda_i r),$$

which is satisfied according to the formal theory developed in Chapter V, §2.5, whenever

$$c_i = \frac{\displaystyle\int_0^1 r\phi(r) J_0(\lambda_i r)\, dr}{\displaystyle\int_0^1 r J_0^2(\lambda_i r)\, dr}. \tag{VI.86}$$

The value of the integral in the denominator of (VI.86) will be found in the next subsection.

The most interesting problem, namely the computation of the frequencies $\lambda_i/2\pi$ of the components of the membrane vibration cannot be solved yet.

We refer the reader to Chapter VII, where we discuss a method by which the eigenvalues of self-adjoint boundary value problems can be found (with patience and skill).

8. The Normalization of the Bessel Functions

The problem which we encountered at the end of the preceding subsection, namely the evaluation of the denominator in (VI.86) is the problem of normalizing the orthogonal system of functions $\sqrt{x}J_\rho(\lambda_i x)$, i.e., of finding—in the terminology of Chapter V—constant factors c_i such that

$$\int_{x_1}^{x_2} x[c_i J_\rho(\lambda_i x)]^2\,dx = 1$$

where x_1, x_2 are the end points of the interval in which the homogeneous boundary conditions are imposed.

We have shown that

$$c_i = \left[\int_{x_1}^{x_2} R(x)y_\rho^2(x)\,dx\right]^{-1/2}$$

Thus, we have to evaluate for our purpose

$$\int_{x_1}^{x_2} xJ_\rho^2(\lambda_i x)\,dx.$$

If λ_i and λ_k denote two continuous variables, $J_\rho(\lambda_i x)$ and $J_\rho(\lambda_k x)$ are solutions of the Bessel equation for any pair of values λ_i, λ_k. Hence,

$$L[J_\rho(\lambda_i x)] + \lambda_i^2 xJ_\rho(\lambda_i x) - \frac{\rho^2}{x}J_\rho(\lambda_i x) = 0,$$

$$L[J_\rho(\lambda_k x)] + \lambda_k^2 xJ_\rho(\lambda_k x) - \frac{\rho^2}{x}J_\rho(\lambda_k x) = 0,$$

where L stands for the self-adjoint operator $L = (d/dx)[x(d/dx).]$ (see lemma V.2). Multiplying the latter equation by $J_\rho(\lambda_i x)$ and subtracting it from the first equation, which is to be multiplied by $J_\rho(\lambda_k x)$, yields

$$J_\rho(\lambda_k x)L[J_\rho(\lambda_i x)] - J_\rho(\lambda_i x)L[J_\rho(\lambda_k x)] + (\lambda_i^2 - \lambda_k^2)xJ_\rho(\lambda_i x)J_\rho(\lambda_k x) = 0.$$

Integrating this equation between x_1 and x_2 and observing lemma V.1 (Lagrange's identity), we obtain

$$\left\{x[\lambda_i J_\rho'(\lambda_i x)J_\rho(\lambda_k x) - \lambda_k J_\rho(\lambda_i x)J_\rho'(\lambda_k x)]\right\}_{x_1}^{x_2}$$
$$= (\lambda_k^2 - \lambda_i^2)\int_{x_1}^{x_2} xJ_\rho(\lambda_i x)J_\rho(\lambda_k x)\,dx. \quad \text{(VI.87)}$$

The left side of (VI.87) can be simplified insofar as the derivatives of J_ρ can be expressed in terms of $J_{\rho-1}$ and $J_{\rho+1}$ respectively.

As the reader can easily verify by substituting the series representations of the Bessel functions and comparing the coefficients, the relations

$$2J'_\rho(x) = J_{\rho-1}(x) - J_{\rho+1}(x) \tag{VI.88}$$

$$\frac{2\rho}{x} J_\rho(x) = J_{\rho-1}(x) + J_{\rho+1}(x) \tag{VI.89}$$

hold (see problem VI.57).

By elimination of $J_{\rho+1}$ and $J_{\rho-1}$, respectively, one obtains the following two pairs of relations:

$$J_{\rho-1}(x) = J'_\rho(x) + \frac{\rho}{x} J_\rho(x), \tag{VI.90}$$

or

$$x^\rho J_{\rho-1}(x) = \frac{d(x^\rho J_\rho(x))}{dx}, \tag{VI.91}$$

and

$$J_{\rho+1}(x) = \frac{\rho}{x} J_\rho(x) - J'_\rho(x), \tag{VI.92}$$

or

$$-x^{-\rho} J_{\rho+1}(x) = \frac{d(x^{-\rho} J_\rho(x))}{dx} \tag{VI.93}$$

(see problem VI.58).

(Note that the relation (VI.89) can be and is used for the computation of tables of Bessel functions.)

In view of (VI.92), the left side of (VI.87) assumes the form

$$\left\{x\left[\lambda_i \frac{\rho}{\lambda_i x} J_\rho(\lambda_i x) - \lambda_i J_{\rho+1}(\lambda_i x)\right] J_\rho(\lambda_k x) - x\left[\lambda_k \frac{\rho}{\lambda_k x} J_\rho(\lambda_k x) - \lambda_k J_{\rho+1}(\lambda_k x)\right] J_\rho(\lambda_i x)\right\}_{x_1}^{x_2}$$

and after cancellation of the appropriate terms we obtain for (VI.87)

$$\left\{x[\lambda_k J_\rho(\lambda_i x) J_{\rho+1}(\lambda_k x) - \lambda_i J_{\rho+1}(\lambda_i x) J_\rho(\lambda_k x)]\right\}_{x_1}^{x_2} = (\lambda_k^2 - \lambda_i^2) \int_{x_1}^{x_2} x J_\rho(\lambda_i x) J_\rho(\lambda_k x)\, dx. \tag{VI.94}$$

This relation as it stands will certainly not furnish any information about

$$\int_{x_1}^{x_2} x J_\rho^2(\lambda_k x)\, dx$$

because of the term $\lambda_k^2 - \lambda_i^2$. The following artifice, however, will enable us to approach the desired result. We differentiate (VI.94) with respect to λ_k and subsequently set $i = k$ (see problem VI.59). Choosing at the same time $x_1 = 0$ and $x_2 = 1$ without too much loss of generality,

$$\lambda_k[J_\rho(\lambda_k)J'_{\rho+1}(\lambda_k) - J'_\rho(\lambda_k)J_{\rho+1}(\lambda_k)] + J_\rho(\lambda_k)J_{\rho+1}(\lambda_k) = 2\lambda_k \int_0^1 xJ_\rho^2(\lambda_k x)\,dx.$$

If we finally specify the boundary conditions such that one becomes $y(1) = 0$, i.e., $J_\rho(\lambda_k) = 0$, we obtain

$$\int_0^1 xJ_\rho^2(\lambda_k x)\,dx = -\frac{1}{2}J'_\rho(\lambda_k)J_{\rho+1}(\lambda_k)$$

or in view of (VI.92)

$$\int_0^1 xJ_\rho^2(\lambda_k x)\,dx = \frac{1}{2}J_{\rho+1}^2(\lambda_k) \tag{VI.95}$$

for all λ_k for which $J_\rho(\lambda_k) = 0$.

We are now in a position to put the solution of the problem of the circular membrane with initial axially symmetric displacement into a closed form:

$$u(r, t) = \sum_{i=1}^{\infty} \frac{\int_0^1 r\phi(r)J_0(\lambda_i r)\,dr}{\frac{1}{2}J_1^2(\lambda_i)} \cos \lambda_i t J_0(\lambda_i r)$$

where the λ_i are the solutions of $J(\lambda_i) = 0$.

Problems VI.57–59

57. Prove relations (VI.88) and (VI.89).

58. Derive the relations (VI.90), (VI.91), (VI.92), and (VI.93) from (VI.88) and (VI.89).

59. Differentiate (VI.94) with respect to λ_k and evaluate the result at $\lambda_i = \lambda_k$.

9. Integral Representation of Bessel Functions of Integral Order

We will develop in this section a representation of the Bessel functions of integral order by a definite integral. We start out with the expansion of the two factors on the right side of the identity

$$e^{[x/2][t-(1/t)]} = e^{xt/2}e^{-x/2t}$$

into power series:

$$e^{xt/2} = \sum_{n=0}^{\infty} \frac{1}{n!}\left(\frac{xt}{2}\right)^n,$$

$$e^{-x/2t} = \sum_{n=0}^{\infty} \frac{(-1)^n}{n!}\left(\frac{x}{2t}\right)^n.$$

Forming the product of these series, we observe that the factor c_n of t^n is obtained by multiplying all terms of $(n+k)$th order of the first series with all terms of kth order in the second series for all integers k:

$$c_n = \sum_{k=0}^{\infty} \frac{1}{(n+k)!}\left(\frac{x}{2}\right)^{n+k} \cdot \frac{(-1)^k}{k!}\left(\frac{x}{2}\right)^k = \sum_{k=0}^{\infty} \frac{(-1)^k (x/2)^{n+2k}}{(n+k)!\,k!}$$

and in view of (VI.58),

$$c_n = J_n(x).$$

Similarly, for the factor c_{-n} of t^{-n} we may write

$$c_{-n} = \sum_{k=0}^{\infty} \frac{(-1)^{n+k}}{k!}\left(\frac{x}{2}\right)^k \frac{1}{(n+k)!}\left(\frac{x}{2}\right)^{n+k}$$

$$= (-1)^n \sum_{k=0}^{\infty} \frac{(-1)^k (x/2)^{n+2k}}{n!\,(n+k)!} = (-1)^n J_n(x)\,.$$

Hence,

$$e^{[x/2][t-(1/t)]} = J_0(x) + \sum_{n=1}^{\infty} J_n(x)\left(t^n + \frac{(-1)^n}{t^n}\right). \tag{VI.96}$$

If we let $t = e^{i\theta}$ where $i = \sqrt{-1}$, and use De Moivre's formula

$$\frac{1}{2}\left(t^n + \frac{(-1)^n}{t^n}\right) = \frac{1}{2}\,(e^{in\theta} + (-1)^n e^{-in\theta}) = \begin{cases} \cos n\theta & \text{for } n = 2\nu, \\ i \sin n\theta & \text{for } n = 2\nu + 1, \end{cases}$$

we find from (VI.96) and

$$e^{[x/2][t-(1/t)]} = e^{xi \sin\theta} = \cos(x \sin\theta) + i \sin(x \sin\theta)$$

the following relation:

$$\cos(x\sin\theta) + i\sin(x\sin\theta) = J_0(x) + 2\sum_{\nu=1}^{\infty} J_{2\nu}(x)\cos 2\nu\theta$$
$$+ 2i\sum_{\nu=0}^{\infty} J_{2\nu+1}(x)\sin(2\nu+1)\theta.$$

Equating real and imaginary parts yields

$$\cos(x\sin\theta) = J_0(x) + 2\sum_{\nu=1}^{\infty} J_{2\nu}(x)\cos 2\nu\theta, \tag{VI.97}$$

$$\sin(x\sin\theta) = 2\sum_{\nu=0}^{\infty} J_{2\nu+1}(x)\sin(2\nu+1)\theta. \tag{VI.98}$$

$\{\sin \mu\theta\}$ and $\{\cos \mu\theta\}$ are orthogonal systems on the interval $0 \leq \theta \leq \pi$ (see problem V.25); i.e.,

$$\int_0^\pi \sin \mu x \sin nx\, dx = \begin{cases} 0 & \text{for } \mu \neq n, \\ \dfrac{\pi}{2} & \text{for } \mu = n, \end{cases}$$
$$\int_0^\pi \cos \mu x \cos nx\, dx = \begin{cases} 0 & \text{for } \mu \neq n, \\ \dfrac{\pi}{2} & \text{for } \mu = n. \end{cases} \tag{VI.99}$$

If we multiply (VI.97) by $\cos \mu\theta$ and integrate from 0 to π, we have, in view of (VI.99),

$$\int_0^\pi \cos (x \sin \theta) \cos \mu\theta\, d\theta = \begin{cases} \pi J_\mu(x) & \text{for } \mu = 2\nu, \\ 0 & \text{for } \mu = 2\nu + 1. \end{cases} \tag{VI.100}$$

Multiplying (VI.98) by $\sin \mu\theta$ and integrating from 0 to π yields, in view of (VI.99),

$$\int_0^\pi \sin (x \sin \theta) \sin \mu\theta\, d\theta = \begin{cases} \pi J_\mu(x) & \text{for } \mu = 2\nu + 1, \\ 0 & \text{for } \mu = 2\nu. \end{cases} \tag{VI.101}$$

If we combine (VI.100) and (VI.101) and divide by π, it follows that

$$\begin{aligned} J_\mu(x) &= \frac{1}{\pi} \int_0^\pi (\cos \mu\theta \cos (x \sin \theta) + \sin \mu\theta \sin (x \sin \theta))\, d\theta \\ &= \frac{1}{\pi} \int_0^\pi \cos (\mu\theta - x \sin \theta)\, d\theta \end{aligned} \tag{VI.102}$$

for all integral values of μ.

One can deduce from (VI.102) without difficulties that

$$|J_\mu(x)| \leq 1 \tag{VI.103}$$

for all integral values of μ and all x (see problem VI.62) and

$$\lim_{x \to \infty} J_\mu(x) = 0 \tag{VI.104}$$

for all integral values of μ (see problem VI.65).

Problems VI.60–65

***60.** Prove that (VI.102) satisfies the Bessel equation for integral μ.

61. Show that it follows from (VI.102) that

$$J_\mu(0) = \begin{cases} 1 & \text{for } \mu = 0, \\ 0 & \text{for } \mu \geq 1, \end{cases} \quad J'_\mu(0) = \begin{cases} 0 & \text{for } \mu \neq 1, \\ \frac{1}{2} & \text{for } \mu = 1. \end{cases}$$

62. Prove (VI.103) for all integral μ and all x.

63. Expand the functions $f(\theta) = \cos(x \sin\theta)$ and $g(\theta) = \sin(x \sin\theta)$ into Fourier series. (*Hint:* Use (VI.100) and (VI.101) for evaluation of the Fourier coefficients.)

64. Prove that $\lim_{\mu\to\infty} J_\mu(x) = 0$ for all x and all integral values of μ. (*Hint:* Use the result in problem VI.63 and the fact that the Fourier coefficients of a convergent Fourier series tend to zero.)

***65.** Show that $\lim_{x\to\infty} J_\mu(x) = 0$ for all integers μ. (*Hint:* Make use of (VI.100) and (VI.101).)

10. The Almost-Periodic Behavior of the Functions $\sqrt{x}J_\rho(x)$

It is clear that we cannot obtain a very accurate picture of the Bessel functions unless we know their zeros. We have seen in Chapter V, §2.3, however, that a Sturm-Liouville boundary value problem has infinitely many eigenvalues under circumstances which are met by Bessel's equation for $x > 0$, and this means that in case of a boundary condition of the type $y(1) = 0$ that $J_\rho(x)$ has to have infinitely many zeros.

We know more than that: We have seen in Chapter V, §1.4 that two consecutive zeros of the solutions of the Bessel equation

$$\frac{d}{dx}(xy') + \left(\lambda x - \frac{\rho^2}{x}\right)y = 0$$

satisfy in an interval $A \le x \le A + d$, where d is sufficiently large but fixed, the inequality

$$\frac{\pi}{\sqrt{\lambda}}\sqrt{\frac{A}{A+d}} \le |x_1 - x_2| \le \frac{\pi}{\sqrt{\lambda}}\sqrt{\frac{A+d}{A-(\rho^2/\lambda A)}}.$$

If we eliminate λ, we find for two consecutive zeros

$$\pi\sqrt{\frac{A}{A+d}} \le |x_1 - x_2| \le \pi\sqrt{\frac{A+d}{A-(\rho^2/A)}}$$

and if $A \to \infty$

$$\pi \le |x_1 - x_2| \le \pi, \quad \text{i.e.:} \quad |x_1 - x_2| = \pi.$$

These facts suggest rather strongly that Bessel functions show a kind of periodic behavior.

This can be seen very easily in the case of $J_{1/2}$ and $J_{-1/2}$. If we carry out the transformation

$$y = \frac{u(x)}{\sqrt{x}}$$

in Bessel's equation, which we assume in the form

$$x^2y'' + xy' + (x^2 - \rho^2)y = 0,$$

we have

$$u'' + \left(1 - \frac{\rho^2 - \frac{1}{4}}{x^2}\right)u = 0. \tag{VI.105}$$

It is readily seen that this equation reduces in the case where $\rho = \pm\frac{1}{2}$ to

$$u'' + u = 0$$

with the solution

$$u = A\cos x + B\sin x,$$

or in terms of y

$$y = A\frac{\cos x}{\sqrt{x}} + B\frac{\sin x}{\sqrt{x}}.$$

If we expand y into a Taylor series for $A = 0$ we obtain

$$y = B\left(\sqrt{x} - \frac{x^{5/2}}{3!} + \cdots\right),$$

and for $B = 0$

$$y = A\left(\frac{1}{\sqrt{x}} - \frac{x^{3/2}}{2!} + \cdots\right).$$

On the other hand

$$J_{1/2}(x) = \frac{\sqrt{x}}{\sqrt{2}\Gamma(\frac{3}{2})} + \cdots$$

and

$$J_{-1/2}(x) = \frac{\sqrt{2}}{\sqrt{x}\Gamma(\frac{1}{2})} + \cdots.$$

Since $\Gamma(\frac{1}{2}) = \sqrt{\pi}$ (see problem VI.66) and in view of (VI.61) $\Gamma(\frac{3}{2}) = \sqrt{\pi}/2$, we find

$$J_{1/2} = \sqrt{2/\pi}\,\frac{\sin x}{\sqrt{x}},$$

$$J_{-1/2} = \sqrt{2/\pi}\,\frac{\cos x}{\sqrt{x}}$$

(see Fig. 27).

If $\rho \neq \pm\frac{1}{2}$, we can see from (VI.105) that the term $(\rho^2 - \frac{1}{4})/x^2$ will become of smaller influence as x increases and hence the solutions of this equation, i.e., the functions $\sqrt{x}J_\rho(x)$, will behave more and more like $\sin x$ and $\cos x$ as x becomes large. Functions of this type are called almost-periodic functions.[16]

[16] See H. Weyl, *Mathem. Annalen*, 97 (1926), p. 338.

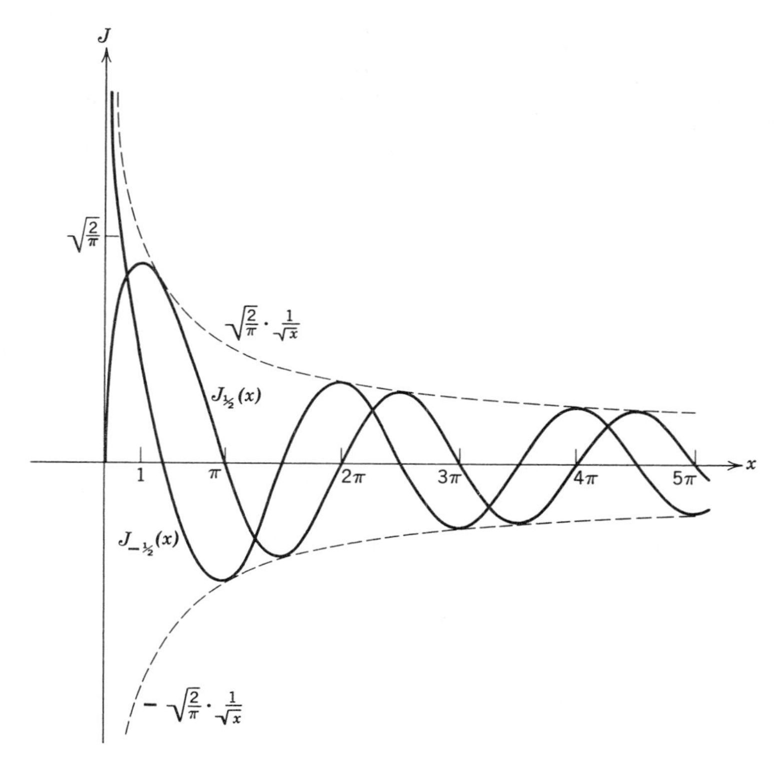

Fig. 27

Problems VI.66–67

***66.** Evaluate $\Gamma(\frac{1}{2})$. (*Hint:* Transform $\int_0^\infty \frac{1}{\sqrt{t}} e^{-t}\, dt$ into $\int_0^\infty e^{-x^2}\, dx$ and evaluate instead of $\int_0^\infty e^{-x^2}\, dx$ the double integral $\int_0^\infty e^{-x^2} \int_0^\infty e^{-y^2}\, dx\, dy$ by introducing polar coordinates.)

***67.** Find the temperature distribution in a homogeneous isotropic right circular cylinder of radius 1 and height h, if the top and bottom are insulated, the lateral surface is kept at constant temperature 0, and the initial temperature distribution is given by a function $\phi(r, z)$, which is consistent with the boundary conditions.

RECOMMENDED SUPPLEMENTARY READING

G. N. Watson: *A Treatise on the Theory of Bessel Functions*, Cambridge University Press, London, 1944.

E. W. Hobson: *The Theory of Spherical and Ellipsoidal Harmonics*, Cambridge University Press, London, 1931.

T. M. MacRobert: *Spherical Harmonics*, Dover Publications, New York, 1948.

R. V. Churchill: *Fourier Series and Boundary Value Problems*, McGraw-Hill Book Co., 1941.

Ph. Frank and R. von Mises: *Die Differential und Integralgleichungen der Mechanik und Physik*, Vol. I, F. Vieweg and Sohn, Braunschweig, 1930.

VII

CHARACTERIZATION OF EIGENVALUES BY A VARIATIONAL PRINCIPLE

§1. INTRODUCTION AND GENERAL EXPOSITION

1. The Sturm-Liouville Equation as Euler-Lagrange Equation of a Certain Isoperimetric Problem

The Euler-Lagrange equation in problems I.25, 26, 27, and 28 turned out to be a self-adjoint differential equation of the second order. We will refresh our memory by restating and solving problem I.28.

We seek a function $y = y(x)$ which satisfies the boundary conditions

$$y(x_1) = 0, \quad y(x_2) = 0 \tag{VII.1}$$

and is such that

$$\int_{x_1}^{x_2} [P(x)y'^2 - Q(x)y^2]\, dx \to \text{minimum} \tag{VII.2}$$

and

$$\int_{x_1}^{x_2} R(x)y^2\, dx = 1, \tag{VII.3}$$

where $P(x)$, $Q(x)$, and $R(x)$ are continuous functions of x, and $P(x)$ is differentiable in $x_1 \leq x \leq x_2$.

With the notation of Chapter I, we have

$$f = Py'^2 - Qy^2, \quad g = Ry^2$$

and thus

$$h = f - \lambda g = Py'^2 - Qy^2 - \lambda Ry^2.$$

(We write for convenience $h = f - \lambda g$ instead of $h = f + \lambda g$.) This function h has to satisfy the Euler-Lagrange equation (I.36)

$$\frac{\partial h}{\partial y} - \frac{d}{dx}\frac{\partial h}{\partial y'} = 0,$$

which turns out to be

$$\frac{d}{dx}[P(x)y'] + [Q(x) + \lambda R(x)]y = 0, \tag{VII.4}$$

a self-adjoint differential equation of the second order, as discussed in Chapter V, §2.

Let us now consider the problem where a function $y(x)$ which satisfies the boundary conditions (VII.1) and is such that

$$\frac{\int_{x_1}^{x_2} (Py'^2 - Qy^2)\, dx}{\int_{x_1}^{x_2} Ry^2\, dx} \to \text{minimum} \tag{VII.5}$$

is sought.

It is readily seen that if $y(x)$ is a solution of the problem (VII.5) under the boundary conditions (VII.1), then it also is, but for a multiplicative constant, a solution of the problem (VII.2) under the constraint (VII.3) and the boundary conditions (VII.1). On the other hand, if $y(x)$ is a solution of the problem (VII.2) under the constraint (VII.3) and the boundary conditions (VII.1), then it is also a solution of (VII.5) under the boundary conditions (VII.1). (See problem VII.2.) Thus, for all practical purposes, we can consider the two variational problems as equivalent, since our major concern is the study of the solutions of the Sturm-Liouville problem, where a multiplicative constant in the solution is immaterial.

We can also see by a direct check that (VII.5) leads to equation (VII.4) and this process will reveal a significant interpretation of the eigenvalue λ which corresponds to the solution y.

Let

$$I[y] = \int_{x_1}^{x_2} (Py'^2 - Qy^2)\, dx$$

and

$$C[y] = \int_{x_1}^{x_2} Ry^2\, dx$$

and let us consider the first variation of the quotient $I[y]/C[y]$; i.e., we introduce for y a family of curves with the parameter ϵ which contains the solution of our minimum problem for $\epsilon = 0$, and form

$$\delta \frac{I}{C} = \frac{\partial}{\partial \epsilon}\left[\frac{I(\epsilon)}{C(\epsilon)}\right]_{\epsilon=0}$$

We obtain

$$\delta \frac{I}{C} = \frac{(\partial I/\partial \epsilon)_{\epsilon=0} C(0) - (\partial C/\partial \epsilon)_{\epsilon=0} I(0)}{C^2(0)}.$$

If we denote the minimum value of the quotient (VII.5) by λ:

$$\frac{I(0)}{C(0)} = \lambda,$$

we can write the first necessary condition $\delta(I/C) = 0$ in the form

$$\left(\frac{\partial I}{\partial \epsilon}\right)_{\epsilon=0} - \lambda\left(\frac{\partial C}{\partial \epsilon}\right)_{\epsilon=0} = 0.$$

If we carry out the indicated operations and invoke the fundamental lemma of the calculus of variations (see Chapter I, §1.2), we are led again to equation (VII.4).

As pointed out before, the preceding discussion revealed a very important fact, namely, that the minimum value of (VII.5) appears to be the eigenvalue λ of the corresponding solution of (VII.4). This fact is far more important for our purpose than the fact that an eigenfunction of (VII.4), (VII.1) can be interpreted as a solution of a variational problem of the type (VII.5), as we will see in the next subsection.

We stumbled upon this apparent equivalence of the Sturm-Liouville problem and the variational problem (VII.5), (VII.1) merely by accident. The reader will understand that this situation warrants closer investigation, which will be undertaken in §2 of this chapter.

Problems VII.1–2

1. Carry out the operations in

$$\left(\frac{\partial I}{\partial \epsilon}\right)_{\epsilon=0} - \lambda\left(\frac{\partial C}{\partial \epsilon}\right)_{\epsilon=0} = 0$$

and show that it yields (I.36) where

$$I = \int_{x_1}^{x_2} f(x, y, y')\, dx, \quad C = \int_{x_1}^{x_2} g(x, y, y')\, dx, \quad \text{and } \lambda \text{ is replaced by } -\lambda.$$

2. Show that if $y(x)$ is a solution of (VII.1), (VII.5), then $ky(x)$ for some k is a solution of (VII.1), (VII,2), (VII.3) and if $y(x)$ is a solution of (VII.1), (VII.2), (VII.3), then it is also a solution of (VII.1), (VII.5).

2. Numerical Examples

While it is not yet established that the variational problem (VII.1), (VII.5) is equivalent to the Sturm-Liouville problem (VII.1), (VII.4), we will nevertheless assume for the time being that this is the case and try to find numerical approximations to eigenvalues on this basis. The necessary condition for a function $y(x)$ to yield a minimum for the quotient (VII.5) is, that it satisfies (VII.4), and we have seen in the preceding subsection that the eigenvalue corresponding to $y(x)$ represents the minimum value of (VII.5). As we have seen in Chapter V, §2, equation (VII.4) has infinitely many solutions corresponding to infinitely many eigenvalues and it appears that the smallest eigenvalue is the minimum of (VII.5).

We will now try to obtain a numerical approximation to the smallest eigenvalue by substituting for y in (VII.5) a function that satisfies the given boundary conditions and resembles the eigenfunction somewhat. We will see that the results will be quite gratifying.

Let us first consider the following Sturm-Liouville problem:

$$y'' + \lambda^2 y = 0,$$
$$y(0) = y(1) = 0,$$

which seems to be equivalent to the variational problem

$$\frac{\int_0^1 y'^2\,dx}{\int_0^1 y^2\,dx} \to \text{minimum}.$$

We know that the solutions of our problem are the functions

$$y = \sin k\pi x$$

with the corresponding eigenvalues $\lambda^2 = k^2\pi^2$. We will now compute an approximation to the smallest eigenvalue $\lambda_1^2 = \pi^2$ by approximating the first eigenfunction $y_1 = \sin \pi x$ by a parabola

$$\bar{y}_1 = x(1 - x).$$

Now, we call the reader's attention to the fact that this constitutes a very rough approximation to the sine function, but we will see nevertheless that the approximation thus obtainable for π is surprisingly good.

We obtain

$$\frac{\int_0^1 \bar{y}_1'^2\,dx}{\int_0^1 \bar{y}_1^2\,dx} = \frac{\int_0^1 (1 - 4x + 4x^2)\,dx}{\int_0^1 (x^2 - 2x^3 + x^4)\,dx} = 10 \cong \lambda_1^2 = \pi^2;$$

i.e., $\pi \cong \sqrt{10} = 3.1622\cdots$, which deviates about 0.6 per cent from the actual value. This example serves as a nice illustration to the contemplated procedure, and this is essentially its only value.

We will now discuss another example which is by no means as trivial as the preceding one. (The preceding example is trivial in the sense that the reader already knows much better methods of approximating π while for the approximation which is sought in the next example, the reader does not yet have at his disposal any better means, or, as a matter of fact, any approach at all.) The example of which we speak constitutes quite a genuine application of the method that we are about to introduce in this chapter.

We will try to find an approximation to the first zero of the Bessel function of zeroth order and the first kind.

We know from Chapter VI, §2 that $y = J_0(\lambda_i x)$ with $J_0(\lambda_i) = 0$ are the solutions of the Sturm-Liouville problem

$$\frac{d}{dx}(xy') + \lambda^2 xy = 0$$

with the boundary conditions $y(1) = 0$, $|y(0)| < \infty$.

Thus, the first zero of $J_0(x)$ appears as the first eigenvalue of this boundary value problem or as the minimum value of

$$\frac{\int_0^1 xy'^2\,dx}{\int_0^1 xy^2\,dx}$$

with the boundary conditions $y(1) = 0$, $|y(0)| < \infty$. We take as an approximation to $J_0(\lambda_1 x)$ in the interval $0 \leq x \leq 1$ the quadratic polynomial $\bar{y}_1 = 1 - x^2$, which satisfies the prescribed boundary conditions, as one can immediately see, and obtain

$$\frac{4\int_0^1 x^3\,dx}{\int_0^1 (x - 2x^3 + x^5)\,dx} = 6 \cong \lambda_1^2,$$

i.e., $\lambda_1 \cong \sqrt{6} = 2.44949\cdots$. This approximation does not compare very favorably with the exact value of the first zero of $J_0(x)$, which is given as $\lambda_1 = 2.40483\cdots$. However, if one considers the roughness of the approximation of $J_0(\lambda_1 x)$ by $1 - x^2$, one should be quite satisfied.

Problems VII.3–4

3. Find an approximation to the first eigenvalue of

$$y'' + \lambda y = 0, \quad y(-1) = y(1) = 0$$

by approximating the true solution by $\bar{y} = 1 - x^2$.

4. (a) Determine a in

$$y = x + a$$

such that

$$\frac{\int_{-1}^{1} (1 - x^2) y'^2 \, dx}{\int_{-1}^{1} y^2 \, dx}$$

yields a maximum, and evaluate the quotient.

(b) Do the same as in (a) with

$$y = a - x^2.$$

3. Minimizing Sequences

It is quite obvious that the numerical method which was employed in the preceding subsection will yield values which are larger than the true eigenvalues. If $y(x)$ is the true solution of our problem, i.e., is such that the quotient (VII.5) assumes its minimum value for $y(x)$, then any other function will yield a larger (or equal, in exceptional cases) value for (VII.5). In both examples, functions which are representable by infinite power series seem to be the true solutions of the problems discussed ($\sin \pi x$, $J_0(\lambda_1 x)$), and in both cases, polynomials of the second order were used to obtain an approximation. It is reasonable to surmise that approximations by polynomials of higher order may have rendered better approximations. In general, the idea is the following one: In solving our minimum problem, we admit all functions with a continuous derivative to compete and one of these functions will yield the minimum, i.e., the smallest value as compared with all other values obtainable from functions with a continuous derivative. If one imposes restrictions on the class of competing functions (e.g., permits only polynomials), then one of these functions will again yield a minimum relative to values which the quotient (VII.5) assumes for functions of this restricted class, and it is quite obvious that this new minimum value will be larger than, or equal to the one obtained before:

Lemma VII.1. *Let $\{y\}$ and $\{\eta\}$ be two classes of functions all members of which satisfy the same boundary conditions (VII.1) and let $\{\eta\} \subset \{y\}$ ("$\subset$" stands for "contained in" or "subclass of").*

Then, if

$$\min_{[y]} \frac{\int_{x_1}^{x_2} f(x, y, y') \, dx}{\int_{x_1}^{x_2} g(x, y, y') \, dx} = \lambda_y$$

(minimum relative to $\{y\}$*) and*

$$\min_{[\eta]} \frac{\int_{x_1}^{x_2} f(x, \eta, \eta')\, dx}{\int_{x_1}^{x_2} g(x, \eta, \eta')\, dx} = \lambda_\eta$$

(minimum relative to $\{\eta\}$*), then*

$$\lambda_y \leq \lambda_\eta.$$

("minimum relative to $\{y\}$*" and "minimum relative to* $\{\eta\}$*" mean, respectively, the smallest value of the quotient (VII.5) obtainable with functions from* $\{y\}$ *and* $\{\eta\}$ *respectively.)*

Proof. Suppose $\lambda_\eta < \lambda_y$, i.e., $\lambda_\eta < \min_{[y]} \dfrac{\int_{x_1}^{x_2} f(x, y, y')\, dx}{\int_{x_1}^{x_2} g(x, y, y')\, dx}$. But this is impossible, because $\eta \epsilon \{y\}$, q.e.d.

This lemma expresses the fundamental principle on which approximative methods that are concerned with finding upper bounds for the minimum values of (VII.5) are based. Let us summarize the procedure: We want to find the minimum of (VII.5) relative to a class of functions $\{y\}$ which have a continuous derivative. In order to find an upper bound to this minimum value, we consider a class $\{\eta_1\} \subset \{y\}$. In order to improve the approximation, we consider still another class $\{\eta_2\} \subset \{y\}$ such that $\{\eta_1\} \subset \{\eta_2\}$, etc. We rightly expect that by considering successively classes of functions $\{\eta_1\} \subset \{\eta_2\} \subset \{\eta_3\} \subset \cdots$ we obtain increasingly better approximations $\lambda^{(1)} \geq \lambda^{(2)} \geq \lambda^{(3)} \geq \cdots$ from above to the minimum value λ of (VII.5).

Clearly, the sequence of values thus obtained converges. Whether it converges actually to the minimum value λ is another question, a partial answer to which will be given in §3, subsections 2 and 4.

Before continuing this line of argument, let us interject a remark. If one obtains upper bounds to the minimum by restricting the class of competing functions, would one not obtain lower bounds by enlarging the class of competing functions?

This idea is basically sound; however, what one means by enlarging the class of competing functions has to be clarified. It would certainly be very impractical to consider, e.g., the class of all functions; besides, the differentiations and integrations which are to be carried out would become senseless. The only practical way of carrying out this idea seems to be to relax the boundary conditions. This was recognized by

A. Weinstein in 1937 and developed into a method now bearing his name.[1]

To return to our problem: Let us suppose that we find functions $\eta_1, \eta_2, \eta_3, \cdots$ in the classes $\{\eta_1\}, \{\eta_2\}, \{\eta_3\}, \cdots$ which are such that the corresponding values of the quotient (VII.5) $\lambda^{(1)}, \lambda^{(2)}, \lambda^{(3)}, \cdots$ converge towards the minimum value λ. In general, we call a sequence of functions which is such that the corresponding values of (VII.5) (or simply an integral) converge to the minimum value of the quotient (VII.5) (or the integral) a *minimizing sequence*. The question arises: Does the minimizing sequence $\eta_1, \eta_2, \eta_3, \cdots$ itself converge towards the function y which yields the minimum of the quotient (integral)?

One can easily make the mistake of answering this question in the affirmative, and yet, a very simple example will show that this does not have to be the case.

We consider for reasons of simplicity instead of a quotient of the type (VII.5) the integral

$$I[u] = \iint_R \sqrt{1 + u_x^2 + u_y^2}\, dx\, dy \to \text{minimum}$$

where R shall designate the circular region $x^2 + y^2 \leq 1$ and where u satisfies the boundary condition

$$u(x, y)\big|_{x^2+y^2=1} = 0.$$

The reader will recognize this as the problem of finding the surface with the smallest area that is bounded by a full circular arc.

It is quite clear that

$$I[u] \geq \pi$$

and that the value π is assumed when $u = 0$. We can nevertheless construct a minimizing sequence u_n such that

$$\lim_{n\to\infty} I[u_n] = \pi, \quad \text{but} \quad \lim_{n\to\infty} u_n \neq 0.$$

Let

$$u_n(x, y) = \begin{cases} 0 \text{ for all } x, y \text{ in } x^2 + y^2 \geq \dfrac{1}{n^2}, \\ \dfrac{1}{2}\cos n\pi\sqrt{x^2 + y^2} + \dfrac{1}{2} \quad \text{for all } x, y \text{ in } x^2 + y^2 < \dfrac{1}{n^2}. \end{cases}$$

[1] A. Weinstein, *Mémor. des sciences math.*, fasc. 88, 1937. See S. M. Gould: *Variational Methods for Eigenvalue Problems*, University of Toronto Press, 1957.

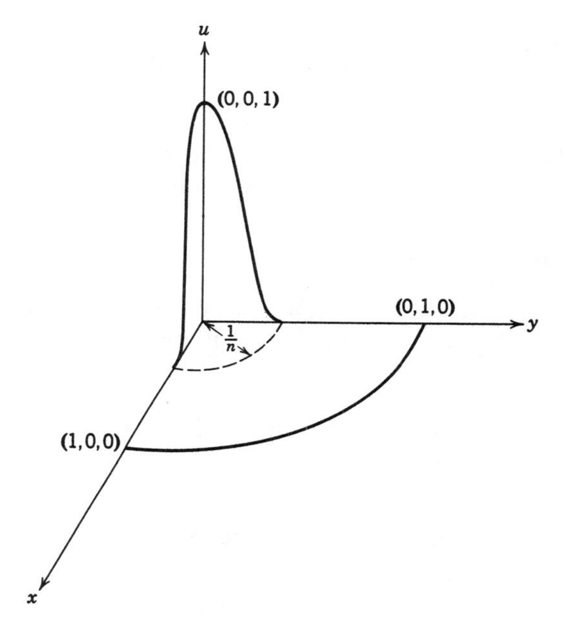

Fig. 28

These functions are continuous, have everywhere continuous derivatives, and satisfy the boundary conditions (see Fig. 28). We obtain

$$I[u_n] = \iint\limits_{1/n^2 \leq x^2+y^2 \leq 1} dx\,dy + \iint\limits_{x^2+y^2<1/n^2} \sqrt{1 + \frac{n^2\pi^2}{4}\sin^2 n\pi\sqrt{x^2+y^2}}\,dx\,dy$$

and if we introduce polar coordinates,

$$I[u_n] = \pi - \frac{\pi}{n^2} + \int_0^{2\pi}\int_0^{1/n} \sqrt{1 + \frac{n^2\pi^2}{4}\sin^2 n\pi r} \cdot r\,dr\,d\theta$$

$$\leq \pi\left(1 - \frac{1}{n^2}\right) + \int_0^{2\pi}\int_0^{1/n} \sqrt{1 + \frac{n^2\pi^2}{4}}\, r\,dr\,d\theta \leq \pi\left(1 - \frac{1}{n^2}\right)$$

$$+ \frac{\pi^2}{2n}\sqrt{1 + \frac{4}{n^2\pi^2}} = \pi + \frac{\pi^2}{2n} + \cdots.$$

It is clear that

$$\lim_{n\to\infty}\left(\pi + \frac{\pi^2}{2n} + \cdots\right) = \pi$$

while

$$\lim u_n = \begin{cases} 1 & \text{for } x = y = 0, \\ 0 & \text{for } (x, y) \neq (0, 0). \end{cases}$$

We see that the sequence as defined above is indeed a minimizing sequence but it does not converge to the function that renders the minimum.

4. The Existence of a Minimum

In the preceding discussion, we frequently referred to the minimum of the quotient (VII.5) and the minimum of the integral $I[u]$, implicitly assuming that such a minimum exists. Are we really justified in such an assumption?

Again, let us consider some examples. First, we again consider the problem

$$I[u] \rightarrow \text{minimum},$$

where $I[u]$ is defined in the preceding subsection, under the boundary conditions

$$u(0, 0) = 1, \quad u(x, y)|_{x^2+y^2=1} = 0.$$

We see that in this case the above sequence is still a minimizing sequence in the sense that the value of the integral approaches π as n becomes large but there does not seem to exist a function satisfying the given boundary conditions which has a continuous derivative, nor a function which is continuous in the entire region of integration and is such that $I[u] = \pi$; i.e., there does not seem to exist a minimum for $I[u]$ under the prescribed boundary conditions. To make the point clearer still, let us consider the following problem:

We seek a function such that

$$L[y] = \int_0^1 \sqrt{1 + y'^2}\, dx$$

becomes a minimum, where

$$y'(0) = 1, \quad y(1) = 0.$$

Clearly $L[y] \geq 1$. The functions of the family

$$y_n(x) = \begin{cases} x + \delta, \quad \text{for } 0 \leq x \leq \dfrac{1}{n}, \\ -\dfrac{(1/n) + \delta}{1 - (1/n)}\, x + \dfrac{n\delta + 1}{n - 1}, \qquad \dfrac{1}{n} < x \leq 1 \end{cases}$$

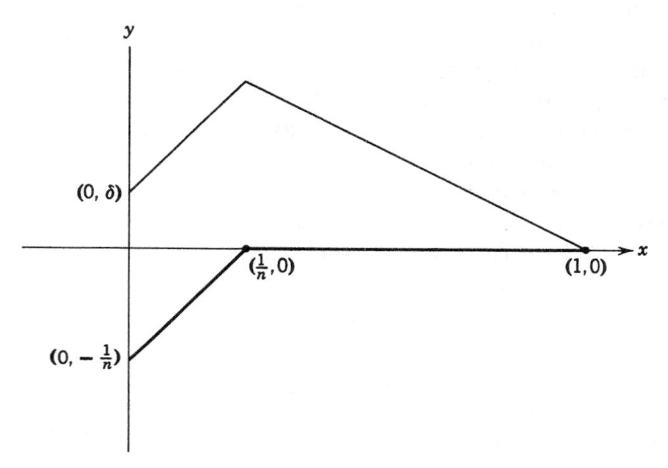

Fig. 29

satisfy the boundary conditions (see Fig. 29) whatever value we assign to δ. It is quite obvious that for any fixed n, the curve with $\delta = -1/n$ will render the shortest path compared to those obtainable from other values of δ. We obtain with this specific value of δ

$$I[y_n] = \sqrt{2}\frac{1}{n} + 1 - \frac{1}{n} = 1 + (\sqrt{2} - 1)\frac{1}{n}$$

and we see that

$$\lim_{n\to\infty} I[y_n] = 1.$$

However,

$$\lim_{n\to\infty} y_n = 0$$

is certainly not the solution of our problem, because it does not satisfy the first boundary condition. (This problem can be interpreted as the problem of choosing the shortest path in walking through the front gate towards a given point (mailbox) down the street.[2] Clearly, such a shortest path does not exist because there is no earliest possible turn after passing through the front gate.) Also in this case, a minimum does not exist.

Up to the time of Weierstrass it was assumed without further investigation that whenever an integral has a lower bound, then there always

[2] This illustration differs from the problem we discussed, insofar as the initial slope here is vertical. In order to formulate this problem mathematically, we would have to use a parametric representation. Then, the initial condition would read $\dot{x} = 0, \dot{y} = 1$ at $t = 0$.

exists a function satisfying given boundary conditions such that it renders the minimum for the integral, and physical evidence seemed to support such an assumption.[3] It was Weierstrass who revealed the unreliability of this principle by constructing counterexamples and thus dealt quite a blow to the mathematical world of his day.

In order to gain some insight into this problem, let us consider an example of a more elementary nature: We consider the function $y = x^2$ in the closed interval $-1 \leq x \leq 1$. Clearly, the minimum of this function in the given interval is 0 and this minimum value is attained for $x = 0$.

However, the function

$$y = \begin{cases} x^2 & \text{for all } x \neq 0, \\ \frac{1}{2} & \text{for } x = 0 \end{cases}$$

does not have a minimum value in the interval $-1 \leq x \leq 1$. It certainly is true that $y \geq 0$ and $\lim_{n\to\infty} x_n^2 = 0$ whereby $\{x_n\}$ stands for any sequence of x-values which approaches zero. It is equally obvious that there exists no point in the interval $-1 \leq x \leq 1$ at which the function as defined above would actually assume the value zero. This is, of course, no surprise to us because Weierstrass's theorem only says that if a function is continuous in a closed interval, then it will attain its minimum value in the interval. The function which is under consideration does not satisfy the requirement of being continuous in the entire interval and this fact is responsible for its deficiency in assuming a minimum value.

However, if we consider the function

$$y = \begin{cases} x^2 & \text{for } x \neq 0, \\ -1 & \text{for } x = 0, \end{cases}$$

then, we obtain the minimum value -1, which is attained by the function for $x = 0$, even though this function too is not continuous.

Continuity at $x = 0$ requires that for any arbitrarily choosen $\epsilon > 0$ there exists a δ_ϵ such that

$$|y(0) - y(x)| < \epsilon \quad \text{for all } |x| < \delta_\epsilon.$$

Clearly, this condition is not satisfied by the function which is defined above, however, if we omit the absolute value signs we obtain a condition

$$y(0) - y(x) = -1 - x^2 < \epsilon \quad \text{for all } |x| < \delta_\epsilon$$

[3] This assumption in regard to the integral $\iint_R (u_x^2 + u_y^2)\, dx\, dy$ is widely referred to as *Dirichlet's principle* and was extensively used by Riemann, one of Dirichlet's most outstanding disciples.

which is satisfied (in this particular case for any δ_ϵ). A function with such a property at $x = 0$ (or any other point) is called *lower semicontinuous* at $x = 0$ (or any other point). It turns out that lower semicontinuity is sufficient for the existence of a minimum. (The reader is challenged to define upper semicontinuity and relate it to the existence of a maximum value.)

Let us now relate this discussion to the problem at hand, namely the minimization of an integral. Here, the independent variable is a function and we have to deal with a definite integral over an expression in the variable function and its derivatives. In the future we will call such an integral a *functional* and we denote its dependence on y by $[y]$, as we have done already in the preceding discussion:

$$I[y] = \int_{x_1}^{x_2} f(x, y, y')\, dx.$$

In dealing with a problem involving the minimum of such a functional, one has to establish the lower semicontinuity of $I[y]$, which was introduced into the calculus of variations by L. Tonelli:[4]

A functional $I[y]$ is lower semicontinuous at y_0 if one can find for any arbitrary small $\epsilon > 0$ a δ_ϵ such that

$$I[y_0] - I[y] < \epsilon \quad \text{for all } y \text{ for which } |y - y_0| < \delta.$$

What does it mean to say "for all functions y for which $|y - y_0| < \delta$"? $|y - y_0|$ certainly cannot stand for the ordinary absolute value since one can construct functions which certainly satisfy the inequality and for which the functional does not even exist. Take, for example, the following problem: A function $y(x)$ satisfies the boundary conditions $y(0) = y(1) = 0$ and has the property that $\int_0^1 y'^2\, dx$ assumes its smallest value. Obviously, $y = 0$ is the solution of this problem. Now, let us take the sequence of functions

$$y_n(x) = \begin{cases} \dfrac{1}{n} & \text{for } x \text{ transcendental,} \\ 0 & \text{for all other values of } x. \end{cases}$$

Clearly, among all functions for which $|y - 0| < \delta$ whatever δ is, there are also infinitely many functions y_n as defined above. Still, $I[y_n]$ does not even exist.

So, it appears that it will largely depend on what class of functions one considers and what definition one gives of the distance $|y - y_0|$ of two

[4] L. Tonelli: "La semicontinuita nel calcolo delle variazioni," *Cir. mat. Palermo Rend.*, Vol. 44, 1920.

functions y, y_0 whether the semicontinuity of a functional at a certain "point" y_0 can be established or not (see problems VII.5, 6, and 7).

While we will deal with the problem of minimizing sequences to some extent in §3 of this chapter, the problem of the existence of a minimum cannot be dealt with in this treatment because of its very nature, which would require the introduction of rather fancy and powerful concepts.[5] Let us point out here that the first proof of a theorem which established a sound basis for Dirichlet's principle was given by D. Hilbert in 1899.[6] This theorem necessarily curtails the class of competing functions and also the structure of the boundary, as one has to expect from the preceding discussion.

Problems VII.5–8

5. Let $I[y] = \int_0^1 f(x, y)\, dx$, where f and $\partial f/\partial x$, $\partial f/\partial y$ are continuous in x, y, be defined on the class C^1 of continuous functions with a continuous derivative on $0 \leq x \leq 1$. Let the distance of two functions y_1, y_2 be defined as follows:

$$d(y_1, y_2) = \max_{[0,1]} |y_1 - y_2|.$$

Show that $I[y]$ is lower semicontinuous for any $y(x) \in C^1$. (*Hint:* Use the mean value theorem.)

6. Let $I[y] = \int_0^1 y'^2\, dx$ be defined on the class C^1 on the interval $0 \leq x \leq 1$ and let

$$d(y_1, y_2) = \max_{[0,1]} |y_1' - y_2'|.$$

Show that the functional is lower semicontinuous for any $y(x) \in C^1$.

***7.** Show that a distance defined as

$$d(y_1, y_2) = \left[\int_0^1 (y_1 - y_2)^2\, dx\right]^{1/2}$$

has the properties of a Euclidean length (IV.9). A functional which is semicontinuous with this definition of a distance is called *semicontinuous on the average.* (If the functions y_i are assumed to be square integrable, then two functions are considered as identical, if they are equal everywhere except on a set of measure zero.)

[5] For a discussion of these problems see Ph. Frank and R. von Mises: *Die Differential- und Integralgleichungen der Mechanik und Physik*, Vol. I, pp. 902–908.

[6] See R. Courant and D. Hilbert: *Methoden der Mathematischen Physik*, Vol. II, Springer, Berlin, 1937; p. 471. R. Courant: *Dirichlet's Principle*, Interscience Publishers, 1950, p. 23. S. H. Gould: *Variational Methods for Eigenvalue Problems*, University of Toronto Press, 1957, pp. 59, 77.

8. Show that a minimizing sequence of continuous functions for the problem

$$I[u] = \iint\limits_{x^2+y^2\leq 1} \sqrt{1 + u_x^2 + u_y^2}\, dx\, dy$$

with the boundary conditions $u|_{x^2+y^2=1} = 0$, $u(0) = 1$ can be obtained by setting a conical frustrum of height 1 with base radius $1/n$ and top radius $1/2n$ with the u-axis as axis onto the x, y–plane and letting $n \to \infty$.

§2. MINIMUM PROPERTIES OF THE EIGENVALUES OF A SELF-ADJOINT BOUNDARY VALUE PROBLEM

1. The Solutions of a Certain Isoperimetric Problem as the Eigenfunctions of a Self-Adjoint Boundary Value Problem

It is our purpose to establish in this section the equivalence of the Sturm-Liouville boundary value problem as discussed in Chapter V, §2 with an isoperimetric problem as discussed in the preceding section.

We will present here a proof which is adaptable for generalizations to more dimensions[7] (actually, all one has to do in a generalization to more dimensions is to interpret the symbols in a more general sense, while the formalism can be largely preserved) and will at the same time give a deep insight into the matter.

In order to state and prove our theorems in as simple as possible a form, let us introduce the following abbreviations:

Let $P(x)$, $Q(x)$, $R(x)$ be continuous; $P(x)$ differentiable: $P(x) > 0$, $R(x) > 0$, and $Q(x) < 0$ in the interval $x_1 \leq x \leq x_2$; and let L stand for the self-adjoint operator

$$L = \frac{d}{dx}\left[P(x)\frac{d}{dx}\cdot\right] + Q(x).$$

Depending on whether we consider the Sturm-Liouville problem with the simple boundary conditions

$$y(x_1) = 0, \quad y(x_2) = 0 \qquad \text{(VII.6)}$$

or with the more general boundary conditions

$$\begin{aligned} a_{11}y(x_1) + a_{12}y'(x_1) &= 0, \qquad a_{12} \neq 0, \\ a_{21}y(x_2) + a_{22}y'(x_2) &= 0, \qquad a_{22} \neq 0, \end{aligned}$$

[7] The same proof for the problem in two dimensions can be found in R. Courant and D. Hilbert: *Methods of Mathematical Physics*, Vol. I, Interscience Publishers, New York, 1953, p. 398.

which we can write after division by a_{12} and a_{22}, respectively, as

$$\begin{aligned} \alpha_1 y(x_1) + y'(x_1) &= 0, \\ \alpha_2 y(x_2) + y'(x_2) &= 0, \end{aligned} \tag{VII.7}$$

we have to distinguish between two different variational problems. In case of the boundary conditions (VII.6), these have to be imposed in the variational problem also, while in the case of the boundary conditions (VII.7) for the Sturm-Liouville problem these boundary conditions will appear as natural boundary conditions of a slightly modified variational problem. (See also Chapter I, §4.2.) Accordingly, we will have to distinguish between the two functionals

$$I_1[y] = \int_{x_1}^{x_2} (Py'^2 - Qy^2)\,dx \tag{VII.8}$$

and

$$I_2[y] = I_1[y] + [\alpha_2 P(x_2) y^2(x_2) - \alpha_1 P(x_1) y^2(x_1)]. \tag{VII.9}$$

Let us further denote by $C[y]$ the functional

$$C[y] = \int_{x_1}^{x_2} Ry^2\,dx. \tag{VII.10}$$

All these functionals are "quadratic forms" in y, y'. If we denote the corresponding "*polar forms*" ("*bilinear forms*") by $I_i[y, \eta]$ and $C[y, \eta]$ respectively, namely

$$\begin{aligned} I_1[y, \eta] &= \int_{x_1}^{x_2} (Py'\eta' - Qy\eta)\,dx, \\ I_2[y, \eta] &= I_1[y, \eta] + [\alpha_2 P(x_2) y(x_2)\eta(x_2) - \alpha_1 P(x_1) y(x_1)\eta(x_1)], \\ C[y, \eta] &= \int_{x_1}^{x_2} Ry\eta\,dx, \end{aligned}$$

it is obvious that the following relations hold true

$$I_i[c_1y_1 + c_2y_2] = c_1^2 I_i[y_1] + 2c_1c_2 I_i[y_1, y_2] + c_2^2 I_i[y_2], \tag{VII.11}$$

$$C[c_1y_1 + c_2y_2] = c_1^2 C[y_1] + 2c_1c_2 C[y_1, y_2] + c_2^2 C[y_2], \tag{VII.12}$$

$$I_i[y_1 + y_2, \eta] = I_i[y_1, \eta] + I_i[y_2, \eta], \tag{VII.13}$$

$$C[y_1 + y_2, \eta] = C[y_1, \eta] + C[y_2, \eta]. \tag{VII.14}$$

As mentioned already, we will have to distinguish between the Sturm-Liouville problem under the boundary conditions (VII.6) and the one under the boundary conditions (VII.7). Let us formulate these problems for further-reference purposes as follows:

Problem IA. To solve the self-adjoint differential equation

$$L[y] + \lambda Ry = 0$$

under the boundary conditions

$$y(x_1) = 0, \; y(x_2) = 0.$$

Problem IB. To solve the self-adjoint differential equation

$$L[y] + \lambda Ry = 0$$

under the boundary conditions

$$\alpha_1 y(x_1) + y'(x_1) = 0,$$
$$\alpha_2 y(x_2) + y'(x_2) = 0.$$

We know from Chapter V, §2 that these problems have infinitely many solutions $y_1, y_2, y_3, \cdots$ (eigenfunctions) which correspond to certain values of λ: $\lambda_1, \lambda_2, \lambda_3, \cdots$ (eigenvalues) and we know further, that the functions $\{\sqrt{R(x)}y_k(x)\}$ form an orthogonal system on $x_1 \leq x \leq x_2$.

It is our aim to establish the equivalence of problems IA and IB, respectively, with the following variational problems:

Problem IIA. To find functions $y_1, y_2, y_3, \cdots$ which satisfy the boundary conditions

$$y_i(x_1) = 0, \quad y_i(x_2) = 0, \qquad i = 1, 2, 3, 4, \cdots n$$

and are such that

$$C[y_i, y_n] = \delta_{in}, \qquad i = 1, 2, 3, \cdots n, \quad n = 1, 2, 3, \cdots$$

where $\delta_{in} = \begin{cases} 0 \text{ for } i \neq n \\ 1 \text{ for } i = n \end{cases}$ and yield a minimum for $I_1[y]$:

$$\min I_1[y] = I_1[y_n].$$

We will call this problem the *variational problem associated with IA.*

Problem IIB. To find functions $y_1, y_2, y_3, \cdots$ which are such that

$$C[y_i, y_n] = \delta_{in}, \qquad i = 1, 2, 3, \cdots, n, \quad n = 1, 2, 3, \cdots$$

and yield a minimum for $I_2[y]$:

$$\min I_2[y] = I_2[y_n].$$

We call this problem the *variational problem associated with IB.*

Let us elaborate a little on the process by which the functions y_n in IIA and IIB are generated.

Let $n = 1$: We have to consider all functions (which satisfy the boundary conditions (VII.6) in case of IA and) which are continuous and have a continuous first-order derivative in $x_1 \leq x \leq x_2$; norm them, i.e., multiply them by a suitable constant such that $C[y] = 1$ for all y; and then single out the one for which $I[y]$ assumes its minimum value. The function which renders the minimum is called y_1.

Now, let $n = 2$: We consider all functions (which satisfy (VII.6) in case of IA and) which are such that $C[y_1, y] = 0$ and $C[y] = 1$ and then single out the one for which $I[y]$ yields a minimum. This function is called y_2; etc.

The reader will notice that with each step one more constraint has to be taken into account; i.e., the class of competing functions is narrowed down with each step. For the nth function y_n, we have to consider besides (the boundary conditions (VII.6) in the case of problem IA and) the normalization condition $C[y] = 1$, the $n - 1$ constraints

$$C[y_i, y_n] = 0 \quad \text{for } i = 1, 2, 3, \cdots, n - 1.$$

Theorem VII.1A. *Provided that there exists a function y_k which is a solution of the associated variational problem IIA, then y_k is also a solution of the Sturm-Liouville boundary value problem IA and*

$$\lambda_k = I_1[y_k] \text{ is the eigenvalue corresponding to } y_k.$$

Theorem VII.1B. *Provided that there exists a function y_k which is a solution of the associated variational problem IIB, then y_k is also a solution of the Sturm-Liouville problem IB and*

$$I_2[y_k] = \lambda_k \text{ is the eigenvalue corresponding to } y_k.$$

Instead of minimizing $I[y]$ under the condition $C[y] = 1$, one can minimize the quotient $I[y]/C[y]$ and dispense with the condition $C[y] = 1$, as we pointed out in §1.1 of this chapter. Then, the eigenfunctions will not appear to be normalized and the eigenvalues will be given by $\lambda = \min I[y]/C[y]$.

Proof of Theorem VII.IB. (For a proof of theorem VII.IA see problem VII.12).

Let $y = y_1$ be the solution of

$$I_2[y] \to \text{minimum}$$
$$C[y] = 1.$$

Then

$$\frac{I_2[y_1]}{C[y_1]} = \lambda_1 \tag{VII.15}$$

is an eigenvalue, as we now prove, and y_1 is the corresponding eigenfunction of IB. Let η_1 be an arbitrary continuous function with a continuous derivative in $x_1 \leq x \leq x_2$—we will call such a function an *admissible function*—and let ϵ be an arbitrary parameter which can assume positive and negative values. Then

$$\frac{I_2[y_1 + \epsilon\eta_1]}{C[y_1 + \epsilon\eta_1]} \geq \lambda_1 \tag{VII.16}$$

holds for any ϵ and all admissible η_1, for, if there were for a certain η_1 an ϵ_0 such that

$$\frac{I_2[y_1 + \epsilon_0\eta_1]}{C[y_1 + \epsilon_0\eta_1]} < \lambda_1,$$

then, the function $\phi = [C[y_1 + \epsilon_0\eta_1]]^{-1/2}(y_1 + \epsilon_0\eta_1)$ would render a smaller value for $I_2[y]$ than y_1, i.e., y_1 would not be the solution of the minimum problem.

From (VII.16) it follows that

$$I_2[y_1 + \epsilon\eta_1] \geq \lambda_1 C[y_1 + \epsilon\eta_1]. \tag{VII.17}$$

In view of (VII.11) and (VII.12) we have

$$I_2[y_1 + \epsilon\eta_1] = I_2[y_1] + 2\epsilon I_2[y_1, \eta_1] + \epsilon^2 I_2[\eta_1],$$
$$C[y_1 + \epsilon\eta_1] = C[y_1] + 2\epsilon C[y_1, \eta_1] + \epsilon^2 C[\eta_1],$$

and since it follows from (VII.15) that

$$I_2[y_1] = \lambda_1 C[y_1],$$

we have

$$2\epsilon\left\{I_2[y_1, \eta_1] - \lambda_1 C[y_1, \eta_1] + \frac{\epsilon}{2}(I_2[\eta_1] - \lambda_1 C[\eta_1])\right\} \geq 0 \tag{VII.18}$$

for any admissible function η_1 and any arbitrary ϵ.

Clearly, this relation can only hold if

$$I_2[y_1, \eta_1] - \lambda_1 C[y_1, \eta_1] = 0 \tag{VII.19}$$

for any admissible η_1. Otherwise, we can always find for any η_1 an ϵ such that the left side of (VII.18) becomes negative (see problem VII.10).

Let us now examine the condition (VII.19): We consider

$$I_2[y_1, \eta_1] = \int_{x_1}^{x_2} (Py_1'\eta_1' - Qy_1\eta_1)\,dx$$
$$+ \alpha_2 P(x_2)y_1(x_2)\eta_1(x_2) - \alpha_1 P(x_1)y_1(x_1)\eta_1(x_1)$$

and apply integration by parts to the first portion under the integral. We obtain

$$I_2[y_1, \eta_1] = Py_1'\eta_1\Big|_{x_1}^{x_2} - \int_{x_1}^{x_2} \eta_1\left[\frac{d}{dx}(Py_1') + Qy_1\right]dx$$
$$+ [\alpha_2 P(x_2)y_1(x_2)\eta_1(x_2) - \alpha_1 P(x_1)y_1(x_1)\eta_1(x_1)]$$
$$= P(x_2)y_1'(x_2)\eta_1(x_2) - P(x_1)y_1'(x_1)\eta_1(x_1) + \alpha_2 P(x_2)y_1(x_2)\eta_1(x_2)$$
$$- \alpha_1 P(x_1)y_1(x_1)\eta_1(x_1) - \int_{x_1}^{x_2} \eta_1\left[\frac{d}{dx}(Py_1') + Qy_1\right]dx$$

or

$$I_2[y_1, \eta_1] = P(x_2)\eta_1(x_2)(\alpha_2 y_1(x_2) + y_1'(x_2))$$
$$- P(x_1)\eta_1(x_1)(\alpha_1 y_1(x_1) + y_1'(x_1)) - \int_{x_1}^{x_2} \eta_1\left[\frac{d}{dx}(Py_1') + Qy_1\right]dx.$$

Hence, (VII.19) yields in view of $C[y_1, \eta_1] = \int_{x_1}^{x_2} \eta_1 y_1 R\,dx$ for any admissible η_1,

$$P(x_2)\eta_1(x_2)(\alpha_2 y_1(x_2) + y_1'(x_2)) - P(x_1)\eta_1(x_1)(\alpha_1 y_1(x_1) + y_1'(x_1))$$
$$- \int_{x_1}^{x_2} \eta_1\left[\frac{d}{dx}(Py_1') + Qy_1 + \lambda_1 R y_1\right]dx = 0. \quad \text{(VII.20)}$$

This relation has to hold for any admissible function η_1. In particular it has to be true for a function η_1 which vanishes at x_1 and x_2; hence

$$\int_{x_1}^{x_2} \eta_1\left[\frac{d}{dx}(Py_1') + (Q + \lambda_1 R)y_1\right]dx = 0$$

for any admissible η_1 which is such that $\eta_1(x_1) = \eta_1(x_2) = 0$ and it follows from the fundamental lemma of the calculus of variations (see Chapter I, §1.2) that

$$\frac{d}{dx}(Py_1') + (Q + \lambda_1 R)y_1 = 0.$$

Consequently it follows from (VII.20) that

$$\eta_1(x_2)P(x_2)(\alpha_2 y_1(x_2) + y_1'(x_2)) - \eta_1(x_1)P(x_1)(\alpha_1 y_1(x_1) + y_1'(x_1)) = 0$$

for any admissible η_1. Therefore, the left member has to vanish in particular for an admissible η_1 for which $\eta_1(x_1) = 0$, $\eta_1(x_2) \neq 0$ and one for which $\eta_1(x_1) \neq 0$, $\eta_1(x_2) = 0$ respectively. Hence

$$\alpha_2 y_1(x_2) + y_1'(x_2) = 0,$$
$$\alpha_1 y_1(x_1) + y_1'(x_1) = 0;$$

i.e.: $y_1(x)$ is a solution of the problem IB and λ_1 is its eigenvalue.

We now proceed to the next following minimum problem:

We assume that y_2 satisfies $C[y_2] = 1$ and $C[y_1, y_2] = 0$ and that

$$I_2[y_2] = \lambda_2$$

is the minimum of $I_2[y]$ relative to the class of functions which is restricted by the above conditions.

Let us now consider a continuous function ζ with a continuous derivative which is such that

$$C[y_1, \zeta] = 0 \qquad \text{(VII.21)}$$

where y_1 is the solution of the preceding minimum problem. In this case, we also have

$$C[y_1, y_2 + \epsilon\zeta] = 0$$

according to (VII.14). We obtain by the same argument which we used before the result that (VII.16) and consequently (VII.17) has to hold for $y_2 + \epsilon\zeta$ and λ_2 where ζ satisfies (VII.21). This has (VI.19) again as a consequence but for a rather restricted class of functions ζ only. In order to obtain (VII.19) for any admissible function, we have to proceed as follows: Let η_2 by any admissible function and let t be such that

$$\zeta = \eta_2 + ty_1,$$

where y_1, the first eigenfunction, satisfies (VII.21). This can be accomplished by choosing $t = -C[y_1, \eta_2]$ (see problem VII.11).

Since the function y_2 is continuous and has a continuous derivative, we can substitute y_2 for η_1 in (VII.19) and obtain

$$I_2[y_1, y_2] - \lambda_1 C[y_1, y_2] = 0$$

and because of $C[y_1, y_2] = 0$ we have

$$I_2[y_1, y_2] = 0. \qquad \text{(VII.22)}$$

Now, let us substitute $\zeta = \eta_2 + ty_1$ into (VII.19) with y_2 and λ_2 and we obtain

$$I_2[y_2, \eta_2 + ty_1] - \lambda_2 C[y_2, \eta_2 + ty_1] = 0$$

or in view of (VII.13) and (VII.14)

$$I_2[y_2, \eta_2] + tI_2[y_2, y_1] - \lambda_2(tC[y_2, y_1] + C[y_2, \eta_2]) = 0,$$

which reduces in view of (VII.22) and $C[y_2, y_1] = 0$ to

$$I_2[y_2, \eta_2] - \lambda_2 C[y_2, \eta_2] = 0$$

for any admissible η_2. From this it follows again that y_2 is an eigenfunction and that λ_2 is the corresponding eigenvalue of problem IB. We can proceed in this fashion and obtain thus an unlimited number of eigenvalues and eigenfunctions of problem IB. Since the class C_k of all functions which are admissible in the kth minimum problem is a subclass of the class C_{k-1} of all functions that are admissible in the $k-1$th problem, $C_k \subset C_{k-1}$, we can apply lemma VII.1 and obtain.

Lemma VII.2. *The process as outlined in theorem VII.1B renders an unlimited number of eigenvalues* $\lambda_1, \lambda_2, \lambda_3, \cdots$ *and* $\lambda_{k-1} \leq \lambda_k$.

Problems VII.9–14

9. Verify the relations (VII.11)–(VII.14).

***10.** Let $I_2[y_1, \eta_1] - \lambda_1 C[y_1, \eta_1] \neq 0$. Show that one can find for any admissible η_1 an ϵ such that

$$2\epsilon\left\{I_2[y_1, \eta_1] - \lambda_1 C[y_1, \eta_1] + \frac{\epsilon}{2}\,[I_2[\eta_1] - \lambda_1 C[\eta_1]]\right\} < 0.$$

11. Show that if

$$\zeta = \eta_2 + ty_1$$

is to satisfy $C[y_1, \zeta] = 0$, where $C[y_1] = 1$, for any admissible η_2, then $t = -C[y_1, \eta_2]$.

12. Prove theorem VII.1A. (*Hint:* Use for admissible functions η such functions which are continuous, have a continuous derivative, and for which $\eta(x_1) = \eta(x_2) = 0$. Otherwise, proceed as in the proof of theorem VII.1B.)

13. Show that

$$\int_{-1}^{1} (1 - x^2)y_n'^2\, dx \rightarrow \text{minimum},$$

where $\int_{-1}^{1} y_n y_i\, dx = \delta_{in}$, $i = 1, 2, 3 \cdots, n$; $n = 1, 2, 3, \cdots$ yields the normalized Legendre polynomials. (See also Chapter V, §2.4.)

14. Show that

$$\int_0^1 (xy_n'^2 + \frac{\mu^2}{x}\, y_n^2)\, dx \rightarrow \text{minimum}$$

with $\int_0^1 xy_i y_n\, dx = \delta_{in}$, $i = 1, 2, 3, \cdots, n$; $n = 1, 2, 3, \cdots$ and the boundary conditions $y(0) = 0$, $y(1) = 0$ yields the normalized Bessel functions of integral order $\mu \geq 1$.

2. The Eigenfunctions of the Sturm-Liouville Boundary Value Problem as Solutions of the Associated Variational Problem

We will show in this subsection that every eigenfunction and eigenvalue of the Sturm-Liouville problems IA and IB is also a solution of the associated variational problems IIA and IIB. Since we have already shown in subsection 1 that every solution of the problems IIA and IIB is also a solution of the problems IA and IB, it will then follow that these problems are equivalent.

Before we can prove a theorem of such content, however, we have to establish first the fact that the sequence of eigenvalues $\lambda_1, \lambda_2, \lambda_3, \cdots$ as obtained from problems IIA and IIB has no upper bound. To prove this is actually very simple, once we have established the following lemmas:

Lemma VII.3 (Courant's maximum-minimum principle). *Let* $\eta_1, \eta_2, \cdots, \eta_{n-1}$ *be a set of* $n - 1$ *functions and let* ζ *be any function such that*

$$C[\eta_i, \zeta] = 0 \quad \text{for } i = 1, 2, 3, \cdots, n-1.$$

If we denote by $m(\zeta;\ \eta_1, \eta_2, \cdots, \eta_{n-1})$ *the minimum of* $I[\zeta]/C[\zeta]$ *relative to a fixed basis* $(\eta_1, \eta_2, \cdots, \eta_{n-1})$:

$$m(\zeta; \eta_1, \eta_2, \cdots, \eta_{n-1}) = \min_{(\eta_i)} \frac{I[\zeta]}{C[\zeta]},$$

then, the nth *eigenvalue* λ_n *of problem II (A or B) is given by*

$$\lambda_n = \max_{[\eta_i]} m(\zeta; \eta_1, \eta_2, \cdots, \eta_{n-1}),$$

where $[\eta_i]$ *means "over all possible systems of* $n - 1$ *functions."*

Remark: To find the nth eigenvalue according to this maximum minimum principle, one has to proceed as follows: One has to consider all possible systems of $n - 1$ functions $\eta_1, \eta_2, \cdots, \eta_{n-1}$ (which satisfy in the case of problem IIA the boundary conditions (VII.6)). For every system of this sort one has to find the function ζ (which satisfies in the case of IIA the boundary conditions (VII.6)) which is such that $C[\zeta, \eta_i] = 0$ and for which $I[\zeta]/C[\zeta]$ assumes its minimum. Among all the minima thus found, the largest is λ_n.

Proof. It is clear from theorems VII.1A and 1B that in the notation of this lemma

$$\lambda_n = m(y_n;\ y_1, y_2, \cdots, y_{n-1}).$$

It remains to be shown that

$$\lambda_n \geq m(\zeta;\ \eta_1, \eta_2, \cdots, \eta_{n-1}), \tag{VII.23}$$

where $\eta_1, \eta_2, \cdots, \eta_{n-1}$ is any system of $n - 1$ functions.

Let us consider such a system $\eta_1, \eta_2, \cdots, \eta_{n-1}$ where the η_i are chosen quite arbitrarily. If we can find a function ζ such that $C[\eta_i, \zeta] = 0$ and

$$\frac{I[\zeta]}{C[\zeta]} \leq \lambda_n \tag{VII.24}$$

and since clearly

$$m(\zeta; \eta_1, \eta_2, \cdots, \eta_{n-1}) \le \frac{I[\zeta]}{C[\zeta]}$$

then, from this and (VII.24) it will follow that (VII.23) is true.

Let us choose

$$\zeta = \sum_{k=1}^{n} c_k y_k$$

where the y_k shall be the first n normed eigenfunctions of problem II (A or B) and let us determine the c_k such that

$$C\left[\sum_{k=1}^{n} c_k y_k, \eta_i\right] = 0, \qquad i = 1, 2, 3, \cdots, (n-1).$$

These are $n - 1$ homogeneous equations for the n unknowns $c_1, c_2, \cdots, c_n$ (see problem VII.15). These unknowns will be determined except for a common multiple and we can therefore norm them such that

$$\sum_{k=1}^{n} c_k^2 = 1,$$

which means simply that $C[\zeta] = 1$ since $C[\zeta] = C\left[\sum_{k=1}^{n} c_k y_k\right] = \sum_{k=1}^{n} c_k^2 C[y_k]$ and the y_k are normed by hypothesis.

Now, we have in view of (VII.11)—or rather a generalization of (VII.11)—

$$I[\zeta] = I\left[\sum_{k=1}^{n} c_k y_k\right] = \sum_{i=1}^{n} \sum_{k=1}^{n} c_i c_k I[y_i, y_k].$$

Since

$$I[y_i, y_k] = \begin{cases} 0 & \text{for } i \neq k \\ \lambda_k & \text{for } i = k, \end{cases} \qquad \text{(see (VII.22) and problem VII.20)},$$

we have

$$I[\zeta] = \sum_{k=1}^{n} c_k^2 \lambda_k.$$

In view of lemma VII.2

$$\lambda_i \le \lambda_n, \qquad i = 1, 2, 3, \cdots, (n-1)$$

and hence

$$I[\zeta] = \sum_{k=1}^{n} c_k^2 \lambda_k \le \lambda_n \sum_{k=1}^{n} c_k^2 = \lambda_n,$$

which proves (VII.24), and this in turn proves our lemma.

Lemma VII.4. *Let λ_n be the* nth *eigenvalue of problem II (A or B) for the functionals $\bar{I}_i[y]$, $\bar{C}[y]$ and let Λ_n be the* nth *eigenvalue of problem II (A or B) for the functionals $I_i[y]$, $C[y]$.*

If

$$\frac{\bar{I}_i[y]}{\bar{C}[y]} \leq \frac{I_i[y]}{C[y]} \tag{VII.25}$$

for any smooth function y, then $\lambda_n \leq \Lambda_n$.

Remark: It would be rather easy to prove a theorem of this kind for the eigenvalues of the Sturm-Liouville problem I by the means which we developed in Chapter V (indeed, we proved a very similar theorem concerning the zeros of the eigenfunctions). However, that would be of no help to us because we have not yet established the equivalence of problems I and II and we need this lemma to establish this equivalence. (See also problem VII.17.)

Proof. Let us consider a system of $n - 1$ functions $\eta_1, \eta_2, \cdots, \eta_{n-1}$ (which satisfy the boundary conditions (VII.6) in case $i = 1$) and let η_0 (which satisfies the boundary condition (VII.6) in case of $i = 1$) be such that

$$C[\eta_0, \eta_i] = 0, \qquad i = 1, 2, 3, \cdots, n - 1$$

and

$$\min_{(\eta_i)} \frac{I_i[\eta]}{C[\eta]} = \frac{I_i[\eta_0]}{C[\eta_0]}.$$

Because of (VII.25) we have

$$\frac{\bar{I}_i[\eta_0]}{\bar{C}[\eta_0]} \leq \frac{I_i[\eta_0]}{C[\eta_0]}$$

and consequently with $\bar{\eta}_i = \dfrac{R}{\bar{R}}\, \eta_i$

$$\min_{(\bar{\eta}_i)} \frac{\bar{I}_i[\eta]}{\bar{C}[\eta]} \leq \frac{I_i[\eta_0]}{C[\eta_0]}.$$

Observing that every conceivable set $\bar{\eta}_1, \bar{\eta}_2, \cdots, \bar{\eta}_{n-1}$ is obtained in this manner, it follows that

$$\max_{[\bar{\eta}_i]} \min_{(\bar{\eta}_i)} \frac{\bar{I}_i[\bar{\eta}]}{\bar{C}[\bar{\eta}]} \leq \max_{[\eta_i]} \min_{(\eta_i)} \frac{I_i[\eta]}{C[\eta]}$$

and therefore in view of lemma VII.3

$$\lambda_n \leq \Lambda_n, \quad \text{q.e.d.}$$

We are now ready to prove the lemma which will guarantee the non-existence of an upper bound for the eigenvalues of problem II:

Lemma VII.5. *If $\lambda_1, \lambda_2, \lambda_3, \cdots$ are the eigenvalues of the variational problem II (A or B) as described in theorems VII.1A and 1B, then*

$$\lim_{k\to\infty} \lambda_k = \infty.$$

Proof. We give the proof of this lemma for the general case of problem IIA and for the special case of problem IIB where $\alpha_2 \geq 0$, $\alpha_1 \leq 0$ only.[8]

We assumed that the functions $P(x)$, $Q(x)$, $R(x)$ are continuous in $x_1 \leq x \leq x_2$ and that $P(x) > 0$, $R(x) > 0$, $Q(x) < 0$.

It follows that they assume in this interval their maximum and minimum values

$$P_m = \min P(x)$$
$$Q_M = \max Q(x),$$
$$R_M = \max R(x).$$

Then, we have in view of $\alpha_2 \geq 0$, $\alpha_1 \leq 0$

$$\frac{\int_{x_1}^{x_2} (P_m y'^2 - Q_M y^2)\,dx + \alpha_2 P_m y^2(x_2) - \alpha_1 P_m y^2(x_1)}{\int_{x_1}^{x_2} R_M y^2\,dx}$$
$$\leq \frac{\int_{x_1}^{x_2} (P y'^2 - Q y^2)\,dx + \alpha_2 P(x_2) y^2(x_2) - \alpha_1 P(x_1) y^2(x_1)}{\int_{x_1}^{x_2} R y^2\,dx}; \qquad \text{(VII.26)}$$

i.e., the condition (VII.25) of lemma VII.4 is satisfied for $i = 2$. Clearly, it is also satisfied for $i = 1$ because in that case the boundary function vanishes.

Therefore, we have according to lemma VII.4

$$\lambda_n \leq \Lambda_n$$

where λ_n is the nth eigenvalue of the minimum problem involving the quotient in the left side of (VII.26) and Λ_n is the nth eigenvalue of the minimum problem involving the quotient on the right side of (VII.26).

Since P_m, Q_M, R_M are constants, we can write the quotient on the left side of (VII.26) as

$$\frac{P_m}{R_M} \cdot \frac{\int_{x_1}^{x_2} y'^2\,dx + \alpha_2 y^2(x_2) - \alpha_1 y^2(x_1)}{\int_{x_1}^{x_2} y^2\,dx} - \frac{Q_M}{R_M}.$$

[8] For the general case see R. Courant and D. Hilbert: *Methods of Mathematical Physics*, Vol. 1, Interscience Publishers, New York, 1953, p. 416.

Let us first consider the problem IIA. Then $y(x_1) = y(x_2) = 0$ and we have

$$\frac{P_m}{R_M} \cdot \frac{\int_{x_1}^{x_2} y'^2 \, dx}{\int_{x_1}^{x_2} y^2 \, dx} - \frac{Q_M}{R_M}.$$

All the eigenvalues of

$$\frac{\int_{x_1}^{x_2} y'^2 \, dx}{\int_{x_1}^{x_2} y^2 \, dx}$$

are in view of theorem VII.1A also eigenvalues of

$$y'' + ly = 0$$

with the boundary conditions $y(x_1) = 0$, $y(x_2) = 0$.
The eigenvalues of this problem of type IA are

$$l_k = \frac{k^2\pi^2}{(x_2 - x_1)^2}, \qquad k = 1, 2, 3, \cdots.$$

Therefore we have for the eigenvalues of the quotient on the left side of (VII.26) in the case of problem IIA

$$\lambda_n = \frac{P_m}{R_M} l_{k_n} - \frac{Q_M}{R_M} = \frac{P_m}{R_M} \frac{k_n^2\pi^2}{(x_2 - x_1)^2} - \frac{Q_M}{R_M},$$

where $k_1, k_2, k_3, \cdots$ is either identical with or a subsequence of $1, 2, 3, \cdots$.
We see from this that $\lim_{n\to\infty} \lambda_n = \infty$ and hence in view of lemma VII.4

$$\lim_{n\to\infty} \Lambda_n = \infty.$$

Next, we have to discuss problem IIB.
All eigenvalues of

$$\frac{\int_{x_1}^{x_2} y'^2 \, dx + \alpha_2 y^2(x_2) - \alpha_1 y^2(x_1)}{\int_{x_1}^{x_2} y^2 \, dx}$$

are in view of theorem VII.1B also eigenvalues of

$$y'' + ly = 0$$

with the boundary conditions

$$\alpha_1 y(x_1) + y'(x_1) = 0,$$
$$\alpha_2 y(x_2) + y'(x_2) = 0.$$

This problem of type IB has in view of theorem V.8 infinitely many eigenvalues l_k and

$$\lim_{k\to\infty} l_k = \infty.$$

The eigenvalues λ_n of the quotient on the left side of (VII.26) are therefore

$$\lambda_n = \frac{P_m}{R_M} l_{k_n} - \frac{Q_M}{R_M},$$

where k_n is either identical with or a subsequence of $1, 2, 3, \cdots$ and it follows as before that

$$\lim_{n\to\infty} \Lambda_n = \infty.$$

(See also problem VII.21.)

We are now ready to prove that every solution of the Sturm-Liouville problem 1A or 1B is also a solution of the associated variational problem IIA or IIB.

Theorem VII.2. *The* nth *eigenfunction* y_n $(n = 1, 2, 3, \cdots)$ *of the Sturm-Liouville problem IA or IB is the solution of the* nth *associated variational problem IIA or IIB and the corresponding eigenvalue* λ_n *is equal to* $I_i[y_n]$ $(i = 1$ *or* $2)$.

Proof (*by contradiction*). Let η with $C[\eta] = 1$ be an eigenfunction of problem I and λ the corresponding eigenvalue and let us assume that η is not among the solutions of the associated variational problem II.

Then,

$$\lambda \neq \lambda_i, \qquad i = 1, 2, 3, \cdots$$

where the λ_i are the eigenvalues of the associated variational problem II. Should there exist a λ_k such that $\lambda_k = \lambda$ but $\eta \neq y_k$, we would obtain a contradiction to the fact that the solution of problem I for a fixed λ is unique.

Since $\lambda \neq \lambda_i$ there exists in view of lemma VII.5 a λ_n such that

$$\lambda < \lambda_n.$$

We will show now that there exists among the first $n - 1$ eigenfunctions $y_1, y_2, \cdots, y_{n-1}$ of the variational problem II at least one y_k such that

$$C[\eta, y_k] \neq 0, \qquad 1 \leq k \leq n - 1. \tag{VII.27}$$

It follows from

$$L[\eta] + \lambda R\eta = 0$$

by multiplying this equation by η and integrating both sides with respect to x over the interval from x_1 to x_2 that

$$\frac{\int_{x_1}^{x_2}\left[\frac{d}{dx}(P\eta')\eta + Q\eta^2\right] dx}{\int_{x_1}^{x_2} R\eta^2\, dx} = -\lambda.$$

If we integrate by parts the first portion in the numerator, we obtain in view of $C[\eta] = 1$

$$\lambda = -P\eta'\eta\Big|_{x_1}^{x_2} + \int_{x_1}^{x_2}(P\eta'^2 + Q\eta^2)\, dx.$$

Since η is a solution of IA or IB, it satisfies either the boundary conditions (VII.6) or the boundary condition (VII.7):

In the case of IA, we have

$$-P\eta'\eta\Big|_{x_1}^{x_2} = 0$$

and in the case of IB,

$$-P\eta'\eta\Big|_{x_1}^{x_2} = \alpha_2 P(x_2)\eta^2(x_2) - \alpha_1 P(x_1)\eta^2(x_1).$$

Hence

$$\lambda = \begin{cases} I_1[\eta] & \text{in the case of problem IA,} \\ I_2[\eta] & \text{in the case of problem IB.} \end{cases}$$

Now, if η were orthogonal to all the $y_1, y_2, \cdots, y_{n-1}$, then, since

$$\lambda = I_i[\eta] < \lambda_n = I_i[y_n],$$

we would obtain a contradiction to the fact that y_n, of all functions which are orthogonal to the $y_1, y_2, \cdots, y_{n-1}$, renders $I_i[y]$ smallest.

η as well as y_k are solutions of problem IA or IB, i.e.,

$$L[\eta] + \lambda R\eta = 0,$$
$$L[y_k] + \lambda_k R y_k = 0.$$

Let us multiply the first equation by y_k and the second equation by η and subsequently subtract the second equation from the first equation:

$$y_k L[\eta] - \eta L[y_k] = (\lambda_k - \lambda) R\eta y_k.$$

If we integrate this equation from x_1 to x_2 and observe Green's lemma (lemma V.5), which is applicable since y_k and η satisfy boundary conditions of the type (V.33), we obtain

$$(\lambda_k - \lambda) C[\eta, y_k] = 0$$

and in view of (VII.27) we have $\lambda_k - \lambda = 0$. This contradicts our assumption that η is not among the y_i's which had $\lambda_k \neq \lambda$ as a consequence, q.e.d.

Combining theorems VII.1A, B and VII.2, we can state:

Theorem VII.3. *The Sturm-Liouville problem IA is equivalent with the associated variational problem IIA and the Sturm-Liouville problem IB is equivalent with the associated variational problem IIB.*

Problems VII.15–19

15. Write the system of $n - 1$ equations for $c_1, c_2, \cdots, c_n$

$$C\left[\sum_{k=1}^{n} c_k y_k, \eta_i\right] = 0, \quad i = 1, 2, 3, \cdots, n - 1$$

explicitly and discuss its solvability.

16. Solve the above equations for $C[y] = \int_{-1}^{1} y^2\,dx$, $y_1 = 1/\sqrt{2}$, $y_2 = \sqrt{\tfrac{3}{2}}x$, $y_3 = \tfrac{1}{2}\sqrt{\tfrac{5}{2}}(3x^2 - 1)$, $\eta_1 = \sin \pi x$, $\eta_2 = \cos \pi x$ and norm the solutions such that $\sum_{k=1}^{3} c_k^2 = 1$.

***17.** Prove lemma VII.4 for the eigenvalues of the Sturm-Liouville problem with coefficients which are related to each other according to $\bar{P} \leq P$, $\bar{R} \geq R$, $\bar{Q} \geq Q$. (*Hint:* Use the x, ϕ–diagram.)

18. Give a simple example for problem 17.

***19.** Establish the validity of lemma VII.3 for the case where only functions from a manifold

$$y = a_1 y_1(x) + a_2 y_2(x) + a_3 y_3(x) + \cdots + a_n y_n(x),$$

where the $y_i(x)$ are fixed functions that satisfy (VII.6) in case of problem IIA, are permitted to compete. (See also R. Courant and D. Hilbert: *Methods of Mathematical Physics*, Vol. I, Interscience Publishers, New York, 1953, p. 23.)

20. Prove that

$$I_n[y_i, y_k] = 0$$

for any two solutions y_i, y_k of problem IIA ($n = 1$) or problem IIB ($n = 2$). (*Hint:* Establish $I[y_i, \eta] - \lambda_i C[y_i, \eta] = 0$ for any admissible η and proceed from there.)

21. Find an explicit representation of the eigenvalues l_k of

$$y'' + ly = 0$$

under the boundary conditions

$$\alpha_1 y(x_1) + y'(x_1) = 0,$$
$$\alpha_2 y(x_2) + y'(x_2) = 0,$$

where $\alpha_1 \leq 0$, $\alpha_2 \geq 0$ and show that

$$\lim_{k \to \infty} l_k = \infty.$$

§3. THE METHOD OF RAYLEIGH AND RITZ

1. Theorems Regarding the Secular Equation

We will have to refer in our subsequent studies to some theorems in matrix theory with which we do not expect the reader to be familiar. Therefore, we will state and prove these theorems in this subsection. Let

$$A = \begin{pmatrix} a_{11} & a_{12} & \cdots & a_{1n} \\ \cdot & & & \\ \cdot & & & \\ \cdot & & & \\ a_{n1} & a_{n2} & \cdots & a_{nn} \end{pmatrix}$$

and

$$B = \begin{pmatrix} b_{11} & b_{12} & \cdots & b_{1n} \\ \cdot & & & \\ \cdot & & & \\ \cdot & & & \\ b_{n1} & b_{n2} & \cdots & b_{nn} \end{pmatrix}$$

be two *symmetric matrices*,

$$A = A^*, \quad B = B^* \tag{VII.28}$$

(where * denotes the transpose), i.e., $a_{ik} = a_{ki}$ and $b_{ik} = b_{ki}$, where the a_{ik} and b_{ik} are constant and real and let $C = (c_1, c_2, \cdots, c_n)$ denote a vector in the n-dimensional Euclidean space E_n.

An equation which plays an important role in various phases of mathematical development ever since *Laplace* used it in 1772 in his determination of the secular perturbations of the planets and which is usually first encountered by the student of mathematics in his dealings with the principal axis problem of quadratic forms is the so-called *secular equation*. We may arrive at this equation by considering the following problem:

To find a vector C in E_n such that the homogeneous linear equations

$$\begin{gathered} c_1(a_{11} - \lambda b_{11}) + \cdots + c_n(a_{1n} - \lambda b_{1n}) = 0 \\ \cdot \\ \cdot \\ \cdot \\ c_1(a_{n1} - \lambda b_{n1}) + \cdots + c_n(a_{nn} - \lambda b_{nn}) = 0 \end{gathered} \tag{VII.29}$$

where the a_{ik} and b_{ik} satisfy (VII.28) and λ is an arbitrary parameter, are satisfied. Clearly, this system of n homogeneous linear equations in

$c_1, c_2, \cdots, c_n$ has a nontrivial solution only if its coefficient determinant vanishes:

$$\begin{vmatrix} a_{11} - \lambda b_{11} \cdots a_{1n} - \lambda b_{1n} \\ \cdot \\ \cdot \\ \cdot \\ a_{n1} - \lambda b_{n1} \cdots a_{nn} - \lambda b_{nn} \end{vmatrix} = 0 \qquad \text{(VII.30)}$$

and this constitutes the *secular equation.*

It is an algebraic equation of nth degree in λ and, according to the fundamental theorem of algebra, we expect n real or complex roots.

We will now prove the following theorems:

Theorem VII.4. *If λ_1, λ_2 are two distinct roots of the secular equation (VII.30), then the corresponding solutions $C_1 = (c_1^{(1)}, c_2^{(1)}, \cdots, c_n^{(1)})$, $C_2 = (c_1^{(2)}, c_2^{(2)}, \cdots, c_n^{(2)})$ of (VII.29) are such that*

$$\sum_{i=1}^{n} \sum_{k=1}^{n} c_i^{(1)} c_k^{(2)} b_{ik} = 0.$$

Remark: For the special case where $b_{ik} = \delta_{ik}$ (*unit matrix*) we obtain two orthogonal vectors C_1, C_2 as solutions, i.e., $(C_1 \cdot C_2) = \sum_{i=1}^{n} c_i^{(1)} c_i^{(2)} = 0.$

Proof. Since λ_1, λ_2 are roots of (VII.30) and C_1, C_2 the corresponding solutions of (VII.29), we have

$$\sum_{i=1}^{n} c_i^{(1)} (a_{ik} - \lambda_1 b_{ik}) = 0,$$

$$\sum_{i=1}^{n} c_i^{(2)} (a_{ik} - \lambda_2 b_{ik}) = 0,$$

for $k = 1, 2, 3, \cdots, n$. Let us multiply the kth equation of the first group by $c_k^{(2)}$ and the kth equation of the second group by $c_k^{(1)}$ and then take in both cases the sum over all k from 1 to n:

$$\sum_{k=1}^{n} \sum_{i=1}^{n} c_i^{(1)} c_k^{(2)} a_{ik} = \lambda_1 \sum_{k=1}^{n} \sum_{i=1}^{n} c_i^{(1)} c_k^{(2)} b_{ik},$$

$$\sum_{k=1}^{n} \sum_{i=1}^{n} c_i^{(2)} c_k^{(1)} a_{ik} = \lambda_2 \sum_{k=1}^{n} \sum_{i=1}^{n} c_i^{(2)} c_k^{(1)} b_{ik}.$$

In view of (VII.28) we obtain from these relations

$$(\lambda_1 - \lambda_2) \sum_{k=1}^{n} \sum_{i=1}^{n} c_i^{(1)} c_k^{(2)} b_{ik} = 0$$

and because of $\lambda_1 \neq \lambda_2$ we can deduce from this our assertion.

A second theorem which is also related to the secular equation and will prove of great value for our subsequent investigations is the following one:

Theorem VII.5. *All roots of the secular equation (VII.30) are real under the conditions (VII.28).*

Proof (by contradiction). We will present the proof only for the case where $b_{ik} = \delta_{ik}$, i.e., where B is the unit matrix

$$B = \begin{pmatrix} 1 & 0 & 0 & \cdot & \cdot & \cdot & 0 \\ 0 & 1 & 0 & \cdot & \cdot & \cdot & 0 \\ \cdot & \cdot & \cdot & \cdot & \cdot & \cdot & \cdot \\ 0 & 0 & 0 & \cdot & \cdot & \cdot & 1 \end{pmatrix}.$$

Thus, we deal with the simpler equation

$$det\,|a_{ik} - \lambda\delta_{ik}| = 0$$

and we will assume that this equation has a complex root

$$\lambda = \alpha + i\beta.$$

Then, it will also have the conjugate complex root

$$\bar{\lambda} = \alpha - i\beta$$

since the a_{ik} are real numbers.

With these two roots λ, $\bar{\lambda}$ there will correspond two solutions $c_i^{(1)}$ and $c_i^{(2)}$ of the system (VII.29):

$$\sum_{i=1}^{n} c_i^{(1)}(a_{ik} - \lambda\delta_{ik}) = 0, \qquad k = 1, 2, 3, \cdots, n \qquad \text{(VII.31)}$$

and because the conjugate number of a product is the product of the conjugate factors and since a_{ik} and δ_{ik} are real, we have

$$\sum_{i=1}^{n} \bar{c}_i^{(1)}(a_{ik} - \bar{\lambda}\delta_{ik}) = 0, \qquad k = 1, 2, 3, \cdots, n, \qquad \text{(VII.32)}$$

i.e., $\bar{c}_i^{(1)} = c_i^{(2)}$, or in words: The solution of (VII.29) that corresponds to the root λ is the complex conjugate of the solution that corresponds to the complex conjugate root $\bar{\lambda}$.

We can write (VII.31) and (VII.32) in the form

$$\sum_{i=1}^{n} c_i a_{ik} = \lambda \sum_{i=1}^{n} c_i\delta_{ik},$$

$$\sum_{i=1}^{n} \bar{c}_i a_{ik} = \bar{\lambda} \sum_{i=1}^{n} \bar{c}_i\delta_{ik}.$$

If we multiply the first system by $\bar{c}_k$ and the second system by c_k, sum over k, and take into account the symmetry of A, then by subtraction

$$(\lambda - \bar{\lambda})\sum_{k=1}^{n}\sum_{i=1}^{n} c_i\bar{c}_k\delta_{ik} = 0.$$

If

$$\sum_{i=1}^{n}\sum_{k=1}^{n} c_i\bar{c}_k\delta_{ik} = \sum_{i=1}^{n} c_i\bar{c}_i\delta_{ii} = \sum_{i=1}^{n}(\gamma_i^2 + \delta_i^2) = 0$$

where $c_k = \gamma_k + i\delta_k$, it would appear that $\gamma_k = \delta_k = 0$ for all k, i.e.: that the solution c_i is a trivial one which contradicts the fact that the coefficient determinant of (VII.29) vanishes. Therefore, $\lambda - \bar{\lambda} = 0$, i.e.: $2i\beta = 0$, and this means that λ cannot be complex, q.e.d.

We wish to indicate how this proof can be applied to the general case where b_{ik} is any real nonsingular symmetric matrix. If B is not *singular*, i.e., *det* $|b_{ik}| \neq 0$ then there exists a matrix inverse to B and the general equation can be reduced to the special case which we treated in the above proof. We do not intend to go into this, however, because the implementation of this idea would involve the use of several theorems in matrix theory whose proofs would take us too far afield.[9]

Problems VII.22–24

22. Find the values of λ for which

$$\sum_{k=1}^{2} c_k(a_{ik} - \lambda b_{ik}) = 0, \qquad i = 1, 2$$

where $A = \begin{pmatrix} 1 & 2 \\ 2 & 1 \end{pmatrix}$, $B = \begin{pmatrix} 1 & -1 \\ -1 & 0 \end{pmatrix}$, has a nontrivial solution.

23. Verify theorem VII.4 for

$$a_{ik} = \begin{pmatrix} 1 & 0 & 1 \\ 0 & 2 & 0 \\ 1 & 0 & 3 \end{pmatrix}, \qquad b_{ik} = \delta_{ik}.$$

24. Given the matrix

$$B = \begin{pmatrix} 1 & 0 & 2 \\ 0 & 1 & -1 \\ 2 & -1 & 2 \end{pmatrix}.$$

Find a matrix B^{-1} such that $B \cdot B^{-1} = \begin{pmatrix} 1 & 0 & 0 \\ 0 & 1 & 0 \\ 0 & 0 & 1 \end{pmatrix}$.

[9] See F. E. Hohn: *Elementary Matrix Algebra*, The Macmillan Company, New York, 1958, p. 225.

2. The Method of Rayleigh and Ritz

The fact that the solutions and eigenvalues of a homogeneous self-adjoint boundary value problem as discussed in Chapter V (and to which we referred in the preceding section as problem I) can be found by solving a variational problem as previously referred to as problem II, was first discovered by *Lord Rayleigh* in the form that the potential and kinetic energy in an elastic system are distributed such that the frequencies (eigenvalues) of the components are a minimum. Lord Rayleigh published this discovery in the *Philosophical Transactions of the Royal Society, London*, A, 161, 77 (1870). This principle is now widely referred to as *Rayleigh's principle*. The numerical procedure which we are going to develop in this section was first proposed by *W. Ritz* in his paper "Ueber eine neue Methode zur Loesung gewisser Variationsprobleme der Mathematischen Physik."[10] This is the reason for speaking of the *method of Rayleigh and Ritz* or simply the *method of Ritz*.

Ritz's original idea is the following one:

In order to find an approximation to the minimum of the functional

$$I[y] = \int_{x_1}^{x_2} f(x, y, y')\, dx \qquad \text{(VII.33)}$$

where y is supposed to satisfy the boundary conditions

$$y(x_1) = 0, \quad y(x_2) = 0, \qquad \text{(VII.34)}$$

he restricts the class of competing functions as follows:

Let $f_1(x), f_2(x), f_3(x), \cdots$ be a system of functions each of which satisfies the boundary conditions (VII.34). Then, the function

$$\phi_n = c_1 f_1(x) + c_2 f_2(x) + \cdots + c_n f_n(x) \qquad \text{(VII.35)}$$

where the c_k are arbitrary constants will also satisfy (VII.34) for any n. Now, instead of admitting the class of all functions with a continuous derivative for the minimization of (VII.33), Ritz permits only functions of the type (VII.35) to compete, where the $f_k(x)$ are chosen once and for all. We will call (VII.35) a *Ritz Manifold*. In other words: The c_k have to be chosen such that (VII.33) becomes a minimum compared with all possible values it may assume for functions of the Ritz manifold (VII.35) for a given n. (An analogy in the theory of extreme values of functions of one variable would be the following one: Instead of asking for the value x—where x may range over all real numbers—for which $f(x)$ assumes an

[10] "On a New Method of Solving Certain Variational Problems of Mathematical Physics," *Journal fuer die reine und angewandte Mathematik* (Crelle), Vol. CXXXV, 1908.

extreme value, one asks for the value c that can be written as a linear combination of a finite number of given algebraic irrational numbers with integral coefficients, for example, for which the function $f(x)$ assumes an extreme value with respect to this rather restricted set.)

Clearly, with this restriction an ordinary extreme value problem for a function of n variables, namely

$$I[c_1 f_1 + c_2 f_2 + \cdots + c_n f_n] = I(c_1, c_2, \cdots, c_n)$$
$$= \int_{x_1}^{x_2} f(x, \phi_n, \phi_n')\, dx \rightarrow \text{minimum}$$

presents itself. Thus we have the n necessary conditions

$$\frac{\partial I(c_1, c_2, \cdots, c_n)}{\partial c_k} = 0, \qquad k = 1, 2, 3, \cdots, n. \tag{VII.36}$$

The question which arises now is the following one: Does one obtain a minimizing sequence with a Ritz manifold of the type (VII.35); i.e., if $I[y_0] = \min I[y]$ is the minimum of the functional (provided such a minimum exists) which is obtained for a function y_0 which is continuous and has a continuous derivative and satisfies the boundary conditions (VII.34), then, under what conditions on $\{f_k(x)\}$ does

$$\lim_{n \to \infty} I[\phi_n] = I[y_0]$$

hold? A partial answer in form of a sufficient condition is given in

Theorem VII.6. *Let $f(x, y, y')$ be continuous and differentiable with respect to y and y'; let the derivatives $\partial f/\partial y$ and $\partial f/\partial y'$ be bounded; and let $\{f_k\}$ be a system of given functions which satisfy the boundary conditions (VII.34).*

If it is possible to find for any function y which is continuous, has a continuous derivative, and satisfies the boundary conditions (VII.34), and for any $\epsilon > 0$ an n and values $c_1, c_2, \cdots, c_n$ such that with

$$\phi_n = c_1 f_1 + c_2 f_2 + \cdots + c_n f_n$$

the inequalities

$$|\phi_n - y| < \epsilon, \quad |\phi_n' - y'| < \epsilon$$

hold for all x in $x_1 \leq x \leq x_2$, then

$$\lim_{n \to \infty} I[\phi_n] = I[y_0].$$

(A system $\{f_k(x)\}$ which satisfies the above conditions is called relatively complete.)

Proof. Since the condition of relative completeness is assumed to hold, we have in particular for the solution y_0 of our minimum problem that for any arbitrary small $\epsilon > 0$ there exists an n and values $c_1, c_2, \cdots, c_n$ such that

$$|\phi_n - y_0| < \epsilon, \quad |\phi_n' - y_0'| < \epsilon$$

for all x in $x_1 \leq x \leq x_2$.

Since

$$f(x, \phi_n, \phi_n') - f(x, y_0, y_0') = \frac{\partial f(x, \bar{y}, \bar{y}')}{\partial y}(\phi_n - y_0) + \frac{\partial f(x, \bar{y}, \bar{y}')}{\partial y'}(\phi_n' - y_0'),$$

where $\bar{y}, \bar{y}'$ are some values between ϕ_n, ϕ_n' and y_0, y_0' respectively, we have

$$\left| f(x, \phi_n, \phi_n') - f(x, y_0, y_0') \right| \leq \left(\left| \frac{\partial f}{\partial y} \right| + \left| \frac{\partial f}{\partial y'} \right| \right) \epsilon.$$

Since $\partial f/\partial y$, $\partial f/\partial y'$ are assumed to be bounded, we can find an M such that

$$\left| \frac{\partial f}{\partial y} \right| + \left| \frac{\partial f}{\partial y'} \right| \leq M$$

and we obtain

$$|I[\phi_n] - I[y_0]| \leq \epsilon M(x_2 - x_1) = \epsilon^*,$$

i.e., $\lim_{n \to \infty} I[\phi_n] = I[y_0]$, q.e.d.

The degree of complexity experienced in solving the system (VII.36) of n equations in $c_1, c_2, \cdots, c_n$ depends on the character of the integrand $f(x, y, y')$. However, we do not wish to indulge in such a general and rather fruitless investigation since our interest is directed towards the approximate solution of a very specific problem, namely, the variational problem of the type IIA. Our problem is to find an approximation to

$$\min \frac{I[y]}{C[y]} = \lambda_r \quad \text{where } C[y, y_i] = 0, \quad i = 1, 2, 3, \cdots, r-1 \qquad \text{(VII.37)}$$

and where the y_i satisfy the boundary conditions (VII.34).

We now restate our problem in Ritz' sense as follows:

To find

$$\min \frac{I[\phi_n]}{C[\phi_n]} = \Lambda_r, \quad \text{where } C[\phi_n, \phi_n^{(i)}] = 0, \quad i = 1, 2, 3, \cdots, r-1 \qquad \text{(VII.38)}$$

and where the ϕ_n and the $\phi_n^{(i)}$ are functions of the Ritz manifold (VII.35) for a given n.

Before becoming involved in the proving of theorems, let us experiment a little and see where it will get us:

We take (after having made a choice of the functions $f_1, f_2, f_3, \cdots$) (VII.35) and substitute it for ϕ_n into (VII.38), defining

$$Q(c_1, c_2, \cdots, c_n) = \frac{I[c_1 f_1 + c_2 f_2 + \cdots + c_n f_n]}{C[c_1 f_1 + c_2 f_2 + \cdots + c_n f_n]}.$$

It is necessary for this quotient to become a minimum that

$$\frac{\partial Q}{\partial c_k} = 0, \qquad k = 1, 2, 3, \cdots, n$$

and since

$$\frac{\partial Q}{\partial c_k} = \frac{\dfrac{\partial I}{\partial c_k} C - \dfrac{\partial C}{\partial c_k} I}{C^2} = \frac{1}{C}\left[\frac{\partial I}{\partial c_k} - \frac{I}{C} \cdot \frac{\partial C}{\partial c_k}\right],$$

we have, if we set

$$\left(\frac{I}{C}\right)_{\min} = \Lambda \tag{VII.39}$$

the condition

$$\frac{\partial I}{\partial c_k} - \Lambda \frac{\partial C}{\partial c_k} = 0, \qquad k = 1, 2, 3, \cdots, n. \tag{VII.40}$$

In view of the definitions of I and C in (VII.8) and (VII.10), it follows that

$$\begin{aligned}\frac{\partial I(c_1, c_2, \cdots, c_n)}{\partial c_k} &= \frac{\partial}{\partial c_k} \int_{x_1}^{x_2} [P(c_1 f_1' + \cdots + c_n f_n')^2 \\ &\qquad - Q(c_1 f_1 + \cdots + c_n f_n)^2]\, dx \\ &= 2\sum_{i=1}^{n} c_i \int_{x_1}^{x_2} (P f_i' f_k' - Q f_i f_k)\, dx\end{aligned}$$

and

$$\begin{aligned}\frac{\partial C(c_1, c_2, \cdots, c_n)}{\partial c_k} &= \frac{\partial}{\partial c_k} \int_{x_1}^{x_2} R(c_1 f_1 + \cdots + c_n f_n)^2\, dx \\ &= 2\sum_{i=1}^{n} c_i \int_{x_1}^{x_2} R f_i f_k\, dx.\end{aligned}$$

The functions f_k and their derivatives f_k' are known; therefore we can evaluate all the definite integrals which are involved in the above expressions. We will denote their values by

$$a_{ik} = \int_{x_1}^{x_2} (P f_i' f_k' - Q f_i f_k)\, dx, \tag{VII.41}$$

$$b_{ik} = \int_{x_1}^{x_2} R f_i f_k\, dx \tag{VII.42}$$

and it is clear that $a_{ik} = a_{ki}$ and $b_{ik} = b_{ki}$.

Thus, (VII.40) becomes

$$\sum_{i=1}^{n} c_i(a_{ik} - \Lambda b_{ik}) = 0, \qquad k = 1, 2, 3, \cdots, n \tag{VII.43}$$

and it is our job to find a nontrivial solution $C = (c_1, c_2, \cdots, c_n)$ of this system of n linear homogeneous equations.

Such a solution will exist whenever Λ satisfies the secular equation

$$det\,|a_{ik} - \Lambda b_{ik}| = 0, \tag{VII.44}$$

which has, according to theorem VII.5, n real roots; and if Λ_l and Λ_m are two distinct roots of (VII.44), then the corresponding solutions $c_i^{(l)}$ and $c_i^{(m)}$ are in view of theorem VII.4 such that

$$\sum_{k=1}^{n} \sum_{i=1}^{n} c_i^{(l)} c_k^{(m)} b_{ik} = 0. \tag{VII.45}$$

We will assume that all the roots of (VII.44) are of multiplicity 1 and arrange them according to increasing values

$$\Lambda_1 < \Lambda_2 < \Lambda_3 < \Lambda_4 < \cdots < \Lambda_n. \tag{VII.46}$$

Now we are ready to prove a theorem regarding the significance of these n roots Λ_k, namely:

Theorem VII.7. *The values $\Lambda_1 < \Lambda_2 < \Lambda_3 < \cdots < \Lambda_n$ are the eigenvalues of the variational problem*

$$\min \frac{I[\phi_n^{(k)}]}{C[\phi_n^{(k)}]} = \Lambda_k$$

where $C[\phi_n^{(k)}, \phi_n^{(s)}] = 0$ for $s = 1, 2, 3, \cdots, (k-1)$ and $k = 1, 2, 3, \cdots$.

Proof. First, let us show that

$$\Lambda_1 = \min \frac{I[\phi_n]}{C[\phi_n]}$$

where $\phi_n = c_1 f_1 + c_2 f_2 + \cdots + c_n f_n$ and the f_i are given functions. Since

$$I[\phi_n] = \sum_{i=1}^{n} \sum_{k=1}^{n} a_{ik} c_i c_k,$$

$$C[\phi_n] = \sum_{i=1}^{n} \sum_{k=1}^{n} b_{ik} c_i c_k,$$

our problem is equivalent to the problem of finding the minimum of $I[\phi_n]$:

$$I[\phi_n] = \sum_{i=1}^{n} \sum_{k=1}^{n} a_{ik} c_i c_k \to \min$$

under the constraint

$$C[\phi_n] = \sum_{i=1}^{n} \sum_{k=1}^{n} b_{ik} c_i c_k = 1.$$

(Incidentally, if $P(x) > 0$, $R(x) > 0$, and $Q(x) < 0$ in $x_1 \leq x \leq x_2$, then, $I[y] > 0$, $C[y] > 0$ if $y \neq 0$, and consequently,

$$\sum_{i=1}^{n} \sum_{k=1}^{n} a_{ik} c_i c_k > 0, \qquad \sum_{i=1}^{n} \sum_{k=1}^{n} b_{ik} c_i c_k > 0$$

for all $(c_1, c_2, \cdots, c_n) \neq (0, 0, \cdots, 0)$. A *quadratic form* with this property is called *positive definite*.)

There exists at least one minimum of the quadratic form $\sum_{i=1}^{n} \sum_{k=1}^{n} a_{ik} c_i c_k$ under the constraint $\sum_{i=1}^{n} \sum_{k=1}^{n} b_{ik} c_i c_k = 1$ according to a theorem of Weierstrass[11] and it is necessary that for this minimum

$$\frac{\partial I}{\partial c_i} - \Lambda \frac{\partial C}{\partial c_i} = 0 \quad \text{for all } i = 1, 2, 3, \cdots, n.$$

This relation can be satisfied as we have seen for $\Lambda_1 < \Lambda_2 < \Lambda_3 < \cdots < \Lambda_n$. Since Λ_1 is the smallest of these values, Λ_1 is the minimum and $\phi_n^{(1)} = c_1^{(1)} f_1 + c_2^{(1)} f_2 + \cdots + c_n^{(1)} f_n$ the corresponding solution.

Now we proceed to prove that Λ_2 is the minimum which is attained under the constraint

$$C[\phi_n^{(1)}, \phi_n] = 0.$$

This is equivalent to

$$\sum_{k=1}^{n} \sum_{i=1}^{n} b_{ik} c_i^{(1)} c_k = 0,$$

which is a linear homogeneous equation for the n unknowns $c_1, c_2, \cdots, c_n$:

$$A_1 c_1 + A_2 c_2 + \cdots + A_n c_n = 0.$$

Since we desire to obtain nontrivial solutions, at least one of the c_i can be assumed to be different from zero. Let $c_1 = 1$. Then we obtain

$$A_2 c_2 + A_3 c_3 + \cdots + A_n c_n = -A_1,$$

a linear nonhomogeneous equation for $c_2, c_3, \cdots, c_n$ which has $n - 1$ linearly independent solutions $c_2^{(2)}, c_3^{(2)}, \cdots, c_n^{(2)}$; $c_2^{(3)}, c_3^{(3)}, \cdots, c_n^{(3)}$;

[11] A continuous function $\left(\sum_{i=1}^{n} \sum_{k=1}^{n} a_{ik} c_i c_k\right)$ in several variables (c_i) in a finite closed domain $\left(\sum_{i=1}^{n} \sum_{k=1}^{n} b_{ik} c_i c_k = 1\right)$ assumes in this domain its minimum. See T. M. Apostol: *Mathematical Analysis*, Addison Wesley, Reading, Mass., 1957, p. 73.

$\cdots c_2^{(n)}, c_3^{(n)}, \cdots, c_n^{(n)}$. Thus, there are $n - 1$ functions $\phi_n^{(2)}, \phi_n^{(3)}, \cdots, \phi_n^{(n)}$ for which

$$C[\phi_n^{(1)}, \phi_n^{(k)}] = 0, \qquad k = 2, 3, \cdots, n.$$

But we have already found all $n - 1$ functions which are orthogonal to $\phi_n^{(1)}$ and are of the type ϕ_n, namely the functions which yield the values $\Lambda_2 < \Lambda_3 < \cdots < \Lambda_n$ for the quotient $\dfrac{I[\phi_n]}{C[\phi_n]}$ and for any two of which (VII.45) holds. Λ_2 is the smallest of these values; hence

$$\Lambda_2 = \min \frac{I[\phi_n]}{C[\phi_n]}$$

where $C[\phi_2^{(1)}, \phi_n] = 0$.

One can proceed analogously in order to show that Λ_3 is the minimum of $\dfrac{I[\phi_n]}{C[\phi_n]}$ under the constraints $C[\phi_n^{(i)}, \phi_n] = 0$, $i = 1, 2$ (see problem VII.25), etc.

Problem VII.25

25. Show that Λ_3 is the third eigenvalue of the variational problem stated in theorem VII.7.

3. The Eigenvalues of Ritz's Problem as Upper Bounds for the Eigenvalues of the Original Problem

We can now proceed to prove a theorem which will establish a relation between the eigenvalues of the original variational problem, IIA, and Ritz's problem (VII.38).

Theorem VII.8. *Let $\lambda_1 \leq \lambda_2 \leq \lambda_3 \leq \cdots \leq \lambda_n \leq \cdots$ be the eigenvalues of problem IIA*

$$\min \frac{I[y]}{C[y]} = \lambda_k, \qquad C[y, y_s] = 0, \qquad s = 1, 2, 3, \cdots, (k-1),$$

where y and the y_s satisfy the boundary conditions (VII.34), and let $\Lambda_1 < \Lambda_2 < \Lambda_3 < \cdots < \Lambda_n$ be the eigenvalues of Ritz's problem (VII.38)

$$\min \frac{I[\phi_n]}{C[\phi_n]} = \Lambda_k, \qquad C[\phi_n, \phi_n^{(s)}] = 0, \qquad s = 1, 2, 3, \cdots, (k-1),$$

where ϕ_n and the $\phi_n^{(s)}$ satisfy the same boundary conditions (VII.34) and are of the Ritz manifold (VII.35). Then, the relation

$$\lambda_k \leq \Lambda_k \quad \text{for } k = 1, 2, 3, \cdots, n \quad \textit{holds.}$$

Remark: This theorem establishes the fact that Ritz's eigenvalues are upper bounds for the eigenvalues of the original problem.

Proof. We have

$$\lambda_k = \min \frac{I[y]}{C[y]}, \qquad C[y, y_s] = 0, \quad s = 1, 2, 3, \cdots, (k-1),$$

and

$$\Lambda_k = \min \frac{I[\phi_n]}{C[\phi_n]}, \qquad C[\phi_n, \phi_n^{(s)}] = 0, \quad s = 1, 2, 3, \cdots, (k-1).$$

Clearly, we cannot immediately compare the values λ_k with the values Λ_k in view of the difference in the orthogonal bases $[y_1, y_2, \cdots, y_{k-1}]$ and $[\phi_n^{(1)}, \phi_n^{(2)}, \cdots, \phi_n^{(k-1)}]$ respectively, except for the case where $k = 1$.

For $k = 1$:

$$\lambda_1 = \min \frac{I[y]}{C[y]},$$

and

$$\Lambda_1 = \min \frac{I[\phi_n]}{C[\phi_n]},$$

whereby $\{\phi_n\} \subset \{y\}$. Hence, we have according to lemma VII.1

$$\lambda_1 \leq \Lambda_1.$$

Let us now proceed to the case $k = 2$. We introduce the auxiliary quantity

$$L_2 = \min \frac{I[\phi_n]}{C[\phi_n]}, \qquad C[\phi_n, y_1] = 0,$$

i.e., L_2 denotes the minimum which is attained for a function ϕ_n of the Ritz manifold (VII.35) which is orthogonal to y_1 (the first eigenfunction that belongs to λ_1). (See also problems VII.30 and 31.)

Clearly, since $\{\phi_n\} \subset \{y\}$, we have in view of lemma VII.1

$$L_2 \geq \lambda_2.$$

It remains to be shown that

$$L_2 \leq \Lambda_2.$$

Let us consider

$$\min \frac{I[\phi_n]}{C[\phi_n]},$$

where $C[\phi_n, z] = 0$ and z is any function. For $z = \phi_n^{(1)}$ we obtain Λ_2 and for $z = y_1$ we obtain L_2. Since

$$\Lambda_2 = \max_{[z]} \min_{(z)} \frac{I[\phi_n]}{C[\phi_n]}$$

according to Courant's maximum-minimum principle (lemma VII.3), we have

$$\Lambda_2 \geq L_2.$$

The proof for the higher eigenvalues is analogous.

Problems VII.26–31

26. Given $I[y] = \int_0^1 y'^2\,dx$, $y(0) = 0$, $y(1) = 0$. Substitute for y the function

$$\phi_3 = c_1 \sin \pi x + c_2 \sin 2\pi x + c_3 \sin 3\pi x$$

and state the necessary condition for $I[\phi_3] \to$ minimum.

27. Given $I[y] = \int_0^\pi y'^2\,dx$, $C[y] = \int_0^\pi y^2\,dx$, $y(0) = 0$, $y(\pi) = 0$. Let $\phi_2 = c_1 x(x - \pi) + c_2 x^2(x - \pi)$. Find c_1, c_2 such that

$$\frac{I[\phi_2]}{C[\phi_2]} \to \text{minimum}.$$

28. Given $I[y] = \int_{-1}^1 (1 - x^2) y'^2\,dx$, $C[y] = \int_{-1}^1 y^2\,dx$. Find c_1, c_2, c_3 such that with

$$\phi_3 = c_1 + c_2 x + c_3 x^2, \qquad \frac{I[\phi_3]}{C[\phi_3]} \to \text{minimum}.$$

***29.** Prove that $\Lambda_3 \geq \lambda_3$, where λ_3 is the third eigenvalue of the original problem and Λ_3 is the third eigenvalue of Ritz's problem (see proof of theorem VII.8).

30. Prove that there are $n - 1$ functions $\phi_n^{(s)}$ $(s = 1, 2, 3, \cdots, n - 1)$ of the Ritz manifold (VII.35) such that

$$C[\phi_n^{(s)}, y_1] = 0$$

where y_1 is the first solution of problem II.

31. Prove that there are $n - r$ functions $\phi_n^{(s)}$ $(s = 1, 2, \cdots, n - r)$ of the Ritz manifold (VII.35) such that

$$C[\phi_n^{(s)}, y_k] = 0, \qquad k = 1, 2, 3, \cdots, r,$$

where the y_k are the first r solutions of problem II.

4. Successive Approximations to the Eigenvalues from Above

As in the preceding sections, we consider the variational problem

$$\min \frac{I[y]}{C[y]} = \lambda_k, \qquad \text{where } C[y, y_s] = 0, \quad s = 1, 2, 3, \cdots, k - 1.$$

and where y and y_s satisfy (VII.34).

According to Ritz, we choose n suitable functions

$$f_1(x), f_2(x), \cdots, f_n(x)$$

each of which satisfies (VII.34) and consider the linear combination

$$\phi_n = c_1 f_1 + c_2 f_2 + \cdots + c_n f_n.$$

As we have seen from the developments in the preceding sections, it seems to be of definite advantage to choose the f_k such that

$$C[f_i, f_k] = 0 \quad \text{for } i \neq k,$$

i.e., to choose for $\{f_k\}$ an orthogonal system on $x_1 \leq x \leq x_2$, because such a choice will make the b_{ik} for $i \neq k$ drop out and simplify the secular equation considerably.

Now, let $n = 1$ and we will obtain by solving one linear equation in Λ an upper bound $\Lambda_1^{(1)}$ for the smallest eigenvalue λ_1 according to theorem VII.8. If we go one step further, i.e., take $n = 2$, we obtain by solving a quadratic equation for Λ an upper bound $\Lambda_1^{(2)}$ for λ_1 and an upper bound $\Lambda_2^{(2)}$ for λ_2, and it is evident that $\Lambda_1^{(2)} \leq \Lambda_1^{(1)}$ because the range of competing functions for $\Lambda_1^{(2)}$ is larger than the range of competing functions for $\Lambda_1^{(1)}$ (lemma VII.1). Clearly, by continuation of this process we secure successive approximations to an increasing number of eigenvalues as indicated in the following table:

Eigenvalue	$n = 1$	$n = 2$	$n = 3$	$n = 4$	$\cdots$	$n = k$	$\cdots$	Original Problem
1st	$\Lambda_1^{(1)} \geq$	$\Lambda_1^{(2)} \geq$	$\Lambda_1^{(3)} \geq$	$\Lambda_1^{(4)}$	$\cdots$	$\geq \Lambda_1^{(k)} \geq$	$\cdots$	$\geq \lambda_1$
2nd		$\Lambda_2^{(2)} \geq$	$\Lambda_2^{(3)} \geq$	$\Lambda_2^{(4)}$	$\cdots$	$\geq \Lambda_2^{(k)} \geq$	$\cdots$	$\geq \lambda_2$
3rd			$\Lambda_3^{(3)} \geq$	$\Lambda_3^{(4)}$	$\cdots$	$\geq \Lambda_3^{(k)} \geq$	$\cdots$	$\geq \lambda_3$
.								
.								
.								
kth						$\Lambda_k^{(k)} \geq$	$\cdots$	$\geq \lambda_k$
.								.
.								.
.								.

It appears from the above table that the sequence of upper bounds $\Lambda_k^{(s)}$, being a monotonically decreasing sequence and being bounded below, is convergent. Whether it actually converges to the kth eigenvalue λ_k is not evident from our discussion. It is clear that the convergence of the upper bounds to the eigenvalues (provided minima exist) will largely depend on the choice of the functions $f_1, f_2, \cdots, f_n, \cdots$.[12]

A sufficient condition for the convergence of the upper bounds to the eigenvalues can easily be obtained from a generalization of theorem VII.6, namely:

Theorem VII.9. *Let the functions* $f(x, y, y')$ *and* $g(x, y, y')$ *be continuous and differentiable with respect to* y, y', *let* $\partial f/\partial y, \partial f/\partial y', \partial g/\partial y, \partial g/\partial y'$ *be bounded, and let* $g > 0$ *in* $x_1 \leq x \leq x_2$. *Let* $\{f_k\}$ *be a given system of functions, all of which satisfy the boundary conditions (VII.34). If it is possible to find for any smooth function* y *which satisfies the boundary conditions (VII.34) and for any* $\epsilon > 0$ *an* n *and values* $c_1, c_2, \cdots, c_n$ *such that with*

$$\phi_n = c_1 f_1 + c_2 f_2 + \cdots + c_n f_n$$

$$|\phi_n - y| < \epsilon, \quad |\phi'_n - y'| < \epsilon$$

holds for all x *in* $x_1 \leq x \leq x_2$, *then*

$$\lim_{n\to\infty} \frac{I[\phi_n]}{C[\phi_n]} = \min \frac{I[y]}{C[y]}$$

where $I[y] = \int_{x_1}^{x_2} f(x, y, y')\, dx, \; C[y] = \int_{x_1}^{x_2} g(x, y, y')\, dx.$

The *proof* is left to the reader (see problems VII.32 and 33).

Problems VII.32–35

32. Prove theorem VII.9. (*Hint:* See proof of theorem VII.6.)

33. Interpret the conditions on f, g and their derivatives in theorem VII.9 for the case where $f = Py'^2 - Qy^2, g = Ry^2$.

34. Given $f_1 = cx$. Determine c such that $\int_0^1 f_1^2\, dx = 1$ and find two polynomials f_2, f_3 of the second and third order respectively such that $\int_0^1 f_i f_k\, dx = \delta_{ik}$.

***35.** Find approximations to the first two eigenvalues of the problem

$$y'' + \lambda y = 0,$$
$$y(0) = 0, \quad y(1) + y'(1) = 0$$

by using Ritz' method. (*Hint:* Use $f_1 = 2x^2 - 3x, f_2 = x^3 - 2x$.)

[12] See E. Trefftz: "Konvergenz und Fehlerabschaetzung beim Ritz'schen Verfahren," *Mathem. Annalen*, Vol. 100, p. 503.

5. Numerical Examples

Let us again consider the problem

$$\min \frac{\int_0^1 y'^2\, dx}{\int_0^1 y^2\, dx} = \lambda^2$$

where $y(0) = y(1) = 0$.

As a basis for our approximation of the eigenvalues, we will choose the two functions

$$f_1 = x(1 - x)$$
$$f_2 = x(1 - x)(1 + ax).$$

Both of these functions apparently satisfy the boundary conditions.

The constant a is determined such that

$$b_{12} = \int_0^1 f_1 f_2\, dx = 0.$$

An elementary computation will show that $a = -2$ does the job. The reader can easily verify the following evaluations of the a_{ik} and the b_{ii}:

$$a_{11} = \int_0^1 f_1'^2\, dx = \tfrac{1}{3},$$

$$a_{22} = \int_0^1 f_2'^2\, dx = \tfrac{1}{5},$$

$$a_{12} = \int_0^1 f_1' f_2'\, dx = 0,$$

$$b_{11} = \int_0^1 f_1^2\, dx = \tfrac{1}{30},$$

$$b_{22} = \int_0^1 f_2^2\, dx = \tfrac{1}{210}.$$

Thus, we obtain for Λ^2 the following equation:

$$\begin{vmatrix} \dfrac{1}{3} - \dfrac{\Lambda^2}{30} & 0 \\ 0 & \dfrac{1}{5} - \dfrac{\Lambda^2}{210} \end{vmatrix} = 0,$$

which has the roots

$$\Lambda_1^2 = 10, \quad \Lambda_2^2 = 42.$$

These roots turn out to be rather close approximations for the first two eigenvalues of our problem, namely

$$\lambda_1^2 = \pi^2 = 9.8696\cdots,$$
$$\lambda_2^2 = 4\pi^2 = 39.477\cdots.$$

In the following example, we attempt to find approximations to the first two zeros of the Bessel function of zeroth order and first kind.

We know that $J_0(\lambda x)$ is a solution of

$$\min \frac{\int_0^1 xy'^2\,dx}{\int_0^1 xy^2\,dx} = \lambda^2$$

with $y(1) = 0$, $|y(0)| < \infty$.

As basis for the application of Ritz's method we choose the two even polynomials

$$f_1 = 1 - x^2,$$
$$f_2 = (1 - x^2)(x^2 + a),$$

both of which satisfy the boundary conditions. Again, we determine a such that

$$\int_0^1 xf_1f_2\,dx = 0;$$

here $a = -\frac{1}{4}$.

Thus, we have

$$f_1' = -2x,$$
$$f_2' = \tfrac{1}{2}(5x - 8x^3)$$

and we compute the a_{ik} and the b_{ii}:

$$a_{11} = 1,\quad a_{12} = \tfrac{1}{12},\quad a_{22} = \tfrac{11}{48},\quad b_{11} = \tfrac{1}{6},\quad b_{22} = \tfrac{1}{160}.$$

Hence, the upper bounds Λ^2 of the first two eigenvalues, i.e., the squares of the first two zeros of J_0 appear as solutions of the quadratic equation

$$\begin{vmatrix} 1 - \dfrac{\Lambda^2}{6} & \dfrac{1}{12} \\[2ex] \dfrac{1}{12} & \dfrac{11}{48} - \dfrac{\Lambda^2}{160} \end{vmatrix} = \frac{1}{12\cdot 480}\begin{vmatrix} 12 - 2\Lambda^2 & 1 \\ 40 & 110 - 3\Lambda^2 \end{vmatrix} = 0,$$

or in simplified form

$$3\Lambda^4 - 128\Lambda^2 + 640 = 0.$$

The roots turn out to be

$$\Lambda_1^2 = 5.86,\quad \Lambda_2^2 = 36.86.$$

If we extract the square roots, we see that

$$\Lambda_1 = 2.42, \quad \Lambda_2 = 6.07,$$

while the exact values of the first two zeros of J_0 are given as

$$\lambda_1 = 2.40483 \cdots,$$
$$\lambda_2 = 5.52008 \cdots.$$

The error of the approximation to λ_1 is roughly 0.8 per cent while the error of the approximation to λ_2 is already 10 per cent.[13]

Problems VII.36–39

36. Find approximations to the first two eigenvalues ($\pi^2/4$, $9\pi^2/4$) of

$$y'' + \lambda y = 0,$$
$$y(-1) = y(1) = 0$$

by the method of Rayleigh and Ritz. Use the approximation $y = c_1(1 - x^2) + c_2(x^2 - x^4)$.

37. Same as in problem 36 for

$$y'' + \lambda xy = 0,$$
$$y(0) = y(1) = 0$$

using the approximation $y = c_1(x - x^2) + c_2(x^2 - x^3)$.

***38.** Show that $y = \sqrt{x}J_{1/3}(\frac{2}{3}\sqrt{\lambda}x^{3/2})$ is the solution of problem 37 if λ is such that $J_{1/3}(\frac{2}{3}\sqrt{\lambda}) = 0$.

39. Use tables for Bessel functions[14] and find the two smallest values of λ for which $J_{1/3}(\frac{2}{3}\sqrt{\lambda}) = 0$ and compare these values with your result in problem 37

RECOMMENDED SUPPLEMENTARY READING

R. Courant and D. Hilbert: *Methods of Mathematical Physics*, Vol. I, Interscience Publishers, New York, 1953.

R. Courant and D. Hilbert: *Methoden der Mathematischen Physik*, Vol. II, Springer Verlag, Berlin, 1937.

[13] For more numerical examples we refer the reader to:

Z. Kopal: *Numerical Analysis*, John Wiley and Sons, New York, 1955, p. 309.

L. Collatz: *Numerische Behandlung von Differentialgleichungen*, Springer Verlag, Berlin, 1951, pp. 132 and 325.

L. Collatz: *Eigenwertproblem und ihre numerische Behandlung*, Academische Verlags Gesellschaft, Leipzig, 1945, pp. 233, 249, 251.

Z. V. Kantorovich and V. I. Krylov: *Approximate Methods of Higher Analysis*, Interscience Publishers, New York, 1958, Chapter IV.

[14] Jahnke and Emde: *Tables of Functions with Formulae and Curves*, Dover Publications, 1943, p. 167.

S. H. Gould: *Variational Methods for Eigenvalue Problems*, University of Toronto Press, 1957.

L. Collatz: *Eigenwertprobleme und ihre numerische Behandlung*, Akademische Verlags Gesellschaft, Leipzig, 1945.

R. Courant: *Dirichlet's Principle*, Interscience Publishers, New York, 1950.

L. V. Kantorovich and V. I. Krylov: *Approximate Methods of Higher Analysis*, translated from the Russian by Curtis D. Benster, Interscience Publishers, New York, 1958.

VIII

SPHERICAL HARMONICS

§1. ASSOCIATED LEGENDRE FUNCTIONS, SPHERICAL HARMONICS, AND LAGUERRE POLYNOMIALS

1. The Equation of Wave Propagation

We have seen in Chapter II, §2.1, that the displacement $u(x, y, t)$ of a stretched membrane under tensile forces satisfies equation (II.33)

$$\alpha^2\left(\frac{\partial^2 u}{\partial x^2} + \frac{\partial^2 u}{\partial y^2}\right) = \frac{\partial^2 u}{\partial t^2},$$

Exactly the same type of equation (from a mathematical viewpoint) governs the propagation of a two-dimensional wave and in general, the equation

$$\boxed{\alpha^2\left(\frac{\partial^2 u}{\partial x^2} + \frac{\partial^2 u}{\partial y^2} + \frac{\partial^2 u}{\partial z^2}\right) = \frac{\partial^2 u}{\partial t^2}} \tag{VIII.1}$$

describes the propagation of a wave in a three-dimensional space, where $u(x, y, z, t)$ stands for the velocity potential and α for the propagation velocity. (See also Chapter II, §3.2.)

We will now show that equation (VIII.1) is valid for *compressional* waves in liquids or gases under certain simplifying restrictions.

We will assume that the velocity field $\mathbf{v}(x, y, z, t)$ which is created in our medium by some disturbance that causes a wave, is *irrotational*, i.e.,

$$\text{curl } \mathbf{v} = 0,$$

and this implies, of course, that there exists a potential function $u(x, y, z, t)$ (the *velocity potential*) such that

$$\mathbf{v} = -\text{grad } u \tag{VIII.2}$$

(see **AI.B7**).

In addition, we will assume that there are neither sources nor sinks, i.e., the amount $f(x, y, z, t)$ of matter which is created or lost per unit volume is zero, and consequently the continuity equation of fluid dynamics (see (II.89)).

$$\frac{\partial \rho}{\partial t} + \operatorname{div}(\rho \mathbf{v}) = f(x, y, z, t)$$

where ρ stands for the density of the medium, reduces to

$$\frac{\partial \rho}{\partial t} + \operatorname{div}(\rho \mathbf{v}) = 0. \tag{VIII.3}$$

Let $P(x, y, z, t)$ stand for the pressure at the space-time point (x, y, z, t). Then, a disturbance at some point will cause a change of the velocity field $\mathbf{v}(x, y, z, t)$ which can be described by Newton's equations of motion (see Chapter I, §1.1) if we consider the medium as consisting of moving particles:

$$\rho \frac{\partial \mathbf{v}}{\partial t} = -\operatorname{grad} P. \tag{VIII.4}$$

In view of (VIII.2), (VIII.4) becomes

$$\rho \frac{\partial}{dt} \operatorname{grad} u = \operatorname{grad} P.$$

Let us dot-multiply this equation by $\mathbf{v}(x, y, z, t)$ and interchange the operations $\partial/\partial t$ and grad. Then, we obtain

$$\rho \operatorname{grad} \frac{\partial u}{\partial t} \cdot \mathbf{v} = \operatorname{grad} P \cdot \mathbf{v}$$

and because of the identity

$$\operatorname{grad} F \cdot \mathbf{v} = \frac{\partial F}{\partial x} \cdot \frac{dx}{dt} + \frac{\partial F}{\partial y} \cdot \frac{dy}{dt} + \frac{\partial F}{\partial z} \frac{dz}{dt} = \frac{dF}{dt}$$

we have

$$\rho \frac{d}{dt}\left(\frac{\partial u}{\partial t}\right) = \frac{dP}{dt}$$

where the total derivatives with respect to the time t have to be understood as derivatives with respect to t, wherever it occurs in x, y, z, while $\partial/\partial t$ denotes the derivative with respect to the explicitly occuring t. If we divide the latter equation by ρ and integrate on both sides, we obtain

$$\int d\left(\frac{\partial u}{\partial t}\right) = \int \frac{dP}{\rho}.$$

Carrying out the integration on the left side and applying the mean value theorem with respect to ρ on the right side yields, with an adequate choice of the integration constant,

$$\frac{\partial u}{\partial t} = \frac{1}{\bar{\rho}} \int dP \tag{VIII.5}$$

where $\bar{\rho}$ stands for the *mean density* and $\int dP$ clearly represents the total change or *variation* of the pressure P which is customarily denoted by δP and called the *excess pressure.*

Hence

$$\frac{\partial u}{\partial t} = \frac{\delta P}{\bar{\rho}}.$$

If we likewise denote the *excess density* by $\delta \rho$,

$$\rho = \bar{\rho} + \delta \rho \tag{VIII.6}$$

and assume the excess pressure to be proportional to the excess density

$$\delta P = \alpha^2 \, \delta \rho$$

and introduce the concept of *condensation* s as defined by the ratio of excess density to mean density

$$s = \frac{\delta \rho}{\bar{\rho}}, \tag{VIII.7}$$

then, we can write (VIII.5) in the form

$$\frac{\partial u}{\partial t} = \alpha^2 s. \tag{VIII.8}$$

It follows from (VIII.6) and (VIII.7) that

$$\rho = \bar{\rho} + \delta \rho = \bar{\rho}\left(1 + \frac{\delta \rho}{\bar{\rho}}\right) = \bar{\rho}(1 + s)$$

and the continuity equation (VIII.3) will appear in the form

$$\bar{\rho} \frac{\partial s}{\partial t} + \operatorname{div} [\bar{\rho}(1 + s)\mathbf{v}] = 0.$$

In view of the linearity of the div-operator and after division by $\bar{\rho}$ we obtain

$$\frac{\partial s}{\partial t} + (1 + s) \operatorname{div} \mathbf{v} + \mathbf{v} \cdot \operatorname{grad} s = 0.$$

(Note that grad $(1 + s) =$ grad s.)

Since

$$\mathbf{v} \cdot \operatorname{grad} s = \frac{\partial s}{\partial x} \cdot \frac{dx}{dt} + \frac{\partial s}{\partial y} \cdot \frac{dy}{dt} + \frac{\partial s}{\partial z} \cdot \frac{dz}{dt} = \frac{ds}{dt},$$

where the total derivative with respect to the time has the same significance as before, we obtain

$$\frac{\partial s}{\partial t} + (1 + s)\operatorname{div}\mathbf{v} + \frac{ds}{dt} = 0.$$

If we assume that $\delta\rho \ll \bar{\rho}$[1] and accordingly $1 + s \cong 1$ (where $\ll$ means "much smaller") and if we assume further that the change of s in space is negligible, i.e., $ds/dt \cong 0$, then, we obtain finally

$$\frac{\partial s}{\partial t} + \operatorname{div}\mathbf{v} = 0$$

and in view of (VIII.8) and (VIII.2)

$$\frac{\partial^2 u}{\partial t^2} = \alpha^2 \Delta u$$

and this is equation (VIII.1).

In order to introduce and explain a few technical terms which we will need in the following subsection, let us assume (for reasons of simplicity) a wave that propagates in one direction, the x-direction. Then, equation (VIII.1) reduces to

$$\frac{\partial^2 u}{\partial t^2} = \alpha^2 \frac{\partial^2 u}{\partial x^2}.$$

Clearly,

$$u = A\cos\frac{2\pi}{\lambda}(x - \alpha t - \delta)$$

is a solution of this equation. It represents a wave of *wave length* λ that propagates with velocity α in the direction of the positive x-axis. (Compare Chapter IV, §3.2.)

But really, what is the propagation velocity of a wave? When it is constant as we assume it to be in this case, then it is the ratio of the distance through which the wave travels to the time required to cover this distance. The distance from one point of the wave to the next following one which is in the same phase is the wave length λ. The time required to go through a complete vibration is the period T. Hence, we have

$$\alpha = \frac{\lambda}{T} = \lambda \cdot \nu \qquad \text{(VIII.9)}$$

where $\nu = 1/T$ is called the *frequency* of the wave. Hence, we can write

$$u = A\cos 2\pi\left(\frac{x}{\lambda} - \nu t - \delta\right) \qquad \text{(VIII.10)}$$

[1] This seems to be justified by experiments.

(where we write again δ for $\lambda\delta$), which is a representation of a wave in which frequency ν and wave length λ appear explicitly.

Problems VIII.1–3

1. Show that

$$u = A \cos \frac{2\pi}{\lambda} (x - \alpha t - \delta)$$

is a solution of the wave equation (VIII.1).

2. Prove that

$$\operatorname{curl} f(\sqrt{x^2 + y^2 + z^2})(x\mathbf{i} + y\mathbf{j} + z\mathbf{k}) = 0.$$

3. Given

$$\mathbf{v} = 2xy\mathbf{i} + (x^2 + z^2)\mathbf{j} + 2yz\mathbf{k}.$$

(a) Show that $\mathbf{v}$ is irrotational.
(b) Find the velocity potential u.

2. Schroedinger's Wave Equation

We wish to point out that the derivation of the wave equation (VIII.1) in the preceding section was guided by two principles which are apparently essentially different, one of which involves the continuity equation which is based upon the concept of a wave as propagated in a continuous medium and does not seem to be consistent with the other one, concerned with Newton's law of motion which presumes a medium consisting of moving particles. However, all that appears to count in physics is the success which a theory enjoys when applied to reality and insofar as this is a criterion there is nothing to worry about.

The situation becomes outright weird, however, if one reverses the process, i.e., if one, instead of attributing corpuscular properties to a wave motion, attributes wave properties to moving corpuscles, as atoms, electrons, and other microscopical manifestations of matter. Nevertheless, this was done by *Louis de Broglie* (in 1924) and thus the basis for *quantum mechanics* was established.

de Broglie postulates that every corpuscular motion can be treated as a wave motion whereby the wave length λ is related to the electronic mass m and the velocity α of the particle concerned as follows:

$$\lambda = \frac{h}{m\alpha} \qquad \text{(VIII.11)}$$

where h stands for *Planck's universal constant* which has the value

$$h = 6.625 \cdot 10^{-27} \text{ erg sec.}$$

The total energy E of the particle is related to the wave frequency ν by

$$E = h\nu. \qquad \text{(VIII.12)}$$

We can now easily set up the energy equation. The kinetic energy is given by

$$\frac{m\alpha^2}{2}.$$

If we denote the potential energy by V we have

$$E = \frac{m\alpha^2}{2} + V$$

or

$$h\nu = \frac{h^2}{2m\lambda^2} + V \tag{VIII.13}$$

according to (VIII.11) and (VIII.12).

Let us now consider a wave function

$$\Psi = A \cos 2\pi\left(\frac{x}{\lambda} - \nu t - \delta\right)$$

and try to find out under what conditions Ψ and its derivatives with respect to t and x satisfy (VIII.13).

We obtain

$$\frac{\partial \Psi}{\partial t} = 2\pi\nu A \sin 2\pi\left(\frac{x}{\lambda} - \nu t - \delta\right),$$

$$\frac{\partial^2 \Psi}{\partial t^2} = -4\pi^2\nu^2 A \cos 2\pi\left(\frac{x}{\lambda} - \nu t - \delta\right),$$

$$\frac{\partial \Psi}{\partial x} = -\frac{2\pi}{\lambda} A \sin 2\pi\left(\frac{x}{\lambda} - \nu t - \delta\right),$$

$$\frac{\partial^2 \Psi}{\partial x^2} = -\frac{4\pi^2}{\lambda^2} A \cos 2\pi\left(\frac{x}{\lambda} - \nu t - \delta\right),$$

or as we can write

$$\frac{\partial^2 \Psi}{\partial t^2} = -4\pi^2\nu^2\Psi,$$

$$\frac{\partial^2 \Psi}{\partial x^2} = -\frac{4\pi^2}{\lambda^2}\Psi.$$

Hence

$$h\nu\Psi = -\frac{h}{4\pi^2\nu} \cdot \frac{\partial^2 \Psi}{\partial t^2} \tag{VIII.14}$$

and

$$\frac{h^2}{2m\lambda^2}\Psi = -\frac{h^2}{8\pi^2 m} \cdot \frac{\partial^2 \Psi}{\partial x^2}. \tag{VIII.15}$$

If we multiply equation (VIII.13) by Ψ and substitute the expressions in (VIII.14) and (VIII.15), we obtain

$$-\frac{h}{4\pi^2\nu}\cdot\frac{\partial^2\Psi}{\partial t^2} = -\frac{h^2}{8\pi^2 m}\cdot\frac{\partial^2\Psi}{\partial x^2} + V\Psi. \tag{VIII.16}$$

Since

$$\frac{h}{4\pi^2\nu}\cdot\frac{8\pi^2 m}{h^2} = \frac{2m}{\nu h}$$

which, in view of (VIII.11) and $\alpha = \lambda\nu$ (see (VIII.9))

$$\frac{2m}{\nu h} = \frac{2}{\nu\lambda\alpha} = \frac{2}{\nu^2\lambda^2},$$

we obtain after a few rearrangements from (VIII.16)

$$\frac{\partial^2\Psi}{\partial x^2} - \frac{8\pi^2 m}{h^2} V\Psi = \frac{2}{\lambda^2\nu^2}\cdot\frac{\partial^2\Psi}{\partial t^2}. \tag{VIII.17}$$

For the case where the wave propagates in a three-dimensional space, the term $\partial^2\Psi/\partial x^2$ is to be replaced by $\Delta\Psi$ and we obtain *Schroedinger's wave equation*

$$\boxed{\Delta\Psi - \frac{8\pi^2 m}{h^2} V\Psi = \frac{2}{\lambda^2\nu^2}\frac{\partial^2\Psi}{\partial t^2}.} \tag{VIII.18}$$

The term containing the time derivative of Ψ in (VIII.18) can be removed immediately by Bernoulli's separation method. Let

$$\Psi = A\cos 2\pi\nu(t - \delta)\psi(x, y, z). \tag{VIII.19}$$

We assume here that the hitherto unknown time factor is of this particular form. Such an assumption is plausible when one stops to consider its record of earlier success in this work. Substitution of (VIII.19) into (VIII.18) yields the equation

$$A\cos 2\pi\nu(t-\delta)\,\Delta\psi - \frac{8\pi^2 m}{h^2} VA\cos 2\pi\nu(t-\delta)\psi$$
$$= -\frac{8\pi^2\nu^2}{\lambda^2\nu^2} A\cos 2\pi\nu(t-\delta)\psi.$$

After cancellation of the time factor and rearrangement of terms, we obtain

$$\Delta\psi + 8\pi^2\left(\frac{1}{\lambda^2} - \frac{m}{h^2} V\right)\psi = 0.$$

In view of (VIII.11)

$$\lambda m\alpha = h$$

and because of (VIII.9) we obtain

$$\lambda^2 m\nu = h,$$

which we multiply by h and use (VIII.12) to introduce the total energy E. Then

$$\lambda^2 mE = h^2$$

and therefore

$$\frac{1}{\lambda^2} = \frac{mE}{h^2}.$$

Substituting this expression into the equation above renders the *time-independent Schroedinger equation*

$$\boxed{\Delta\psi + \frac{8\pi^2 m}{h^2}(E - V)\psi = 0} \qquad \text{(VIII.20)}$$

in its most common form.

We require that the solution $\psi(x, y, z)$ vanishes of sufficiently high order at infinity such that the volume integral $\iiint_{-\infty}^{\infty} \psi^2\, dV$ over the entire space exists and norm ψ such that

$$\iiint_{-\infty}^{\infty} \psi^2\, dV = 1. \qquad \text{(VIII.21)}$$

This condition comes in quite naturally if one interprets ψ as a *probability wave*, namely, a function which is such that

$$\psi^2\, dV$$

represents the probability that a particular particle is found in the volume element dV. (In case ψ is complex, ψ^2 has to be understood as $|\psi|^2$.) The relative frequency interpretation of this probability density ψ^2 opens up a possibility of confronting the solution of Schroedinger's equation with reality and certain experiments not only seem to indicate the validity of equation (VIII.20) but enhance the acceptability of the rather far-fetched concept of a probability wave.

Problem VIII.4

4. Transform $\psi(x, y, z)$ into spherical coordinates, separate the variables, and write condition (VIII.21) as an iterated integral with the integration variables r, θ, ϕ.

3. The Hydrogen Atom

We will now make an attempt to solve equation (VIII.20) for the hydrogen atom. The hydrogen atom consists of one nucleus of charge e and one electron of charge $-e$ with the electronic mass m. Hence, the potential energy is given by

$$V = -\frac{e^2}{r},$$

where r is the distance between nucleus and electron, and equation (VIII.20) will appear in the form

$$\Delta\psi + \frac{8\pi^2 m}{h^2}\left(h\nu + \frac{e^2}{r}\right)\psi = 0.$$

We will choose our coordinate system such that its origin coincides with the nucleus and introduce—as appears to be indicated by the nature of our problem—spherical coordinates.

We obtain for equation (VIII.20) the following equation in spherical coordinates (see AI.B6):

$$\frac{1}{r^2}\frac{\partial}{\partial r}\left(r^2\frac{\partial\psi}{\partial r}\right) + \frac{1}{r^2\sin\phi}\cdot\frac{\partial}{\partial\phi}\left(\sin\phi\frac{\partial\psi}{\partial\phi}\right) + \frac{1}{r^2\sin^2\phi}\cdot\frac{\partial^2\psi}{\partial\theta^2} + \frac{8\pi^2 m}{h^2}\left(h\nu + \frac{e^2}{r}\right)\psi = 0$$

whereby $\psi = \psi(r, \theta, \phi)$.

Again, we use Bernoulli's separation method and let

$$\psi(r, \theta, \phi) = R(r)\Theta(\theta)\Phi(\phi).$$

Then, because of

$$\frac{\partial}{\partial r}(r^2R'\Theta\Phi) = r^2R''\Phi\Theta + 2rR'\Theta\Phi,$$

$$\frac{\partial}{\partial\phi}(\sin\phi R\Theta\Phi') = \cos\phi R\Phi'\Theta + \sin\phi R\Phi''\Theta,$$

we arrive at the following equation:

$$\frac{1}{r^2}(r^2R''\Theta\Phi + 2rR'\Theta\Phi) + \frac{\cot\phi}{r^2}R\Theta\Phi' + \frac{R\Theta\Phi''}{r^2} + \frac{1}{r^2\sin^2\phi}R\Theta''\Phi + \frac{8\pi^2 m}{h^2}\left(h\nu + \frac{e^2}{r}\right)R\Theta\Phi = 0,$$

which, on division by $\psi = R\Theta\Phi$, becomes

$$\frac{R''}{R} + \frac{2R'}{rR} + \frac{\cot\phi}{r^2}\frac{\Phi'}{\Phi} + \frac{1}{r^2}\frac{\Phi''}{\Phi} + \frac{1}{r^2\sin^2\phi}\frac{\Theta''}{\Theta} + \frac{8\pi^2 m}{h^2}\left(h\nu + \frac{e^2}{r}\right) = 0$$

or

$$\frac{r^2R''}{R} + \frac{2rR'}{R} + r^2\frac{8\pi^2 m}{h^2}\left(h\nu + \frac{e^2}{r}\right) = -\cot\phi\frac{\Phi'}{\Phi} - \frac{\Phi''}{\Phi} - \frac{1}{\sin^2\phi}\cdot\frac{\Theta''}{\Theta}. \tag{VIII.22}$$

Both sides have to be equal to a constant which for reasons of convenience (which will appear later), we write in the form $n(n + 1)$. Thus, we have

$$r^2R'' + 2rR' - n(n + 1)R + \frac{8\pi^2 m}{h^2}r^2\left(h\nu + \frac{e^2}{r}\right)R = 0. \tag{VIII.23}$$

Separating θ and ϕ in (VIII.22) and taking the constant to be $-\mu^2$, the following two equations become evident:

$$\Theta'' + \mu^2\Theta = 0 \tag{VIII.24}$$

and

$$\sin^2\phi\Phi'' + \sin\phi\cos\phi\Phi' + n(n + 1)\sin^2\phi\Phi - \mu^2\Phi = 0. \tag{VIII.25}$$

The latter equation can be simplified as follows: We substitute (as we did in Chapter II, §3.4) for ϕ a new variable ξ according to

$$\xi = \cos\phi,$$

$$\Phi(\phi) = P(\xi).$$

Then

$$\frac{d}{d\xi}\left[(1 - \xi^2)\frac{dP}{d\xi}\right] + \left(n(n + 1) - \frac{\mu^2}{1 - \xi^2}\right)P = 0, \tag{VIII.26}$$

which would be Legendre's equation were it not for the term $\mu^2/(1 - \xi^2)$.

We will postpone the discussion of this equation to the next subsection. It is called *Legendre's associated equation* and there are certain solutions for integral values of n and μ which are called the *associated Legendre functions.*

For the time being, let us call these solutions of (VIII.26)

$$P_n^\mu(\xi).$$

Then, the solutions of (VIII.25) will be

$$\Phi = P_n^\mu(\cos\phi).$$

The solutions of (VIII.24) are clearly

$$\Theta = A \cos \mu\theta + B \sin \mu\theta \tag{VIII.27}$$

and because we require continuity in (x, y, z) of the solution of Schroedinger's equation, we have to insist that

$$\Theta(\theta + 2\pi) = \Theta(\theta),$$

which has

$$\mu = 0, \pm 1, \pm 2, \pm 3, \cdots \tag{VIII.28}$$

as a consequence (see Chapter II, §2.4). At a later time, we will find an additional restriction for the μ. Next and finally, let us attack equation (VIII.23). Division by r^2 leaves

$$R'' + \frac{2R'}{r} + \left(\frac{8\pi^2 m}{h^2} h\nu + \frac{8\pi^2 m e^2}{h^2 r} - \frac{n(n+1)}{r^2}\right) R = 0. \tag{VIII.29}$$

Let us introduce some abbreviating notation to write this equation in a more useful form.

We set

$$a = \frac{8\pi^2 m}{h^2} h\nu, \quad b = a \frac{e^2}{h\nu}.$$

Then, (VIII.29) will appear in the form

$$R'' + \frac{2R'}{r} + \left(a + \frac{b}{r} - \frac{n(n+1)}{r^2}\right) R = 0.$$

Now, we perform the following transformation of the constants

$$K = \frac{b}{2\sqrt{-a}}$$

and introduce a new variable z according to

$$z = 2\sqrt{-a}r.$$

Then, we see that

$$\frac{d^2R}{dz^2} + \frac{2}{z}\frac{dR}{dz} + \left(-\frac{1}{4} + \frac{K}{z} - \frac{n(n+1)}{z^2}\right) R = 0.$$

Multiplication by z^2 and collection of appropriate terms gives

$$\frac{d}{dz}(z^2 R') - \frac{z^2 - 4Kz + 4n(n+1)}{4} R = 0, \tag{VIII.30}$$

which is a differential equation of second order that is closely related to Laguerre's equation; we deal with it in subsection 6.

4. Legendre's Associated Equation

In this section we will discuss equation (VIII.26). In order to conform to the notation in the literature, let us rewrite (VIII.26) in the form:

$$\frac{d}{dx}[(1-x^2)y'] + \left[n(n+1) - \frac{\mu^2}{1-x^2}\right]y = 0 \qquad \text{(VIII.31)}$$

and compare it with Legendre's equation

$$\frac{d}{dx}[(1-x^2)y'] + n(n+1)y = 0. \qquad \text{(VIII.32)}$$

(See (V.14).)

Merely looking at these equations, one feels intuitively that there is some connection between them. The question is, what is the connecting link? Well, to make a long story short, let us make the substitution

$$y = (1-x^2)^{\mu/2}u \qquad \text{(VIII.33)}$$

in (VIII.31) and differentiate equation (VIII.32) m times and we will see that we arrive in both cases at the same equation. (See also problems VI.13 and 14.) If we differentiate (VIII.32) we have to assume that the solution $y = y(x)$ is differentiable in the entire interval $|x| \leq 1$ since $x = \pm 1$ corresponds to $\phi = 0, \pi$, which represents the z-axis. The only solutions of (VIII.32) which are differentiable in $|x| \leq 1$ are the Legendre polynomials $y = P_n(x)$, since all series solutions of the Legendre equation are not even continuous in $|x| \leq 1$. (See theorem VI.2.) Hence, we have to assume that n is an integer. Further, we have to assume that $\mu \leq n$ if we do not wish to obtain a trivial result since the $(n+1)$th derivative of a polynomial of nth degree vanishes.

Beginning with (VIII.31), substituting (VIII.33), and dividing by $(1-x^2)^{\mu/2}$ (see problem VIII.5) leads to

$$(1-x^2)u'' - 2(\mu+1)xu' + (n-\mu)(n+\mu+1)u = 0. \qquad \text{(VIII.34)}$$

If we differentiate Legendre's equation (VIII.32) μ times, assuming that $y = y(x)$ is a solution, we obtain (see problem VI.13)

$$(1-x^2)y^{(\mu+2)} - 2(\mu+1)xy^{(\mu+1)} + (n-\mu)(n+\mu+1)y^{(\mu)} = 0, \qquad \text{(VIII.35)}$$

which is identical with (VIII.34) for $u = y^{(\mu)}$. Hence, if n is an integer, and $y = P_n(x)$, the Legendre polynomials, then the solutions of (VIII.34) are

$$u = \frac{d^\mu}{dx^\mu} P_n(x)$$

and consequently, the solutions of (VIII.31) are (in view of (VIII.33)) given by

$$y = P_n^\mu(x) = (1 - x^2)^{\mu/2} \frac{d^\mu}{dx^\mu} P_n(x). \qquad \text{(VIII.36)}$$

These functions are called the *associated Legendre functions* of degree n and order μ.

Problems VIII.5–7

5. Substitute for y in (VIII.31) as follows

$$y = (1 - x^2)^{\mu/2} u$$

and show that one obtains (VIII.34).

6. The functions

$$y = (1 - x^2)^{\mu/2} \frac{d^\mu}{dx^\mu} P_n^\mu(x)$$

are not always real-valued for $|x| > 1$. What substitution is to be carried out in (VIII.31) in order to obtain for all μ in $|x| > 1$ real-valued solutions of (VIII.31) in terms of the Legendre polynomials?

7. Find by two independent processes, series solutions of (VIII.31) which converge in $|x| < 1$.

5. Spherical Harmonics

Let us outline in this subsection a few general principles which will help to explain the appearance of Legendre's equation and Legendre's associated equation, respectively, whenever we deal with a partial differential equation that involves the Laplace operator Δ and spherical coordinates.

We give the following definitions:

A function $\Phi(x, y, z)$ is called *homogeneous of degree n*, if with any constant factor λ

$$\Phi(\lambda x, \lambda y, \lambda z) = \lambda^n \Phi(x, y, z)$$

holds.

$\Phi(x, y, z)$ is called a *solid spherical harmonic* (the *harmonics* being the solutions of the Laplace equation) of degree n if

$$\Delta\Phi(x, y, z) = 0$$

and if $\Phi(x, y, z)$ is homogeneous of degree n.

Let us transform the equation $\Delta\Phi = 0$ into spherical coordinates and multiply through by r^2. Then, we obtain

$$r \frac{\partial^2(r\Phi)}{\partial r^2} + \frac{1}{\sin\phi} \cdot \frac{\partial}{\partial\phi}\left(\sin\phi \frac{\partial\Phi}{\partial\phi}\right) + \frac{1}{\sin^2\phi} \cdot \frac{\partial^2\Phi}{\partial\theta^2} = 0. \qquad \text{(VIII.37)}$$

In our attempt to solve the problem of stationary temperature distribution in space (see Chapter II, §3.4) we saw that Bernoulli's separation method led to Legendre's equation and Cauchy-Euler's equation, the latter having a solution of the type r^n. Hence, we will try to reduce the above equation by setting

$$\Phi = r^n \Psi_n. \qquad \text{(VIII.38)}$$

Let us interject at this point the following definition:

If Φ is a solid spherical harmonic of degree n, then the function Ψ_n which is related to Φ by

$$\Phi = r^n \Psi_n$$

is called a *spherical surface harmonic*.

(The term "surface" is chosen for quite obvious reasons, because for a constant value of r, Φ reduces to a function defined on the surface of a sphere with radius r and Ψ_n appears thus as Φ on the surface of the unit sphere.) If we go with (VIII.38) into (VIII.37) and divide by r^n, the following equation for the spherical surface harmonics becomes evident:

$$\frac{1}{\sin\phi}\cdot\frac{\partial}{\partial\phi}\left(\sin\phi\,\frac{\partial\Psi_n}{\partial\phi}\right) + n(n+1)\Psi_n + \frac{1}{\sin^2\phi}\cdot\frac{\partial^2\Psi_n}{\partial\theta^2} = 0. \qquad \text{(VIII.39)}$$

If we let, as usual,

$$x = \cos\phi,$$

we obtain

$$\frac{\partial}{\partial x}\left[(1-x^2)\frac{\partial\Psi_n}{\partial x}\right] + n(n+1)\Psi_n + \frac{1}{1-x^2}\cdot\frac{\partial^2\Psi_n}{\partial\theta^2} = 0.$$

In order to find an explicit representation of the spherical surface harmonics we put

$$\Psi_n = X(x)\Theta(\theta)$$

and by proceeding in the usual way, we arrive at

$$\Theta'' + \mu^2\Theta = 0$$

and the associated Legendre equation (VIII.31).

Hence, from (VIII.36) the following representation of the spherical surface harmonics may be written:

$$\Psi_n^{(\mu)}(\theta, \phi) = (A\cos\mu\theta + B\sin\mu\theta)P_n^\mu(\cos\phi) \qquad \text{(VIII.40)}$$

and because of (VIII.38) we obtain the following representation of the solid spherical harmonics of degree n:

$$\Phi^{(n)}(r, \theta, \phi) = r^n(A\cos\mu\theta + B\sin\mu\theta)P_n^\mu(\cos\phi). \qquad \text{(VIII.41)}$$

Before concluding this subsection, we wish to point out that the spherical surface harmonics of degree n appear as the coefficients of r^n in the expansion of the potential Φ in (VI.29) if one lets m_2 in Fig. 23 move on the surface of a sphere of radius r with its center in C. (See Chapter VIII, §2.4.)

Problems VIII.8–9

8. State explicit forms of all the spherical surface harmonics as represented in (VIII.40) for $n = 0, 1, 2$. Observe that $\mu \leq n$.

9. Same as in problem 8 for the solid spherical harmonics as represented in (VIII.41).

6. Laguerre Polynomials and Associated Functions

We are going to deal in this section with the so-called *Laguerre equation*

$$xy'' + (1 - x)y' + ly = 0 \tag{VIII.42}$$

and will assume—at least for the time being—that l is a positive integer. Of course, we could retrace our steps from equation (VIII.30) in subsection 3 of this section and arrive at equation (VIII.42) via some substitutions, transformations and cumbersome manipulations. We feel, however, that the reader should be motivated sufficiently by now and we can spare ourselves from such tedious procedure.

If we divide equation (VIII.42) by x, we see that

$$y'' + \frac{1 - x}{x}\, y' + \frac{l}{x}\, y = 0,$$

which is an equation with a regular singular point at $x = 0$, since $x\,\dfrac{1 - x}{x}$ and $x^2\,\dfrac{l}{x}$ are analytic in a neighborhood of $x = 0$ (which is in this case the entire real axis) (see Chapter VI, §2.2, theorem VI.3).

Hence, at least one solution will be of the form

$$y = \sum_{k=0}^{\infty} c_k x^{\lambda + k} \tag{VIII.43}$$

and, after having carried out the indicated differentiations and substitutions, the following identity appears:

$$\sum_{k=0}^{\infty} c_k(\lambda + k)(\lambda + k - 1)x^{\lambda+k-1} + \sum_{k=0}^{\infty} c_k(\lambda + k)x^{\lambda+k-1}$$
$$- \sum_{k=0}^{\infty} c_k(\lambda + k)x^{\lambda+k} + l\sum_{k=0}^{\infty} c_k x^{\lambda+k} = 0.$$

Changing the summation index in the two latter sums from k to $k - 1$ and leaving the limits unchanged with the understanding that $c_{-1} = 0$, we have for the coefficient of $x^{\lambda+k-1}$

$$c_k(\lambda + k)(\lambda + k - 1) + c_k(\lambda + k) - c_{k-1}(\lambda + k - 1) + lc_{k-1} = 0. \quad \text{(VIII.44)}$$

For $k = 0$,

$$c_0\lambda^2 = 0$$

and consequently $\lambda = 0$ or $c_0 = 0$ (see problem VIII.10). Let $\lambda = 0$.

Then, from (VIII.44) with $k \geq 1$ the following recursion formula becomes evident:

$$c_k = c_{k-1}\frac{k - l - 1}{k^2}. \quad \text{(VIII.45)}$$

If l is a positive integer, we can see that c_{l+1} and all the following coefficients vanish and we obtain polynomials of lth degree as solutions. As we did in the case of the Legendre polynomials, we express the coefficients c_k in terms of the highest nonvanishing coefficient c_l and without difficulties,

$$c_{l-k} = (-1)^k \frac{l^2(l - 1)^2 \cdots (l - k + 1)^2}{k!} c_l. \quad \text{(VIII.46)}$$

Choosing $c_l = (-1)^l$, the so-called *Laguerre polynomials* L_l arise, a few of which we will list here:

$L_0(x) = 1,$ $\qquad L_1(x) = 1 - x,$

$L_2(x) = 2 - 4x + x^2,$ $\qquad L_3(x) = 6 - 18x + 9x^2 - x^3,$

$L_4(x) = 24 - 96x + 72x^2 - 16x^3 + x^4$, etc., (See problem VIII.11.)

In general, we have

$$L_n(x) = \sum_{k=0}^{n}(-1)^k \frac{n^2(n - 1)^2 \cdots (n - k + 1)^2}{k!} x^{n-k} \quad \text{(VIII.47)}$$

(see problem VIII.12).

If we differentiate Laguerre's equation (VIII.42) σ times, assuming that $y = L_l(x)$ and $\sigma \leq l$, it follows that (see problem VIII.13)

$$xy^{(\sigma+2)} + (1 + \sigma - x)y^{(\sigma+1)} + (l - \sigma)y^{(\sigma)} = 0,$$

which, for

$$u = y^{(\sigma)}$$

becomes

$$xu'' + (1 + \sigma - x)u' + (l - \sigma)u - 0 \quad \text{(VIII.48)}$$

with the solutions

$$u = L_l^\sigma(x) = \frac{d^\sigma}{dx^\sigma} L_l(x), \tag{VIII.49}$$

the so-called *associated Laguerre* functions. If we introduce in (VIII.48) a new function R according to

$$u = x^{(1-\sigma)/2} e^{x/2} R, \tag{VIII.50}$$

after tedious manipulations the following equation (see problem VIII.14) is obtained:

$$x^2 R'' + 2xR' + \left[\frac{1-\sigma^2}{4} - \frac{x^2}{4} + \left(\frac{\sigma+1}{2} + (l-\sigma)\right)x\right] R = 0,$$

or in a more convenient form and by writing z instead of x

$$\frac{d}{dz}(z^2 R') - \frac{z^2 + [2(\sigma-1) - 4l]z + \sigma^2 - 1}{4} R = 0, \qquad \sigma \leq l. \tag{VIII.51}$$

This equation is, but for a different formal appearance of the constants, identical with equation (VIII.30).

The solutions of (VIII.51) are, in view of (VIII.49) and (VIII.50),

$$R(z) = z^{(\sigma-1)/2} e^{-z/2} L_l^\sigma(z). \tag{VIII.52}$$

Equation (VIII.51) has a regular singular point at $z = 0$, as one can easily see (see problem VIII.15). Hence, a solution will exist according to theorem VI.3 in $|z| < \rho$ where ρ is the radius of convergence of $a(z) = 2$, $b(z) = -z^2 - [2(\sigma - 1) - 4]z - \sigma^2 + 1$. Since $a(z)$, $b(z)$ are polynomials in z, $\rho = \infty$ and infinite series solutions of (VIII.51) will be of the type $z^\lambda \sum_{n=0}^{\infty} c_n z^n$ where $\sum_{n=0}^{\infty} c_n z^n$ converges everywhere.

However, it can be shown that the solutions of (VIII.48) for nonintegral values of $l > 0$ approach infinity of at least the order $e^{x/\alpha}$, where $1 < \alpha < 2$ as x approaches infinity (see problem VIII.19). It will become apparent in the next section that for our particular purpose we will have to rule out such solutions for nonintegral values of l on these very grounds.

Problems VIII.10–16

10. Let $c_0 = 0$ in (VIII.44) and determine λ such that (VIII.44) is satisfied for $k = 1$. Investigate the solutions which are thus obtained in relation to the solution which is obtained for $c_0 \neq 0$ and $\lambda = 0$.

11. Give the explicit form of the Laguerre polynomials for $l = 0, 1, 2, 3, 4, 5$.

12. Verify (VIII.47) on the basis of (VIII.46).

13. Differentiate (VIII.42) σ times where $\sigma \leq l$ assuming that $y = y(x)$ is σ times differentiable in the interval which is under consideration.

14. Substitute

$$x = z \qquad u = x^{(1-\sigma)/2}e^{x/2}R(x)$$

into (VIII.48) and show that (VIII.51) is obtained.

15. Show that $z = 0$ is a regular singular point of (VIII.51).

16. Find the series solutions of (VIII.51).

7. Solution of Schroedinger's Equation for the Hydrogen Atom

After all the preparations in the preceding three subsections, we are finally in a position to write down a solution of Schroedinger's equation for the hydrogen atom in a closed form, which will reveal information about the discrete energy spectrum and the various quantum states of the atom.

We recall that we assumed the solution to be of the form

$$\psi(r, \theta, \phi) = R(r)\Theta(\theta)\Phi(\phi) \tag{VIII.53}$$

where

$$\Theta(\theta) = A \cos \mu\theta + B \sin \mu\theta \tag{VIII.54}$$

with $\mu = 1, 2, 3, 4, \cdots$ and $\Phi(\phi)$ satisfies equation (VIII.25) which is equivalent to equation (VIII.26) via the substitution $\xi = \cos \phi$, $\Phi(\phi) = P(\xi)$. The latter equation has, according to (VIII.36), the solutions

$$P(\xi) = P_n^\mu(\xi), \qquad \mu \leq n$$

and hence, as solutions of (VIII.25) we have

$$\Phi(\phi) = P_n^\mu(\cos \phi), \qquad \mu \leq n, \quad \text{where } n \text{ has to be an integer.} \tag{VIII.55}$$

Finally, we have seen that $R(r)$ has to satisfy equation (VIII.23), which we have transformed into (VIII.30) via the substitution

$$K = \frac{b}{2\sqrt{-a}} = \frac{2\pi me^2}{h\sqrt{-2mh\nu}} \tag{VIII.56}$$

and the transformation

$$z = 2\sqrt{-a}r = 2\sqrt{-8\pi^2 m\nu/h} \cdot r. \tag{VIII.57}$$

We found in subsection 6 that the solutions of (VIII.30) with

$$-4K = 2(\sigma - 1) - 4l$$

$$4n(n + 1) = \sigma^2 - 1 \tag{VIII.58}$$

are

$$R = z^{(\sigma-1)/2}e^{-z/2}L_l^\sigma(z).$$

From (VIII.58) it follows that

$$\sigma = 2n + 1$$

$$l = K + n.$$

Hence, we verify as solutions of (VIII.30)

$$R = z^n e^{-z/2} L_{K+n}^{2n+1}(z) \tag{VIII.59}$$

where n is an integer and K might be any number, for all we know. However, we imposed on our solution ψ the condition (VIII.21), i.e.:

$$\iiint_{-\infty}^{\infty} \psi^2 \, dx \, dy \, dz = 1 \text{ [2]}$$

which, in spherical coordinates, may be expressed as an iterated integral

$$\int_{r=0}^{\infty} r^2 R^2 \, dr \int_{\theta=0}^{2\pi} \Theta^2 \, d\theta \int_{\phi=0}^{\pi} \Phi^2 \sin \phi \, d\phi = 1 \tag{VIII.60}$$

(see problem VIII.4).

The R-part of this integral exists only if R approaches zero of sufficiently high order as r approaches infinity. We recall the remark we made at the end of the preceding subsection according to which the solutions of (VIII.48) approach infinity of at least the order $e^{z/\alpha}$ if the subscript of L is not an integer. In that case, the R-part of the iterated integral (VIII.60) cannot possibly exist and we have to reject the possibility of K assuming nonintegral values. (See problem VIII.21.) Hence, we impose the condition

$$K = \frac{2\pi m e^2}{h\sqrt{-2mh\nu}} = N, \quad \text{integer.}$$

If we square this relation and solve for the total energy $E = h\nu$, we see that

$$E = -\frac{2\pi^2 e^4 m}{h^2 N^2}, \qquad N = 1, 2, 3, \cdots$$

are the possible values of the energy. This represents a discrete energy spectrum with a limit point at $E = 0$. For $N = 1$, we obtain the *normal quantum state* or *ground state* (as it is often called) of the atom, which is the most *stable state* that is ordinarily occupied by the electron. For $N > 1$, we obtain a manifold of higher quantum states, as shall be demonstrated below:

[2] This condition simply asserts that the electron is somewhere in space.

Since we had to assume for the solutions of (VIII.51) that $m \leq l$, we have

$$2n + 1 \leq N + n$$

or

$$n \leq N - 1,$$

i.e.: for every N, there are N possibilities for n, namely

$$n = 0, 1, 2, 3, \cdots, N - 1$$

and because of $\mu \leq n$, we have

$$\mu = 0, 1, 2, 3, \cdots, n$$

for every n. Hence, for every N, there are $[N(N + 1)]/2$ independent quantum states.

Finally, the solutions ψ are according to (VIII.53), (VIII.54), (VIII.55), (VIII.52), (VIII.57), and (VIII.59),

$$\psi_{N,n,\mu} = \frac{h^2 N}{4\pi^2 m e^2} (A \cos \mu\theta + B \sin \mu\theta) P_n^\mu(\cos \phi) r^n \cdot e^{-(2\pi^2 m e^2/h^2 N) r} L_{N+n}^{2n+1}\left(\frac{4\pi^2 m e^2}{h^2 N} r\right),$$

where $\mu \leq n \leq N - 1$.

In terms of the spherical surface harmonics $\Psi_n^{(\mu)}(\theta, \phi)$, we can write this also in the form

$$\psi_{N,n,\mu} = \frac{h^2 N}{4\pi^2 m e^2} r^n e^{-(2\pi^2 m e^2/h^2 N) r} L_{N+n}^{2n+1}\left(\frac{4\pi^2 m e^2}{h^2 N} r\right) \Psi_n^{(\mu)}(\theta, \phi)$$

(see (VIII.40)).

N is called the *principal quantum number*, n the *azimuthal quantum number*, and μ the *magnetic quantum number*.

Problems VIII.17–21

17. Find a recursive formula for the coefficients c_n of the series solution of (VIII.48) ($l > 0$).

***18.** Show that with c_n from problem 17,

$$|c_n| > \frac{\gamma}{n!\alpha^n}$$

where γ is a constant and $1 < \alpha < 2$.

***19.** Show that for nonintegral values of $l > 0$, the function L_l^σ approaches infinity of at least the order $e^{x/\alpha}$, where $1 < \alpha < 2$, as x approaches infinity. (*Hint:* Use the result in problem 18.)

20. Show that the solution

$$R(z) = z^{(\sigma-1)/2} e^{-z/2} L_l^\sigma(z)$$

of (VIII.51) for nonintegral values of l approaches infinity of at least the order $\gamma e^{\beta z}/\sqrt{z}$ as z approaches infinity, where $\beta > 0$.

21. Show that $\int_0^\infty r^2 R^2\, dr$ does not exist if R approaches infinity of at least the order $\gamma e^{\beta r}/\sqrt{r}$ where $\beta > 0$.

§2. SPHERICAL HARMONICS AND POISSON'S INTEGRAL

1. Stationary Temperature Distribution Generated by a Spherical Stove

In Chapter II, §3.4, we dealt with the problem of finding the stationary temperature distribution generated by a spherical stove of radius 1, where the temperature on the surface of the unit sphere was kept constant and was assumed to be a function of the elevation ϕ alone. We will now formulate the same problem in greater generality insofar as we will assume the temperature on the unit sphere to be a function of the elevation ϕ and the polar angle θ. If $u(r, \theta, \phi)$ represents the temperature at the point (r, θ, ϕ) and if $f(\theta, \phi)$ represents the given temperature on the surface of the unit sphere at the point (θ, ϕ) we have to impose the boundary condition

$$u(1, \theta, \phi) = f(\theta, \phi). \tag{VIII.61}$$

Since heat cannot accumulate at infinity, we have the additional condition

$$\left|\lim_{r\to\infty} u(r, \theta, \phi)\right| < \infty \tag{VIII.62}$$

and if we are interested in the interior of the sphere, we have to impose the condition

$$|u(0, \theta, \phi)| < \infty \tag{VIII.63}$$

for quite obvious reasons.

According to Chapter II, §3.1, the stationary temperature distribution u in space has to satisfy the potential equation

$$\Delta u(x, y, z) = 0,$$

which, in spherical coordinates and after multiplication by r^2, assumes the form

$$r\frac{\partial^2(ru)}{\partial r^2} + \frac{1}{\sin\phi}\frac{\partial}{\partial\phi}\left(\sin\phi\,\frac{\partial u}{\partial\phi}\right) + \frac{1}{\sin^2\phi}\frac{\partial^2 u}{\partial\theta^2} = 0. \tag{VIII.64}$$

A solution of this equation is, according to (VIII.38),

$$u_n = r^n \Psi_n, \tag{VIII.65}$$

where Ψ_n is a spherical surface harmonic of degree n.

Theorem VIII.1 (theorem of Kelvin). *If Φ_n is a solid spherical harmonic of degree n, then $(1/r^{2n+1})\Phi_n$ is a solid spherical harmonic of degree $-n-1$.*

(With regard to our problem: If $r^n\Psi_n$ is a solution of the potential equation in terms of spherical surface harmonics, then $(1/r^{n+1})\Psi_n$ is also a solution of the potential equation in terms of spherical surface harmonics.)

Proof. If Φ_n is a solid spherical harmonic of degree n, then

$$\Delta\Phi_n = 0 \tag{VIII.66}$$

and

$$\Phi_n(\lambda x, \lambda y, \lambda z) = \lambda^n \Phi_n(x, y, z). \tag{VIII.67}$$

Let us now consider the function $\Phi = r^m\Phi_n$ and examine what value m has to have in order to make Φ a solid spherical harmonic.

Since $r = \sqrt{x^2 + y^2 + z^2}$, we have $\partial r/\partial x = x/r$ and hence

$$\frac{\partial \Phi}{\partial x} = mr^{m-2}x\Phi_n + r^m \frac{\partial \Phi_n}{\partial x},$$

$$\frac{\partial^2 \Phi}{\partial x^2} = m(m-2)r^{m-4}x^2\Phi_n + mr^{m-2}\Phi_n + 2mr^{m-2}x\frac{\partial \Phi}{\partial x} + r^m \frac{\partial^2 \Phi_n}{\partial x^2},$$

with analogous expressions for $\partial^2\Phi/\partial y^2$ and $\partial^2\Phi/\partial z^2$.

Hence

$$\Delta\Phi = r^m\Delta\Phi_n + 2mr^{m-2}\left(x\frac{\partial \Phi_n}{\partial x} + y\frac{\partial \Phi_n}{\partial y} + z\frac{\partial \Phi_n}{\partial z}\right) + m(m+1)\Phi_n r^{m-2}. \tag{VIII.68}$$

On differentiating (VIII.67) with respect to λ, we obtain

$$x\frac{\partial \Phi_n}{\partial x} + y\frac{\partial \Phi_n}{\partial y} + z\frac{\partial \Phi_n}{\partial z} = n\lambda^{n-1}\Phi_n(x, y, z),$$

which, for $\lambda = 1$, yields *Euler's identity* for homogeneous functions:

$$x\frac{\partial \Phi_n}{\partial x} + y\frac{\partial \Phi_n}{\partial y} + z\frac{\partial \Phi_n}{\partial z} = n\Phi_n. \tag{VIII.69}$$

Now, in view of (VIII.66) and (VIII.69), we can write (VIII.68) in the form

$$\Delta\Phi \equiv r^{m-2}(2mn + m(m+1))\Phi_n$$

and from the condition that Φ satisfies the potential equation, the following condition on m is evident:

$$2mn + m(m+1) = 0$$

or

$$m_1 = -2n - 1, \quad m_2 = 0.$$

Thus, $\Phi = r^{-2n-1}\Phi_n$ is a solution of the potential equation. That Φ is homogeneous of degree $-n - 1$ can be seen as follows:

$$\begin{aligned}\Phi(\lambda x, \lambda y, \lambda z) &= (\lambda^2 x^2 + \lambda^2 y^2 + \lambda^2 z^2)^{-\frac{1}{2}(2n+1)}\Phi_n(\lambda x, \lambda y, \lambda z)\\ &= \lambda^{-2n-1}r^{-2n-1}\lambda^n\Phi_n(x, y, z) = \lambda^{-n-1}\Phi(x, y, z), \quad \text{q.e.d.}\end{aligned}$$

According to theorem VIII.1—as we already remarked parenthetically—with

$$u_n = r^n\Psi_n,$$

also

$$u_n = \frac{1}{r^{n+1}}\Psi_n$$

occurs as solution of (VIII.64) in terms of spherical harmonics. Hence, a more general solution is represented by

$$u_n = \left(c_1 r^n + \frac{c_2}{r^{n+1}}\right)\Psi_n \qquad \text{(VIII.70)}$$

where Ψ_n is a spherical surface harmonic.

(The reader will remember that we arrived at the same r-factor in Chapter II, §3.4, in solving the Euler-Cauchy equation (II.102) which resulted from Bernoulli's separation method applied to the potential equation in r and ϕ.)

If we take the second boundary condition, (VIII.62) or (VIII.63), into account, we either have to choose $c_1 = 0$ if we are interested in the solution outside the unit sphere or we have to choose $c_2 = 0$ if we are interested in the interior of the unit sphere. Thus we have

$$u_n = cr^n\Psi_n \quad \text{for } 0 \leq r < 1, \qquad \text{(VIII.71)}$$

and

$$u_n = c\frac{1}{r^{n+1}}\Psi_n \quad \text{for } r > 1. \qquad \text{(VIII.72)}$$

We will deal in the following subsections with the case (VIII.71) only, i.e., with the boundary value problem (VIII.61), (VIII.63), because the result that will be obtained can be easily adjusted so that it will represent then a solution of (VIII.61), (VIII.62).

$\Psi_n(\theta, \phi)$ is a spherical surface harmonic which can be represented in terms of trigonometric functions and the associated Legendre functions, as was done in subsection 5 of the preceding section (formula (VIII.40)) and consequently the superposition principle can be used in order to satisfy the boundary condition (VIII.61). This approach will be taken in the following subsection. Another alternative is to apply the superposition principle directly to (VIII.71) and use this as a point of departure, as will be done in subsection 4 of this section.

Problems VIII.22–23

22. Verify by substitution that

$$u = \left(c_1 r^n + \frac{c_2}{r^{n+1}}\right)\Psi_n,$$

where Ψ_n is defined in (VIII.40), is a solution of the potential equation.

23. Verify Euler's identity for the functions

$$\Phi = \sqrt{x^2 + y^2},$$
$$\Phi = xy^2 + x^3,$$
$$\Phi = (x + y)^2.$$

2. Fourier Expansion in Terms of Associated Legendre Functions

We have seen in subsection 5 of §1 that

$$\Psi_n^{(m)} = (A \cos m\theta + B \sin m\theta)P_n^{(m)}(\cos \phi)$$

is a spherical surface harmonic, where the P_n^m are associated Legendre functions (see (VIII.40)).

Substituting this representation into (VIII.71) and applying the superposition principle, we construct the function

$$u(r, \theta, \phi) = \sum_{n=0}^{\infty} \sum_{m=0}^{n} r^n(a_{nm} \cos m\theta + b_{nm} \sin m\theta)P_n^m(\cos \phi). \tag{VIII.73}$$

In order to satisfy the boundary condition (VIII.61), we have to determine the coefficients a_{nm} and b_{nm} such that

$$u(1, \theta, \phi) = f(\theta, \phi) = \sum_{n=0}^{\infty} \sum_{m=0}^{n} (a_{nm} \cos m\theta + b_{nm} \sin m\theta)P_n^m(\cos \phi). \tag{VIII.74}$$

Suppose that such an expansion is possible and the series on the right side of (VIII.74) converges uniformly to $f(\theta, \phi)$.

Multiplication of (VIII.74) by $\cos k\theta$ ($k = 0, 1, 2, 3, \cdots$) and subsequent integration from $-\pi$ to π yields, in view of (IV.11) and (IV.12),

$$\int_{-\pi}^{\pi} f(\theta, \phi) \cos k\theta \, d\theta = \pi \sum_{n=0}^{\infty} a_{nk} P_n^k(\cos \phi). \tag{VIII.75}$$

Similarly we obtain by using $\sin k\theta$ instead of $\cos k\theta$, in view of (IV.10) and (IV.11),

$$\int_{-\pi}^{\pi} f(\theta, \phi) \sin k\theta \, d\theta = \pi \sum_{n=0}^{\infty} b_{nk} P_n^k(\cos \phi). \tag{VIII.76}$$

Since P_n^k, P_m^k are solutions of Legendre's associated equation (VIII.31), we have

$$\frac{d}{dx}\left[(1 - x^2)\frac{dP_n^k}{dx}\right] + \left[n(n + 1) - \frac{k^2}{1 - x^2}\right]P_n^k = 0$$

and

$$\frac{d}{dx}\left[(1 - x^2)\frac{dP_m^k}{dx}\right] + \left[m(m + 1) - \frac{k^2}{1 - x^2}\right]P_m^k = 0.$$

We multiply the first equation by P_m^k, the second equation by P_n^k and subtract one from the other

$$P_m^k \frac{d}{dx}\left[(1 - x^2)\frac{dP_n^k}{dx}\right] - P_n^k \frac{d}{dx}\left[(1 - x^2)\frac{dP_m^k}{dx}\right]$$
$$= [m(m + 1) - n(n + 1)]P_n^k P_m^k.$$

Integration of the left side from -1 to 1 yields zero in view of (V.16) with $p(x) = 1 - x^2$. Hence,

$$\int_{-1}^{1} P_n^k P_m^k \, dx = 0 \quad \text{if } n \neq m. \tag{VIII.77}$$

If we substitute $x = \cos \phi$, then we obtain from (VIII.77)

$$\int_0^{\pi} P_n^k(\cos \phi) P_m^k(\cos \phi) \sin \phi \, d\phi = 0 \quad \text{for } n \neq m. \tag{VIII.78}$$

Thus, multiplying (VIII.75) and (VIII.76) by $\sin \phi P_i^k(\cos \phi)$ and integrating subsequently from 0 to π:

$$\frac{1}{\pi}\int_0^{\pi}\int_{-\pi}^{\pi} f(\theta, \phi) \cos k\theta P_i^k(\cos \phi) \sin \phi \, d\theta \, d\phi = a_{ik}\int_0^{\pi} (P_i^k(\cos \phi))^2 \sin \phi \, d\phi$$

and

$$\frac{1}{\pi}\int_0^{\pi}\int_{-\pi}^{\pi} f(\theta, \phi) \sin k\theta P_i^k(\cos \phi) \sin \phi \, d\theta \, d\phi = b_{ik}\int_0^{\pi} (P_i^k(\cos \phi))^2 \sin \phi \, d\phi.$$

One can show by various methods[3] that

$$\int_0^{\pi} (P_i^k(\cos \phi))^2 \sin \phi \, d\phi = \int_{-1}^{1} [P_i^k(x)]^2 dx = \frac{2}{2i + 1} \cdot \frac{(i + k)!}{(i - k)!}, \qquad k \leq i.$$

[3] See T. M. MacRobert: *Spherical Harmonics*, Dover Publications, New York, 1948.

Hence,

$$a_{ik} = \frac{(2i+1)(i-k)!}{2\pi(i+k)!}\int_0^\pi\int_{-\pi}^\pi f(\theta,\phi)\cos k\theta P_i^k(\cos\phi)\sin\phi\,d\theta\,d\phi \tag{VIII.79}$$

and

$$b_{ik} = \frac{(2i+1)(i-k)!}{2\pi(i+k)!}\int_0^\pi\int_{-\pi}^\pi f(\theta,\phi)\sin k\theta P_i^k(\cos\phi)\sin\phi\,d\theta\,d\phi. \tag{VIII.80}$$

For $k = 0$ we obtain in (VIII.75) the factor 2π instead of π and hence

$$a_{i0} = \frac{2i+1}{4\pi}\int_0^\pi\int_{-\pi}^\pi f(\theta,\phi)P_i(\cos\phi)\sin\phi\,d\theta\,d\phi. \tag{VIII.81}$$

If we substitute these expressions for a_{nm} and b_{nm} into (VIII.73), there results the solution of the potential equation which satisfies (VIII.61) and (VIII.63). Before we write it down, using α and β as integration variables instead of θ, ϕ, we make use of $\cos\mu\theta\cos\mu\alpha + \sin\mu\theta\sin\mu\alpha = \cos\mu(\theta-\alpha)$. Then

$$u(r,\theta,\phi) = \sum_{n=0}^\infty r^n\frac{2n+1}{4\pi}\int_0^\pi\int_{-\pi}^\pi f(\alpha,\beta)P_n(\cos\beta)\sin\beta\,d\alpha\,d\beta P_n(\cos\phi)$$
$$+\sum_{n=1}^\infty\sum_{m=1}^n r^n\frac{2n+1}{2\pi}\frac{(n-m)!}{(n+m)!}\int_0^\pi\int_{-\pi}^\pi f(\alpha,\beta)P_n^m(\cos\beta)$$
$$\cdot\sin\beta\cos\mu(\theta-\alpha)\,d\alpha\,d\beta P_n^m(\cos\phi),$$

or, after interchange of integration and summation, provided it is permissible:[4]

$$u(r,\theta,\phi) = \int_0^\pi\int_{-\pi}^\pi\left\{\sum_{n=0}^\infty r^n\frac{2n+1}{4\pi}P_n(\cos\beta)f(\alpha,\beta)\sin\beta P_n(\cos\phi)\right.$$
$$+\sum_{n=1}^\infty r^n\sum_{m=1}^n\frac{2n+1}{2\pi}\cdot\frac{(n-m)!}{(n+m)!}P_n^m(\cos\beta)f(\alpha,\beta)$$
$$\left.\cdot\sin\beta\cos m(\theta-\alpha)P_n^m(\cos\phi)\right\}d\alpha\,d\beta. \tag{VIII.82}$$

Problems VIII.24–27

24. Carry out all the necessary substitutions that lead from (VIII.73) to (VIII.82).

25. Find a solution $u(r,\theta,\phi)$ of the potential equation in the form (VIII.82) if the boundary function f on the unit sphere is a function of ϕ alone.

[4] The permissibility of interchange of integration and summation here and on the following pages is not discussed. We refer the reader to our discussion on p. 320.

26. Write down the explicit representation of all the $P_i^k(x)$ for $k \leq i \leq 3$.
27. Let $f(\theta, \phi) = \theta$. Find $a_{00}, a_{01}, b_{10}, b_{11}$.

3. Poisson's Integral Representation of the Solution of the Potential Equation

We will now carry out the summation under the integral sign in (VIII.82) for the specific choice of the reference point $\theta = \phi = 0$:

$$u(r, 0, 0) = \int_0^\pi \int_{-\pi}^\pi \Bigg\{ \sum_{n=0}^\infty r^n \frac{2n+1}{4\pi} P_n(\cos\beta) f(\alpha, \beta) \sin\beta P_n(1)$$

$$+ \sum_{n=1}^\infty \sum_{m=1}^n \frac{2n+1}{2\pi} \frac{(n-m)!}{(n+m)!} P_n^m(\cos\beta) f(\alpha, \beta)$$

$$\cdot \sin\beta \cos m\alpha P_n^m(1) \Bigg\} \, d\alpha \, d\beta.$$

It follows from the definition of P_n^m (see (VIII.36)) that

$$P_n^m(1) = 0 \quad \text{for } n \geq 1.$$

Hence, the second part of the above sum vanishes.

Since $P_n(1) = 1$ (see problem VI.15)), we verify that for $u(r, 0, 0)$:

$$u(r, 0, 0) = \frac{1}{4\pi} \int_0^\pi \int_{-\pi}^\pi \left(\sum_{n=0}^\infty (2n+1) r^n P_n(\cos\beta) f(\alpha, \beta) \sin\beta \right) d\alpha \, d\beta. \tag{VIII.83}$$

The sum $\sum_{n=0}^\infty (2n+1) r^n P_n(x)$ can be easily represented in a closed form by again using the relation between the Legendre polynomials and their generating function (see (VI.29)). If we differentiate (VI.29) with respect to r and subsequently multiply the result by $2r$, it follows that

$$\sum_{n=0}^\infty 2n r^n P_n(x) = \frac{2rx - 2r^2}{(1 - 2rx + r^2)^{3/2}}. \tag{VIII.84}$$

Adding this expression to (VI.29), we obtain

$$\sum_{n=0}^\infty (2n+1) r^n P_n(x) = \frac{1 - r^2}{(1 - 2rx + r^2)^{3/2}}. \tag{VIII.85}$$

Now we substitute the expression on the right side of (VIII.85) into (VIII.83) to get

$$u(r, 0, 0) = \frac{1 - r^2}{4\pi} \int_0^\pi \int_{-\pi}^\pi \frac{f(\alpha, \beta) \sin\beta \, d\alpha \, d\beta}{(1 - 2r\cos\beta + r^2)^{3/2}}. \tag{VIII.86}$$

We note that the expression

$$\rho = (1 - 2r \cos \beta + r^2)^{1/2},$$

which occurs to the third power in the denominator of (VIII.86) represents the distance between the reference point (1, 0, 0) and the integration point (r, α, β). We note further that (VIII.86) has to hold for any point (ϕ, θ)

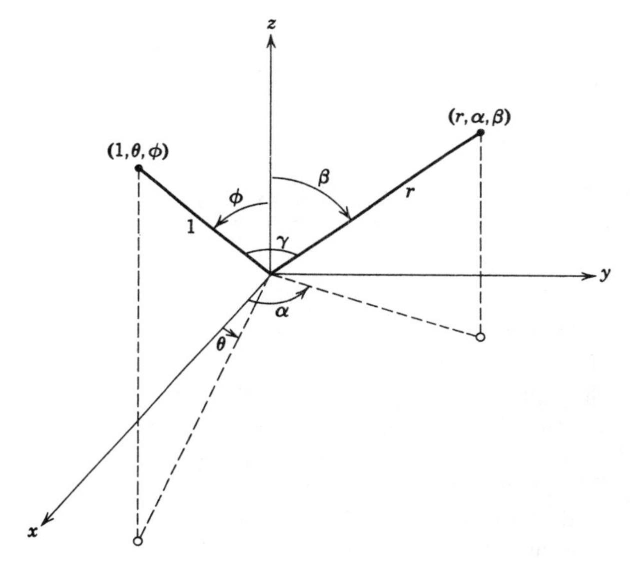

Fig. 30

on the sphere with radius r because it is immaterial where we choose the north pole $\phi = \theta = 0$. Hence, we have

$$u(r, \theta, \phi) = \frac{1 - r^2}{4\pi} \int_0^{\pi} \int_{-\pi}^{\pi} \frac{f(\alpha, \beta) \sin \beta \, d\alpha \, d\beta}{\rho^3}$$

where ρ stands for the distance between integration point and reference point. Now, let γ be the angle between the line L_1 joining the reference point $(1, \theta, \phi)$ with the origin and the line L_2 joining the integration point (r, α, β) with the origin. The direction cosines of these lines are, as the reader can easily see from Fig. 30,

$$L_1: \quad (\sin \phi \cos \theta, \sin \phi \sin \theta, \cos \phi),$$

$$L_2: \quad (\sin \beta \cos \alpha, \sin \beta \sin \alpha, \cos \beta).$$

Hence,

$$\cos \gamma = \cos \phi \cos \beta + \sin \phi \sin \beta \cos (\theta - \alpha). \qquad \text{(VIII.87)}$$

Thus,

$$u(r, \theta, \phi)$$

$$= \frac{1 - r^2}{4\pi} \int_0^{\pi} \int_{-\pi}^{\pi} \frac{f(\alpha, \beta) \sin \beta \, d\alpha \, d\beta}{[1 - 2(\cos \phi \cos \beta + \sin \phi \sin \beta \cos (\theta - \alpha))r + r^2]^{3/2}},$$

$$0 \leq r < 1, \qquad \text{(VIII.88)}$$

The reader can easily verify for himself that for the exterior of the unit sphere:

$$u(r, \theta, \phi)$$

$$= \frac{r^2 - 1}{4\pi} \int_0^{\pi} \int_{-\pi}^{\pi} \frac{f(\alpha, \beta) \sin \beta \, d\alpha \, d\beta}{[1 - 2(\cos \phi \cos \beta + \sin \phi \sin \beta \cos (\theta - \alpha))r + r^2]^{3/2}},$$

$$r > 1. \qquad \text{(VIII.89)}$$

(VIII.88) and (VIII.89) are called the *Poisson integral representation* of the solution of the potential equation in terms of the function $f(\theta, \phi)$ which represents the values of the potential on the surface of the unit sphere. The reader with some background in complex variables will see that (VIII.88) is in a sense a generalization of *Cauchy's integral formula*, which represents the values of an analytic function in the interior of a region in terms of its values on the boundary of the region.

Before we discuss this representation (VIII.88) any further, let us derive it by a different method, as was already indicated on pp. 310-311.

Problems VIII.28–30

28. Find $u(r, \theta, 0)$ for $f(\theta, \phi) = \theta(\theta - 2\pi)$, $0 \leq \theta < 2\pi$.

29. Establish (VIII.89) as the formal solution of the potential equation with the boundary conditions (VIII.61) and (VIII.62) for $r > 1$.

30. Explain the fact that $u(r, \theta, 0) = u(r, 0, 0)$.

4. Expansion in Terms of Laplace Coefficients

We will now try to build up a solution to our problem by beginning again with (VIII.71)

$$u_n = cr^n \Psi_n$$

but we will use a different representation of the spherical surface harmonics. We recall the remark we made at the end of §1.5 about the generation of the spherical surface harmonics by an expansion analogous to the one used in generating the Legendre polynomials, where m_2 instead of being restricted to a circle, is now allowed to move freely on the surface of a sphere. We will place the unit mass m_1 at the fixed point $(1, \alpha, \beta)$ and let the unit mass m_2 move within the surface of a sphere of radius r and its center at the origin, where $0 < r \leq R < 1$ (see Fig. 31). If z is the distance between

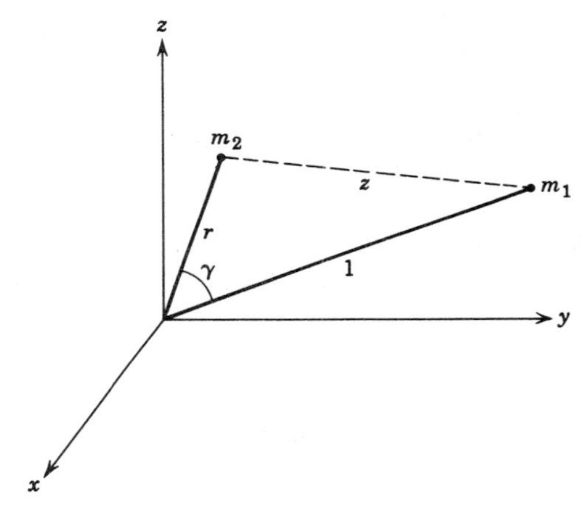

Fig. 31

m_1 and m_2, then the potential of m_2 with respect to m_1 is but for a multiplicative factor

$$\Phi(z) = \frac{1}{z},$$

where z is a function of the angle γ between the line $\overline{0m_1}$ and the line $\overline{0m_2}$. According to (VIII.87),

$$\cos \gamma = \cos \phi \cos \beta + \sin \phi \sin \beta \cos (\theta - \alpha),$$

if (r, θ, ϕ) are the coordinates of m_2. Hence,

$$z = (1 - 2r \cos \gamma + r^2)^{1/2}$$

and

$$\Phi(z) = (1 - 2r \cos \gamma + r^2)^{-1/2}. \qquad \text{(VIII.90)}$$

As in Chapter VI, §1.4, we expand $\Phi(z)$ in a power series in r and obtain

$$\Phi(z) = (1 - 2r \cos \gamma + r^2)^{-1/2} = \sum_{n=0}^{\infty} r^n P_n(\cos \gamma) \qquad \text{(VIII.91)}$$

where the P_n are formally the Legendre polynomials for $\cos \gamma = x$. However, $P_n(\cos \gamma)$ is now a function of two variables in view of (VIII.87).

We will show that these coefficients—called the *Laplace coefficients*—are spherical surface harmonics of degree n, i.e., $r^n P_n(\cos\gamma)$ is a solution of the Laplace equation and is homogeneous of degree n.

r^n is already homogeneous of degree n and one can see easily that $P_n(\cos\gamma)$ is homogeneous of zeroth degree, since $\cos\gamma$ is homogeneous of zeroth degree (see problem VIII.32).

It remains to be shown that $r^n P_n(\cos\gamma)$ is a solution of the Laplace equation. It follows from (VIII.91) that

$$\Delta(1 - 2r\cos\gamma + r^2)^{-1/2} = \Delta \sum_{n=0}^{\infty} r^n P_n(\cos\gamma)$$

where Δ stands for the Laplace operator in spherical coordinates (see AI.B6). In order to show that the left side is zero, it is best to transform the expression under the Δ-operator to Cartesian coordinates:

$$\Phi = (1 - 2r\cos\gamma + r^2)^{-1/2} = [(x - x_1)^2 + (y - y_1)^2 + (z - z_1)^2]^{-1/2}$$

where (x, y, z) are the Cartesian coordinates of the point (r, α, β) and (x_1, y_1, z_1) are the Cartesian coordinates of the point $(1, \theta, \phi)$.

Since

$$\frac{\partial^2\Phi}{\partial x^2} = \frac{(x - x_1)^2 + (y - y_1)^2 + (z - z_1)^2 - 3(x - x_1)^2}{[(x - x_1)^2 + (y - y_1)^2 + (z - z_1)^2]^{5/2}}$$

and the expressions for $\partial^2\Phi/\partial y^2$ and $\partial^2\Phi/\partial z^2$ are obtained by interchanging x, y, z, it follows that

$$\Delta\Phi = \frac{\partial^2\Phi}{\partial x^2} + \frac{\partial^2\Phi}{\partial y^2} + \frac{\partial^2\Phi}{\partial z^2} = 0$$

and consequently

$$\Delta \sum_{n=0}^{\infty} r^n P_n(\cos\gamma) = 0. \tag{VIII.92}$$

Interchange of Δ and summation and carrying out the required operations according to (VIII.37) renders

$$\sum_{n=0}^{\infty} r^n \left[(n + 1)nP_n(\cos\gamma) + \frac{1}{\sin\phi} \cdot \frac{\partial}{\partial\phi}\left(\sin\phi \frac{\partial P_n}{\partial\phi}\right) + \frac{1}{\sin^2\phi} \frac{\partial^2 P_n}{\partial\theta^2} \right] = 0$$

for all r and consequently, the coefficient of r^n for all n has to vanish:

$$\frac{1}{\sin\phi} \frac{\partial}{\partial\phi}\left(\sin\phi \frac{\partial P_n}{\partial\phi}\right) + (n + 1)nP_n(\cos\gamma) + \frac{1}{\sin^2\phi} \cdot \frac{\partial^2 P_n}{\partial\theta^2} = 0.$$

From this equation, which is identical with (VIII.39), we see that $\Phi = r^n P_n(\cos\gamma)$ is a solution of the Laplace equation.

Thus, we can write (VIII.71) in the form

$$u_n(r, \theta, \phi) = cr^n P_n(\cos \gamma)$$

and the solution of our boundary value problem amounts to the determination of coefficients A_n such that

$$f(\theta, \phi) = \sum_{n=0}^{\infty} A_n P_n(\cos \gamma). \tag{VIII.93}$$

Having done this, the solution u will appear in the form

$$u(r, \theta, \phi) = \sum_{n=0}^{\infty} r^n A_n P_n(\cos \gamma). \tag{VIII.94}$$

It can be shown that

$$\int_0^{\pi}\int_{-\pi}^{\pi} P_n(\cos \gamma')P_m(\cos \gamma') \sin \phi' \, d\theta' \, d\phi' = 0, \quad \text{for } n \neq m \tag{VIII.95}$$

and

$$\int_0^{\pi}\int_{-\pi}^{\pi} P_n^2(\cos \gamma') \sin \phi' \, d\theta' \, d\phi' = \frac{4\pi}{2n + 1} \tag{VIII.96}$$

(see problems VIII.33 and 37), where

$$\cos \gamma' = \cos \phi' \cos \beta + \sin \phi' \sin \beta \cos (\theta' - \alpha).$$

Hence, by the usual procedure,

$$A_n = \frac{2n + 1}{4\pi}\int_0^{\pi}\int_{-\pi}^{\pi} P_n(\cos \gamma')f(\theta', \phi') \sin \phi' \, d\theta' \, d\phi'$$

and consequently

$$u(r, \theta, \phi) = \frac{1}{4\pi}\int_0^{\pi}\int_{-\pi}^{\pi} \sum_{n=0}^{\infty} (2n+1)r^n P_n(\cos \gamma')f(\theta', \phi') \sin \phi' \, d\theta' \, d\phi' \, P_n(\cos \gamma).$$

According to (VIII.85) we have

$$\sum_{n=0}^{\infty} (2n + 1)r^n P_n(x) = \frac{1 - r^2}{(1 - 2rx + r^2)^{3/2}},$$

(note that this relation is independent of what x represents) and thus we obtain for $\theta = \alpha$, $\phi = \beta$

$$u(r, \alpha, \beta) = \frac{1 - r^2}{4\pi}\int_0^{\pi}\int_{-\pi}^{\pi} \frac{f(\theta', \phi') \sin \phi' \, d\theta' \, d\phi'}{(1 - 2r \cos \gamma' + r^2)^{3/2}},$$

which is identical with (VIII.88).

Poisson's integral representation of the solution of the potential equation with given boundary values on the surface of the unit sphere was found by assuming that the expansions (VIII.74) and (VIII.93), respectively, are possible and uniformly convergent. One can now establish (VIII.88) directly as a solution of the stated boundary value problem and establish the convergence of (VIII.74) and (VIII.93), respectively, *a posteriori*.

One can show that $u(r, \theta, \phi)$ as represented by (VIII.88) is a solution of the potential equation and that

$$\lim_{r\to 1} u(r, \theta, \phi) = f(\theta, \phi)$$

where $f(\theta, \phi)$ is the given boundary function. This fact can be established for every point on the surface of the sphere, where f is continuous. It can even be shown that this limit exists at points (θ^*, ϕ^*) of discontinuity of f, provided that

$$\lim_{\rho\to 0} \frac{1}{2\pi\rho} \int_C f(\theta, \phi)\, ds$$

exists, where C is a parallel circle with radius ρ on the unit sphere with center at (θ^*, ϕ^*). Then

$$\lim_{r\to 1} u(r, \theta, \phi) = \lim_{\rho\to 0} \frac{1}{2\pi\rho} \int_C f(\theta, \phi)\, ds$$ [5]

Once these facts are established, one can reverse the procedure which was followed in the preceding subsections by representing the kernel of (VIII.88)

$$\frac{1 - r^2}{(1 - 2r\cos\gamma + r^2)^{3/2}}, \qquad |r| \le R < 1$$

by the uniformly convergent series representation in (VIII.85) and is thus led to the series representation of u:

$$u(r, \theta, \phi) = \frac{1}{4\pi} \int_0^{\pi} \int_{-\pi}^{\pi} \sum_{n=0}^{\infty} f(\alpha, \beta) \sin\beta (2n + 1) r^n P_n(\cos\gamma)\, d\alpha\, d\beta. \quad \text{(VIII.97)}$$

In order to obtain the series representation (VIII.73) with the coefficients (VIII.79), (VIII.80), and (VIII.81) one has to make use of a representation of the Laplace coefficients $P_n(\cos\gamma)$ in terms of the associated Legendre functions and the trigonometric functions, namely

$$P_n(\cos\gamma) = P_n(\cos\phi) P_n(\cos\beta) + 2 \sum_{m=1}^{n} \frac{(n - m)!}{(n + m)!} P_n^m(\cos\phi) \cdot P_n^m(\cos\beta) \cos m(\theta - \alpha). \quad \text{(VIII.98)}$$ [6]

[5] See Ph. Frank and R. von Mises: *Die Differential und Integralgleichungen der Mechanik und Physik*, Vol. 1, F. Vieweg und Sohn, Braunschweig, 1930, pp. 748, 749. See also problem VIII.31.

[6] See T. M. MacRobert: *Spherical Harmonics*, Dover Publications, New York, 1948, p. 138.

Problems VIII.31-37

31. Find $\lim_{\rho \to 0} \frac{1}{2\pi\rho} \int_C f(\theta, \phi)\, ds$ for $f = \theta(\theta - 2\pi)$ and compare it with the result in problem 28.

***32.** Show that $P_n(\cos \gamma)$ is homogeneous of zeroth degree. (*Hint:* Note that $\cos \gamma = \cos \phi.(z/r) + \sin \phi \sin \theta.(x/r) + \sin \phi \sin \theta.(y/r)$.)

***33.** Prove (VIII.95). (*Hint:* Note that $P_n(\cos \gamma)$ satisfies (VIII.39). Multiply (VIII.39) by $\sin \phi$ and proceed as in Chapter V, §2.4. Apply the corollary of lemma V.1 to

$$\frac{\partial}{\partial \phi}\left(\sin \phi \frac{\partial P_n}{\partial \phi}\right) P_m - \frac{\partial}{\partial \phi}\left(\sin \phi \frac{\partial P_m}{\partial \phi}\right) P_n$$

and to $\frac{\partial^2 P_n}{\partial \theta^2} P_m - \frac{\partial^2 P_m}{\partial \theta^2} P_n$ and observe in the latter case that $\cos \gamma|_{\theta=-\pi} = \cos \gamma|_{\theta=\pi}$.)

34. Verify (VIII.98) for $n = 0, 1, 2$.

***35.** Establish Poisson's integral representation for the solution of the potential equation in two dimensions where the values of the potential are prescribed on the circumference of the unit circle.

36. Show that (VIII.82) can be obtained from (VIII.97) via (VIII.98).

37. Prove (VIII.96). (*Hint:* Observe that this relation is independent of the choice of the north pole $\theta = \phi = 0$.)

RECOMMENDED SUPPLEMENTARY READING

R. Courant and D. Hilbert: *Methods of Mathematical Physics*, Vol. 1, Interscience Publishers, New York, 1953.

E. W. Hobson: *The Theory of Spherical and Ellipsoidal Harmonics*, Cambridge University Press, London, 1931.

T. M. MacRobert: *Spherical Harmonics*, Dover Publications, New York, 1948.

Ph. Frank and R. von Mises: *Die Differential und Integralgleichungen der Mechanik und Physik*, Vols. I and II, F. Vieweg und Sohn, Braunschweig, 1930, 1935.

IX

THE NONHOMOGENEOUS BOUNDARY VALUE PROBLEM

§1. THE INFLUENCE FUNCTION (GREEN'S FUNCTION)

1. The Nonhomogeneous String Equation

In Chapter IV, §3, the problem of the stretched string under tensile forces was discussed and a solution was obtained in full generality. Except for the tensile force, no other external forces have been taken into account and this fact found its mathematical expression in the homogeneity of the string equation (II.3). We will now proceed to discuss the case where other external forces in addition to the tension have to be taken into consideration. We have seen in Chapter II, §1, that if a force density $f(x, t)$ (i.e., force per unit length) were to act upon the string at the point x at the time t, then the displacement $u(x, t)$ would satisfy equation (II.4)

$$\frac{\partial^2 u}{\partial t^2} = \frac{\tau}{\rho} \frac{\partial^2 u}{\partial x^2} + \frac{1}{\rho} f(x, t).$$

We do not make an attempt to solve this problem in such general form and consider instead the special case of the equilibrium position of the string under a time-independent force $f(x)$ and assume further that there is neither an initial displacement nor an initial drive. Thus, we have with $\tau = 1$ the following boundary value problem

$$\frac{\partial^2 u}{\partial x^2} = -f(x)$$

with the boundary conditions

$$u(x_1) = 0,$$
$$u(x_2) = 0.$$

While the problem in its original form appeared too complicated for a first attempt at a solution, now the problem appears too trivial to bother with. Therefore, we will generalize it insofar as we replace the left side by a self-adjoint differential expression of the second order and the boundary conditions by some of a more general nature as we have done in Chapters V and VII:

$$L[y] = \frac{d}{dx}[p(x)y'] + q(x)y = -f(x), \tag{IX.1}$$

$$\begin{aligned} a_{11}y(x_1) + a_{12}y'(x_1) &= 0, \\ a_{21}y(x_2) + a_{22}y'(x_2) &= 0. \end{aligned} \tag{IX.2}$$

We will assume that $f(x)$ is continuous and sectionally smooth in $x_1 \leq x \leq x_2$ in order to avoid difficulties that would turn up at a later stage of our development. Some insight into the problem and an opening for a constructive solution is obtained from the following heuristic argument:

Instead of considering $f(x)$ as a continuous force, we will view it as a number of point forces $f(\xi_1), f(\xi_2), \cdots, f(\xi_n)$ exerted upon the string at the points $\xi_1, \xi_2, \cdots, \xi_n$. Let $f^0(\xi_1), f^0(\xi_2), \cdots, f^0(\xi_n)$ denote the corresponding unit forces acting upon the points $\xi_1, \xi_2, \cdots, \xi_n$ and let $G(x, \xi_k)$ denote the displacement of the string at the point x due to the exertion of the unit force $f^0(\xi_k)$ at $x = \xi_k$. Then, the displacement of the string at x due to the force $f(\xi_k)$ would appear to be $G(x, \xi_k)f(\xi_k)$. In view of the linearity of our problem, the displacement of the string at the point x due to all point forces $f(\xi_1), f(\xi_2), \cdots, f(\xi_n)$ at $\xi_1, \xi_2, \cdots, \xi_n$ will appear as

$$y_n(x) = \sum_{k=1}^{n} G(x, \xi_k)f(\xi_k).$$

If we let the point forces become force densities, we expect to obtain for the displacement at x due to the external force $f(x)$

$$y(x) = \int_{x_1}^{x_2} G(x, \xi)f(\xi)\, d\xi. \tag{IX.3}$$

Hence, it would appear that the knowledge of the function $G(x, \xi)$ which is for obvious reasons called the *influence function* and often referred to as *Green's function*, is essential for the successful solution of our problem and we will devote the following subsections to an extensive discussion of the same.

2. Determination of Green's Function

Let us investigate in this subsection what properties the function $G(x, \xi)$—provided it exists—has to have in order to make $y(x)$ as represented in

(IX.3) the solution of the problem (IX.1), (IX.2). This consideration will furnish constructive ideas as to the establishment of Green's function.

Let us assume that $G(x, \xi)$ exists and that (IX.3) is a solution of (IX.1), (IX.2). By substituting the expression (IX.3) into (IX.1) we have to take some care when differentiating $G(x, \xi)$ with respect to x at the point $x = \xi$, because $G(x, \xi)$ corresponds to a force 0 at all points $x \neq \xi$. While $G(x, \xi)$, being a displacement function, has to be continuous for a fixed ξ in $x_1 \leq x \leq x_2$, the same certainly does not have to be true of its derivative. Hence, we interrupt the integration at $\xi = x$ when computing the derivatives y' and y''.

$$y'(x) = \int_{x_1}^{x} \frac{\partial G(x, \xi)}{\partial x} f(\xi)\, d\xi + G(x, x - 0) f(x) + \int_{x}^{x_2} \frac{\partial G(x, \xi)}{\partial x} f(\xi)\, d\xi - G(x, x + 0) f(x).$$

Because of the assumed continuity of $G(x, \xi)$ in ξ we have

$$G(x, x - 0) = G(x, x + 0)$$

and therefore

$$y'(x) = \int_{x_1}^{x_2} \frac{\partial G(x, \xi)}{\partial x} f(\xi)\, d\xi.$$

Hence,

$$\begin{aligned} y''(x) &= \int_{x_1}^{x} \frac{\partial^2 G(x, \xi)}{\partial x^2} f(\xi)\, d\xi + \frac{\partial G(x, x - 0)}{\partial x} f(x) \\ &\quad + \int_{x}^{x_2} \frac{\partial^2 G(x, \xi)}{\partial x^2} f(\xi)\, d\xi - \frac{\partial G(x, x + 0)}{\partial x} f(x) \\ &= \int_{x_1}^{x_2} \frac{\partial^2 G(x, \xi)}{\partial x^2} f(\xi)\, d\xi - f(x)\left[\frac{\partial G(x, x + 0)}{\partial x} - \frac{\partial G(x, x - 0)}{\partial x}\right]. \end{aligned}$$

Substitution of these expressions into (IX.1) leads to

$$\begin{aligned} & p(x)\left\{\int_{x_1}^{x_2} \frac{\partial^2 G(x, \xi)}{\partial x^2} f(\xi)\, d\xi - f(x)\left[\frac{\partial G(x, x + 0)}{\partial x} - \frac{\partial G(x, x - 0)}{\partial x}\right]\right\} \\ &\qquad + p'(x) \int_{x_1}^{x_2} \frac{\partial G(x, \xi)}{\partial x} f(\xi)\, d\xi + q(x) \int_{x_1}^{x_2} G(x, \xi) f(\xi)\, d\xi \\ &= \int_{x_1}^{x_2} f(\xi) L[G(x, \xi)]\, d\xi - p(x) f(x)\left[\frac{\partial G(x, x + 0)}{\partial x} - \frac{\partial G(x, x - 0)}{\partial x}\right] \\ &= -f(x). \end{aligned}$$

Thus, we obtain the following sufficient condition for $G(x, \xi)$:

(IX.3) is a solution of the differential equation (IX.1) if $G(x, \xi)$ satisfies the homogeneous equation corresponding to (IX.1)—which we will call in future the *associated homogeneous equation*—

$$L[G(x, \xi)] = 0$$

everywhere except at $x = \xi$ and at this point the relation

$$\frac{\partial G(x, x+0)}{\partial x} - \frac{\partial G(x, x-0)}{\partial x} = \lim_{\epsilon \to 0} \frac{\partial G(x, \xi)}{\partial x}\bigg|_{\xi = x-\epsilon}^{\xi = x+\epsilon} = \frac{1}{p(x)} \tag{IX.4}$$

holds.

Substitution of (IX.3) for y in the boundary conditions (IX.2) yields

$$\int_{x_1}^{x_2} f(\xi)\left[a_{11}G(x_1, \xi) + a_{12}\frac{\partial G(x_1, \xi)}{\partial x}\right] d\xi = 0,$$

$$\int_{x_1}^{x_2} f(\xi)\left[a_{21}G(x_2, \xi) + a_{22}\frac{\partial G(x_2, \xi)}{\partial x}\right] d\xi = 0,$$

which is certainly satisfied if $G(x, \xi)$ satisfies the boundary conditions (IX.2). Thus, we can state the following:

Theorem IX.1. *If a continuous function $G(x, \xi)$ which has a continuous derivative everywhere except at $x = \xi$ exists and is such that it satisfies the boundary conditions (IX.2) and the associated homogeneous equation $L[G] = 0$ everywhere except at the point $x = \xi$ and if (IX.4) holds, then $y(x) = \int_{x_1}^{x_2} G(x, \xi)f(\xi)\, d\xi$ is a solution of (IX.1), (IX.2).*

For reasons of convenience we will frequently refer to condition (IX.4) in the following equivalent form:

$$\lim_{\epsilon \to 0} \frac{\partial G(x, \xi)}{\partial x}\bigg|_{x = \xi-\epsilon}^{x = \xi+\epsilon} = -\frac{1}{p(\xi)}. \tag{IX.5}$$

That (IX.4) and (IX.5) are indeed equivalent can be seen from the following argument: Let

$$G(x, \xi) = \begin{cases} G_1(x, \xi) & \text{for } x_1 \le \xi < x, \\ G_2(x, \xi) & \text{for } x \le \xi \le x_2, \end{cases}$$

where $G_1(x, \xi)$, $G_2(x, \xi)$ are continuous functions and $G_1(x, x) = G_2(x, x)$. Then

$$\frac{\partial G(x, \xi)}{\partial x}\bigg|_{\xi = x-0}^{\xi = x+0} = \frac{\partial G_2(x, \xi)}{\partial x}\bigg|_{\xi = x} - \frac{\partial G_1(x, \xi)}{\partial x}\bigg|_{\xi = x} = \frac{1}{p(x)}.$$

The discontinuity of $\partial G/\partial x$ occurs at $\xi = x$ or, as we may state it, at $x = \xi$. In one case we consider x as the variable and ξ as the parameter and in the other case vice versa. Hence, we can also write

$$G(x, \xi) = \begin{cases} G_1(x, \xi) & \text{for } \xi < x \leq x_2, \\ G_2(x, \xi) & \text{for } x_1 \leq x \leq \xi, \end{cases} \qquad \text{where } G_1(\xi, \xi) = G_2(\xi, \xi)$$

and it follows that

$$\left.\frac{\partial G(x, \xi)}{\partial x}\right|_{x=\xi-0}^{x=\xi+0} = \left.\frac{\partial G_1(x, \xi)}{\partial x}\right|_{x=\xi} - \left.\frac{\partial G_2(x, \xi)}{\partial x}\right|_{x=\xi} = -\frac{1}{p(\xi)}$$

in view of (IX.4). Retracing the steps in this deduction will establish the equivalence of (IX.4) and (IX.5).

Problems IX.1-2

1. Verify (IX.4) and (IX.5) for

$$G(x, \xi) = \begin{cases} \dfrac{x^3\xi}{2} + \dfrac{x\xi^3}{2} - \dfrac{9x\xi}{5} + x, & 0 \leq x < \xi, \\[2ex] \dfrac{x^3\xi}{2} + \dfrac{x\xi^3}{2} - \dfrac{9x\xi}{5} + \xi, & \xi \leq x \leq 1, \end{cases}$$

for $p(x) = 1$.

2. Same as in problem 1 for

$$G(x, \xi) = \begin{cases} -\frac{1}{2} \log |1 - x|\,|1 + \xi| + \log 2 - \frac{1}{2}, & -1 \leq x < \xi, \\ -\frac{1}{2} \log |1 + x|\,|1 - \xi| + \log 2 - \frac{1}{2}, & \xi \leq x \leq 1, \end{cases}$$

for $p(x) = 1 - x^2$.

3. Examples

In order to gain some insight into the possibility of constructing Green's function, let us discuss two simple examples:

First, we deal with the equation

$$y'' + y = 1 \tag{IX.6}$$

and the boundary conditions

$$y(0) = 0 \tag{IX.7}$$

$$y\left(\frac{\pi}{2}\right) = 0. \tag{IX.8}$$

According to theorem IX.1, Green's function has to be a solution of the associated homogeneous equation

$$G'' + G = 0 \tag{IX.9}$$

in $0 \leq x < \xi$ and $\xi < x \leq \pi/2$ and has to satisfy the boundary conditions (IX.7) and (IX.8) besides having to satisfy (IX.4). A solution of (IX.9) in $0 \leq x < \xi$ that satisfies (IX.7) is

$$y_0(x) = \sin x$$

and a solution of (IX.9) in $\xi < x \leq \pi/2$ that satisfies (IX.8) is

$$y_1(x) = \cos x.$$

Hence, Green's function appears to be of the form

$$G(x, \xi) = \begin{cases} \sin x \cdot \phi(\xi) & \text{for } 0 \leq x < \xi, \\ \cos x \cdot \psi(\xi) & \text{for } \xi \leq x \leq \dfrac{\pi}{2}, \end{cases}$$

where ϕ and ψ have to be chosen such that (IX.5) is satisfied:

$$\left.\frac{\partial G(x, \xi)}{\partial x}\right|_{x=\xi+0} - \left.\frac{\partial G(x, \xi)}{\partial x}\right|_{x=\xi-0} = -\sin \xi \cdot \psi(\xi) - \cos \xi \cdot \phi(\xi) = -\frac{1}{p(\xi)} = -1,$$

($p = 1$ in our case). Clearly, the following choice for ϕ and ψ will prove satisfactory:

$$\phi(\xi) = \cos \xi = y_1(\xi),$$

$$\psi(\xi) = \sin \xi = y_0(\xi).$$

Hence, Green's function appears in the form

$$G(x, \xi) = \begin{cases} y_0(x)y_1(\xi) = \sin x \cos \xi & \text{for } 0 \leq x < \xi, \\ y_1(x)y_0(\xi) = \cos x \sin \xi & \text{for } \xi \leq x \leq \dfrac{\pi}{2}. \end{cases}$$

Therefore, we obtain as solution of (IX.6) with the boundary conditions (IX.7) and (IX.8) according to (IX.3)

$$y(x) = \int_0^{\pi/2} G(x, \xi)f(\xi)\, d\xi = -\cos x \int_0^x \sin \xi\, d\xi - \sin x \int_x^{\pi/2} \cos \xi\, d\xi$$

$$= 1 - \cos x - \sin x.$$

The reader can easily verify by substitution that this is indeed a solution. Thus, we succeeded in finding the solution of a nonhomogeneous self-adjoint differential equation with homogeneous boundary conditions by constructing Green's function.

To show how lucky we were in this case, let us now consider another example. We leave the differential equation (IX.6) unchanged, however change the boundary conditions slightly:

$$y(0) = 0, \tag{IX.10}$$

$$y(\pi) = 0. \tag{IX.11}$$

We see quite easily that it is impossible in this case to follow the same pattern as before, because any solution of the homogeneous equation (IX.9) that satisfies one of the two boundary conditions will also satisfy the other boundary condition. In other words, the associated homogeneous problem has a nontrivial solution, while the associated homogeneous problem of (IX.6), (IX.7), (IX.8) has the trivial solution only.

We shall see that the construction of Green's function is doomed to failure in this case and not only because of the reasons stated above, but also because the problem does not have a solution at all:

The solution of the associated homogeneous problem is

$$y_0(x) = \sin x.$$

Let us assume that there exists a solution of (IX.6), (IX.10), (IX.11) and call it $y(x)$. We substitute $y(x)$ into (IX.6), multiply both sides by $y_0(x)$ and integrate from 0 to π:

$$\int_0^\pi y_0(x)[y''(x) + y(x)]\,dx = \int_0^\pi y_0(x)\,dx. \tag{IX.12}$$

Since $L[y] = y'' + y$ is a self-adjoint differential expression and y_0, y both satisfy the same homogeneous boundary conditions, we obtain according to Green's lemma (lemma V.5)

$$\int_0^\pi y_0[y'' + y]\,dx = \int_0^\pi y[y_0'' + y_0]\,dx$$

which in turn is zero since y_0 is a solution of the homogeneous equation. It follows that

$$\int_0^\pi \sin x\,dx = 0$$

which is obviously not the case. Thus, the assumption that our problem has a solution leads to a contradiction. We will see later in §2.1 that whenever the associated homogeneous problem has a nontrivial solution, then in order for the nonhomogeneous problem to have a solution it is necessary that the disturbing force $f(x)$ satisfy the orthogonality condition $\int_{x_1}^{x_2} y_0(x)f(x)\,dx = 0.$

Problems IX.3–6

3. Find the solution of

$$y'' + y = \sin x, \qquad y(0) = 0, \quad y\left(\frac{\pi}{2}\right) = 0.$$

4. Solve problem 3 by Lagrange's method of variation of the parameters.
5. Find the solution of

$$y'' + y = \cos x, \qquad y(0) = 0, \quad y(\pi) = 0,$$

using Lagrange's method of variation of the parameters.

6. Show that the argument put forth at the end of subsection 3 does not apply to problem 5.

4. A General Discussion of the Nonhomogeneous and the Associated Homogeneous Boundary Value Problem

Before going any further in our investigation of Green's function, let us state and prove a general theorem with regard to the solvability of the nonhomogeneous and the associated homogeneous boundary value problem in relation to each other.

Let $\Lambda[y]$ stand for a linear differential expression of the second order with continuous coefficients

$$\Lambda[y] = a_0(x)y'' + a_1(x)y' + a_2(x)y$$

and let $B_1[y]$ and $B_2[y]$ denote the boundary functions

$$B_1[y] = a_{11}y(x_1) + a_{12}y'(x_1) + b_{11}y(x_2) + b_{12}y'(x_2),$$

$$B_2[y] = a_{21}y(x_1) + a_{22}y'(x_1) + b_{21}y(x_2) + b_{22}y'(x_2).$$

(See also (V.34).)

We consider the following nonhomogeneous boundary value problem

$$\begin{aligned} \Lambda[y] &= f(x), \\ B_1[y] &= c_1, \\ B_2[y] &= c_2 \end{aligned} \qquad \text{(IX.13)}$$

where c_1, c_2 are constants, and the associated homogeneous boundary value problem

$$\begin{aligned} \Lambda[\eta] &= 0, \\ B_1[\eta] &= 0, \\ B_2[\eta] &= 0. \end{aligned} \qquad \text{(IX.14)}$$

There are two possibilities to be considered as already indicated in the preceding section: Either (IX.14) has a nontrivial solution or (IX.14) has the trivial solution only. The effect of this on the solvability of (IX.13) is stated in

Theorem IX.2. *If the homogeneous problem (IX.14) has a nontrivial solution, then the nonhomogeneous problem (IX.13) has a solution if* $f(x)$, c_1, c_2 *satisfy certain conditions. If this is the case, then (IX.13) has infinitely many solutions. If, however, the homogeneous problem (IX.14) has the trivial solution only, then (IX.13) has a unique solution for any* $f(x)$, c_1, c_2.

Proof. $\Lambda[\eta] = 0$ has two linearly independent solutions η_1, η_2. Hence, the general solution can be written in the form

$$\eta = \alpha_1\eta_1 + \alpha_2\eta_2$$

and $\Lambda[y] = f(x)$ will have the general solution

$$y = y_0 + \alpha_1\eta_1 + \alpha_2\eta_2,$$

where y_0 is a particular solution of $\Lambda[y] = f(x)$. In order to satisfy the boundary conditions of (IX.13), we have to solve the following system of equations

$$B_1[y] = B_1[y_0] + \alpha_1 B_1[\eta_1] + \alpha_2 B_1[\eta_2] = c_1,$$

$$B_2[y] = B_2[y_0] + \alpha_1 B_2[\eta_1] + \alpha_2 B_2[\eta_2] = c_2.$$

These two linear nonhomogeneous equations for α_1, α_2 have the form

$$\begin{aligned} \alpha_1 B_1[\eta_1] + \alpha_2 B_1[\eta_2] &= c_1 - B_1[y_0], \\ \alpha_1 B_2[\eta_1] + \alpha_2 B_2[\eta_2] &= c_2 - B_2[y_0] \end{aligned} \tag{IX.15}$$

with the coefficient determinant

$$\Delta = \begin{vmatrix} B_1[\eta_1] & B_1[\eta_2] \\ B_2[\eta_1] & B_2[\eta_2] \end{vmatrix}.$$

We have to distinguish now between the case where $\Delta \neq 0$ and the case where $\Delta = 0$.

(1) $\Delta \neq 0$. In this case, (IX.15) has a unique solution α_1, α_2 and the solution of (IX.13) exists and is uniquely determined. For (IX.14) we obtain in this case

$$\eta = \alpha_1\eta_1 + \alpha_2\eta_2$$

and

$$\begin{aligned} \alpha_1 B_1[\eta_1] + \alpha_2 B_1[\eta_2] &= 0, \\ \alpha_1 B_2[\eta_1] + \alpha_2 B_2[\eta_2] &= 0. \end{aligned} \tag{IX.16}$$

Since $\Delta \neq 0$, this system has the trivial solution only and consequently, (IX.14) has the trivial solution $\eta = 0$ only.

(2) $\Delta = 0$. In this case (IX.16) has a nontrivial solution α_1, α_2 which is determined but for a common multiple. Hence (IX.14) has a nontrivial solution. Then, (IX.15) has in general no solution. However, there are values of $c_1 - B_1[y_0]$ and $c_2 - B_2[y_0]$ for which the system may have a solution and whenever a nonhomogeneous system of linear equations with a vanishing coefficient determinant has a nontrivial solution (i.e., if the equations are linearly dependent) then it has infinitely many solutions, q.e.d.

Remark: The solvability of (IX.13) in the latter case depends on the boundary values $c_1 - B_1[y_0]$, $c_2 - B_2[y_0]$ and since y_0 depends on $f(x)$, the solvability hinges on the function $f(x)$ as we pointed out in the preceding subsection.

Problem IX.7

7. Discuss the problem

$$\begin{aligned} y'' + y &= f(x), \\ y(0) &= 1, \\ y(\pi) &= 1 \end{aligned}$$

and the associated homogeneous problem in the light of theorem IX.2. Demonstrate all the possibilities stated in theorem IX.2 by suitable choices for $f(x)$.

5. The Existence of Green's Function

We have seen in subsection 3 that one seems to succeed in constructing Green's function in case the associated homogeneous problem has the trivial solution only. Let us state this as a general theorem:

Theorem IX.3. *Green's function, as defined in theorem IX.1, exists if the associated homogeneous boundary value problem has the trivial solution only.*

Proof. Let

$$y = c_1 y_1 + c_2 y_2$$

be the general solution of the homogeneous equation

$$L[y] = 0$$

where L denotes a self-adjoint operator of the second order. We determine the constants c_1, c_2 such that the first boundary condition in (IX.2)

$$a_{11} y(x_1) + a_{12} y'(x_1) = 0$$

is satisfied and hence

$$\bar{y}_1 = c_1^{(1)} y_1 + c_2^{(1)} y_2.$$

If we determine the constants c_1, c_2 such that the second condition in (IX.2)

$$a_{21}y(x_2) + a_{22}y'(x_2) = 0$$

is satisfied, we see that

$$\bar{y}_2 = c_1^{(2)}y_1 + c_2^{(2)}y_2.$$

Now, there are two possibilities: Either $\bar{y}_1 = \bar{y}_2$ or $\bar{y}_1 \neq \bar{y}_2$.

In the first case, $\bar{y}_1$ is a nontrivial solution of the associated homogeneous boundary value problem. (It is to be understood that the constants c_1, c_2 are only determined to within a constant factor and hence, y_1 and y_2 are to be considered as identical if they differ by a multiplicative factor only.) In the second case, a nontrivial solution of the associated homogeneous problem does not exist and we can construct the following function:

$$G(x, \xi) = \begin{cases} -\dfrac{\bar{y}_1(x)\bar{y}_2(\xi)}{p(\xi)[\bar{y}_1(\xi)\bar{y}_2'(\xi) - \bar{y}_1'(\xi)\bar{y}_2(\xi)]} & \text{for } x < \xi, \\[2ex] -\dfrac{\bar{y}_1(\xi)\bar{y}_2(x)}{p(\xi)[\bar{y}_1(\xi)\bar{y}_2'(\xi) - \bar{y}_1'(\xi)\bar{y}_2(\xi)]} & \text{for } \xi \leq x. \end{cases} \tag{IX.17}$$

The denominator does not vanish because

$$\bar{y}_1(\xi)\bar{y}_2'(\xi) - \bar{y}_1'(\xi)\bar{y}_2(\xi) = W[\bar{y}_1, \bar{y}_2](\xi)$$

and $\bar{y}_1, \bar{y}_2$ are assumed to be linearly independent, i.e., $W[\bar{y}_1, \bar{y}_2] \neq 0$. We can show further that

$$p(\xi)[\bar{y}_1\bar{y}_2' - \bar{y}_1'\bar{y}_2] = c \text{ (constant)}.$$

Since we have in view of (V.16)

$$\bar{y}_1L[\bar{y}_2] - \bar{y}_2L[\bar{y}_1] = \frac{d}{dx}[p(\bar{y}_1\bar{y}_2' - \bar{y}_1'\bar{y}_2)],$$

and because $\bar{y}_1$, as well as $\bar{y}_2$, are solutions of $L[y] = 0$, the left side vanishes and

$$\frac{d}{dx}[p(\bar{y}_1\bar{y}_2' - \bar{y}_1'\bar{y}_2)] = 0;$$

consequently

$$p(\bar{y}_1\bar{y}_2' - \bar{y}_1'\bar{y}_2) = c.$$

Now, we will show that the function $G(x, \xi)$ as defined in (IX.17) is Green's function of the nonhomogeneous problem

$$L[y] = -f(x)$$

with the boundary conditions (IX.2).

(1) $G(x, \xi)$ satisfies $L[G] = 0$ in $x_1 \leq x < \xi$ and $\xi < x \leq x_2$ because $\bar{y}_1, \bar{y}_2$ are solutions of $L[y] = 0$.

(2) $G(x, \xi)$ satisfies the boundary conditions (IX.2) because $\bar{y}_1, \bar{y}_2$ satisfy one boundary condition each,

and finally

$$(3) \qquad \frac{\partial G(x, \xi)}{\partial x}\Bigg|_{\xi = x-0}^{\xi = x+0} = -\frac{1}{c}(\bar{y}_2\bar{y}_1' - \bar{y}_1\bar{y}_2') = \frac{1}{p(x)}$$

and this completes the proof.

Looking at (IX.17) we can see that $G(x, \xi)$ is symmetric in x and ξ. However, we will give an independent proof of this fact in view of later developments, when we try to establish the symmetry of a generalized Green function where simple inspection would not suffice.

We state:

Lemma IX.1. *If a continuous function $G(x, \xi)$ exists, has a continuous derivative everywhere except at $x = \xi$, is such that it satisfies the boundary conditions (IX.2) and the homogeneous equation $L[G] = 0$ everywhere except at $x = \xi$, and if it satisfies the condition (IX.4) (or the equivalent condition (IX.5)), then*

$$G(x, \xi) = G(\xi, x),$$

i.e., Green's function is symmetric in the variable x and the parameter ξ.

Proof. Let us consider the following two functions

$$\Phi = G(x, \xi), \quad \Psi = G(x, \eta), \qquad \xi < \eta$$

where Φ and Ψ satisfy the conditions as stated in lemma IX.1 and in particular Φ satisfies (IX.5) at $x = \xi$ and Ψ satisfies (IX.5) at $x = \eta$. Since L is a self-adjoint operator, (V.16) holds

$$\Phi L[\Psi] - \Psi L[\Phi] = \frac{d}{dx}[p(\Psi'\Phi - \Phi'\Psi)].$$

Since G satisfies $L[G] = 0$ except at $x = \xi$, we have

$$L[\Psi] = L[\Phi] = 0$$

and consequently

$$\frac{d}{dx}[p(\Psi'\Phi - \Phi'\Psi)] = 0.$$

Let us integrate this expression over the interval $x_1 \leq x \leq x_2$ with interruptions at $x = \xi$ and $x = \eta$ due to the discontinuities of Φ' and Ψ', respectively, at these points:

$$\begin{aligned} p(\Psi'\Phi - \Psi\Phi')\Big|_{x_1}^{\xi} + p(\Psi'\Phi - \Psi\Phi')\Big|_{\xi}^{\eta} + p(\Psi'\Phi - \Psi\Phi')\Big|_{\eta}^{x_2} \\ = p(\xi)[\Psi'(\xi - 0, \eta)\Phi(\xi, \xi) - \Psi(\xi, \eta)\Phi'(\xi - 0, \xi)] \\ + p(\eta)[\Psi'(\eta - 0, \eta)\Phi(\eta, \xi) - \Psi(\eta, \eta)\Phi'(\eta - 0, \xi)] \\ - p(\xi)[\Psi'(\xi + 0, \eta)\Phi(\xi, \xi) - \Psi(\xi, \eta)\Phi'(\xi + 0, \xi)] \\ - p(\eta)[\Psi'(\eta + 0, \eta)\Phi(\eta, \xi) - \Psi(\eta, \eta)\Phi'(\eta + 0, \xi)] \\ + \{p(x)(\Psi'\Phi - \Psi\Phi')\}_{x_1}^{x_2} = 0. \end{aligned}$$

The last term, in braces, drops out because Φ and Ψ satisfy the same homogeneous boundary conditions (see lemma V.5). The remaining terms can be rearranged as follows

$$\begin{aligned} p(\xi)\Phi(\xi, \xi)[\Psi'(\xi - 0, \eta) - \Psi'(\xi + 0, \eta)] \\ + p(\xi)\Psi(\xi, \eta)[\Phi'(\xi + 0, \xi) - \Phi'(\xi - 0, \xi)] \\ + p(\eta)\Phi(\eta, \xi)[\Psi'(\eta - 0, \eta) - \Psi'(\eta + 0, \eta)] \\ + p(\eta)\Psi(\eta, \eta)[\Phi'(\eta + 0, \xi) - \Phi'(\eta - 0, \xi)] = 0. \end{aligned}$$

Since Ψ' and Φ' are continuous everywhere except at $x = \xi$ and $x = \eta$ respectively, we have

$$\Psi'(\xi - 0, \eta) - \Psi'(\xi + 0, \eta) = 0, \quad \Phi'(\eta + 0, \xi) - \Phi'(\eta - 0, \xi) = 0.$$

In view of (IX.5), we have

$$\Psi'(\eta - 0, \eta) - \Psi'(\eta + 0, \eta) = \frac{1}{p(\eta)},$$

and

$$\Phi'(\xi + 0, \xi) - \Phi'(\xi - 0, \xi) = -\frac{1}{p(\xi)}.$$

Hence, we obtain

$$-\Psi(\xi, \eta) + \Phi(\eta, \xi) = 0$$

or in view of the definitions of Ψ and Φ

$$G(\xi, \eta) = G[\eta, \xi], \quad \text{q.e.d.}$$

Since Green's function represents the effect at the point x, of a unit force applied at the point ξ we can interpret this result as follows: The unit force applied at the point ξ produces at the point x the same result as the unit force applied at x produces at ξ. (Such a reciprocity is frequently found in physics.)

Problems IX.8–11

8. Given $G(x, \xi) = (1 - \xi)x$ for $x \leq \xi$. Find $G(x, \xi)$ for $x > \xi$.
9. Same as in problem 8 with $G(x, \xi) = -\log \xi$ for $x \leq \xi$.
10. Given $G(x, \xi) = -\frac{1}{2} \log [(1 + x)(1 - \xi)] + c$ for $x > \xi$. Find $G(x, \xi)$ for $x \leq \xi$.
11. Given $G(x, \xi) = (x/2\xi)(1 - \xi^2)$ for $x < \xi$. Find $p(x)$.

§2. THE GENERALIZED GREEN FUNCTION

1. Construction of Green's Function in the Case Where the Associated Homogeneous Boundary Value Problem Has a Nontrivial Solution

Let us now deal with the case where $\bar{y}_1 = \bar{y}_2$, i.e., where the associated homogeneous boundary value problem has a nontrivial solution. As indicated by the second example in subsection 3 of §1, we expect that a theorem of the following type will hold:

Lemma IX.2. *If the homogeneous boundary value problem $L[y] = 0$ with the boundary conditions (IX.2) has a nontrivial solution $y_0(x)$, then, the associated nonhomogeneous boundary value problem $L[y] = -f(x)$ with the same boundary conditions (IX.2) has a solution if and only if*

$$\int_{x_1}^{x_2} f(x)y_0(x)\, dx = 0.$$

Proof. We will prove at this time only that this condition is necessary. That it is also sufficient will follow later when we have proved that some sort of influence function can be constructed under this condition (see theorem IX.4). Let us assume that $y(x)$ is a solution of the nonhomogeneous equation

$$L[y] = -f(x)$$

and that it satisfies (IX.2). Let us multiply this equation by the nontrivial solution $y_0(x)$ of the associated homogeneous problem and let us integrate from x_1 to x_2:

$$\int_{x_1}^{x_2} y_0 L[y]\, dx = -\int_{x_1}^{x_2} f(x)y_0(x)\, dx.$$

In view of (V.16) and lemma V.5,

$$\int_{x_1}^{x_2} y_0 L[y]\, dx = 0.$$

Hence

$$\int_{x_1}^{x_2} f(x)y_0(x)\, dx = 0, \quad \text{q.e.d.}$$

We see from this lemma that it is necessary that $f(x)$ be orthogonal to the solution of the associated homogeneous problem, in order for the nonhomogeneous boundary value problem to have a solution. As soon as we have clarified the existence and construction of Green's function for this case, we have solved our problem.

In order to pave the way for the construction of what is frequently referred to as the *generalized Green function*, let us prove the following:

Lemma IX.3. *If $L[y] = 0$ has a nontrivial solution y_0 that satisfies the boundary conditions (IX.2), then $L[y] = y_0(x)y_0(\xi)$ cannot have a nontrivial solution that satisfies the same boundary conditions.*

Proof (by contradiction). Let y be a solution of $L[y] = y_0(x)y_0(\xi)$ which satisfies (IX.2). We multiply this equation by $y_0(x)$ and integrate over the interval $x_1 \leq x \leq x_2$:

$$\int_{x_1}^{x_2} y_0 L[y]\, dx = y_0(\xi)\int_{x_1}^{x_2} y_0^2(x)\, dx.$$

It follows again in view of (V.16), $L[y_0] = 0$, and lemma V.5 that

$$\int_{x_1}^{x_2} y_0 L[y]\, dx = \int_{x_1}^{x_2} y L[y_0]\, dx = 0 \text{ and consequently}$$

$$y_0(\xi)\int_{x_1}^{x_2} y_0^2(x)\, dx = 0$$

and therefore $y_0(x) = 0$, which contradicts our assumption that $y_0(x)$ is a nontrivial solution, q.e.d.

Let us now assume that $y_0(x)$ is a nontrivial solution of $L[y] = 0$ and satisfies the boundary conditions (IX.2). According to lemma IX.3,

$$L[y] = y_0(x)y_0(\xi) \tag{IX.18}$$

cannot have a nontrivial solution that satisfies the same boundary conditions (IX.2). The general solution of (IX.18) will be

$$y = \eta(x)y_0(\xi) + a(\xi)y_1(x) + b(\xi)y_2(x) \tag{IX.19}$$

where $\eta(x)$ is a particular solution of $L[y] = y_0(x)$ and y_1, y_2 are linearly independent solutions of $L[y] = 0$.

Although a, b cannot be determined such that (IX.19) satisfies (IX.2), we can nevertheless determine a in terms of b such that (IX.19) satisfies the first of the boundary conditions (IX.2) at x_1. Let us call the value thus obtained $a_1(\xi, b_1(\xi))$. Next, we determine a such that the boundary condition (IX.2) at x_2 is satisfied and we will obtain a value which we will call $a_2(\xi, b_2(\xi))$.

Now, we consider the two functions

$$\begin{aligned} G_1(x, \xi) &= \eta(x)y_0(\xi) + a_1(\xi, b_1(\xi))y_1(x) + b_1(\xi)y_2(x), \\ G_2(x, \xi) &= \eta(x)y_0(\xi) + a_2(\xi, b_2(\xi))y_1(x) + b_2(\xi)y_2(x), \end{aligned} \tag{IX.20}$$

where $b_1(\xi)$, $b_2(\xi)$ still are arbitrary functions of ξ. Let us determine $b_1(\xi)$ in terms of $b_2(\xi)$ such that

$$G_1(\xi, \xi) = G_2(\xi, \xi).$$

That leaves us with one arbitrary function $b_2(\xi)$. The reader will realize by this time that we are trying to construct an influence function in analogy to Green's function by making use of the functions $G_1(x, \xi)$ and $G_2(x, \xi)$. Our final goal will be to show that in this case

$$\mathbf{G}(x, \xi) = \begin{cases} G_1(x, \xi) & \text{for } x_1 \le x < \xi, \\ G_2(x, \xi) & \text{for } \xi \le x \le x_2 \end{cases} \tag{IX.21}$$

is Green's function in a generalized sense, i.e., $\int_{x_1}^{x_2} \mathbf{G}(x, \xi)f(\xi)\, d\xi$ shall turn out to be a solution of the nonhomogeneous boundary value problem with the boundary conditions (IX.2) in case the associated homogeneous boundary value problem has a nontrivial solution.

Thus far, the function as defined in (IX.21) is continuous, satisfies the equation $L[\mathbf{G}] = y_0(x)y_0(\xi)$ except at the point $x = \xi$, and satisfies the boundary conditions (IX.2). We know from our previous discussion that the function $\mathbf{G}(x, \xi)$ also has to satisfy the condition (IX.5) and we will see that this condition is automatically satisfied by (IX.21) if $y_0(x)$ is normalized:

Lemma IX.4. *If* $\int_{x_1}^{x_2} y_0^2(x)\, dx = 1$ *and* $\mathbf{G}(x, \xi)$ *is defined as in (IX.21), then*

$$\left.\frac{\partial \mathbf{G}(x, \xi)}{\partial x}\right|_{x=\xi-0}^{x=\xi+0} = -\frac{1}{p(\xi)}.$$

Proof. L is a self-adjoint operator and $L[\mathbf{G}] = y_0(x)y_0(\xi)$ and $L[y_0] = 0$. Hence,

$$y_0L[\mathbf{G}] - \mathbf{G}L[y_0] = y_0^2(x)y_0(\xi) = \frac{d}{dx}[p(\mathbf{G}'y_0 - y_0'\mathbf{G})].$$

If we integrate this relation from x_1 to x_2 with the proper break at $x = \xi$ because of the discontinuity of $\mathbf{G}'$ at this point, we obtain

$$y_0(\xi)\int_{x_1}^{x_2} y_0^2(x)\, dx = [p(\mathbf{G}'y_0 - \mathbf{G}y_0')]_{x_1}^{\xi} + [p(\mathbf{G}'y_0 - \mathbf{G}y_0')]_{\xi}^{x_2}$$

and in view of

$$\int_{x_1}^{x_2} y_0^2(x)\,dx = 1:$$

$$\begin{aligned} y_0(\xi) = p(\xi)\mathbf{G}'(\xi - 0, \xi)y_0(\xi) - p(\xi)\mathbf{G}(\xi - 0, \xi)y_0'(\xi) \\ - p(\xi)\mathbf{G}'(\xi + 0, \xi)y_0(\xi) + p(\xi)\mathbf{G}(\xi + 0, \xi)y_0'(\xi) \\ + [p(x)(\mathbf{G}'y_0 - \mathbf{G}y_0')]_{x_1}^{x_2}. \end{aligned}$$

Since y_0 and **G** both satisfy (IX.2), the last term vanishes in view of lemma V.5. Tne second and fourth term cancel each other in view of the continuity of **G** and there remains

$$y_0(\xi) = p(\xi)y_0(\xi)[\mathbf{G}'(\xi - 0, \xi) - \mathbf{G}'(\xi + 0, \xi)].$$

From this follows immediately that

$$\frac{\partial \mathbf{G}(x, \xi)}{\partial x}\Bigg|_{x=\xi-0}^{x=\xi+0} = -\frac{1}{p(\xi)}, \quad \text{q.e.d.}$$

The symmetry in x and ξ of the generalized Green function as defined in (IX.21) can be established easily because we have still one arbitrary constant, namely $b_2(\xi)$, available. If we can determine this constant such that

$$\int_{x_1}^{x_2} (\Phi L[\Psi] - \Psi L[\Phi])\,dx = 0, \tag{IX.22}$$

where Φ and Ψ are defined as follows:

$$\Phi = \mathbf{G}(x, \xi), \quad \Psi = \mathbf{G}(x, \eta),$$

we can apply the proof of lemma IX.1 and obtain $\mathbf{G}(x, \xi) = \mathbf{G}(\xi, x)$.

Since

$$L[\mathbf{G}] = y_0(x)y_0(\xi),$$

and consequently

$$L[\Phi] = y_0(x)y_0(\xi),$$

and

$$L[\Psi] = y_0(x)y_0(\eta),$$

(IX.22) becomes

$$y_0(\eta)\int_{x_1}^{x_2} \mathbf{G}(x, \xi)y_0(x)\,dx - y_0(\xi)\int_{x_1}^{x_2} \mathbf{G}(x, \eta)y_0(x)\,dx = 0,$$

which is certainly satisfied if we determine $b_2(\xi)$ such that

$$\int_{x_1}^{x_2} \mathbf{G}(x, \xi)y_0(x)\,dx = 0. \tag{IX.23}$$

Now, we can investigate whether the function $\mathbf{G}(x, \xi)$ thus constructed enables us to find a solution of the nonhomogeneous boundary value problem $L[y] = -f(x)$ with the boundary conditions (IX.2).

Let

$$y = \int_{x_1}^{x_2} \mathbf{G}(x, \xi) f(\xi)\, d\xi.$$

Then,

$$\begin{aligned} L[y] &= \int_{x_1}^{x_2} f(\xi) L[\mathbf{G}(x, \xi)]\, d\xi + p(x) f(x)\left[\frac{\partial \mathbf{G}(x, x-0)}{\partial x} \right. \\ &\qquad \left. - \frac{\partial \mathbf{G}(x, x+0)}{\partial x}\right] \\ &= \int_{x_1}^{x_2} f(\xi) y_0(x) y_0(\xi)\, d\xi - \frac{f(x)p(x)}{p(x)} \\ &= y_0(x) \int_{x_1}^{x_2} f(\xi) y_0(\xi)\, d\xi - f(x). \end{aligned}$$

According to lemma IX.2 which states that the nonhomogeneous boundary value problem has a solution in this case only if the solution of the associated homogeneous problem is orthogonal to the disturbing force $f(x)$, we have to require that

$$\int_{x_1}^{x_2} f(\xi) y_0(\xi)\, d\xi = 0$$

and consequently

$$L[y] = -f(x).$$

This proves that we can obtain a solution by constructing Green's function in the generalized sense in the indicated way and shows at the same time that the orthogonality condition in lemma IX.2 is also sufficient. Let us now summarize the results which we obtained in the preceding discussion in the form of a theorem:

Theorem IX.4. *Let L be a self-adjoint operator. If the homogeneous boundary value problem $L[y] = 0$ with the boundary conditions (IX.2) has a nontrivial solution $y_0(x)$ which we consider normalized, i.e., $\int_{x_1}^{x_2} y_0^2(x)\, dx = 1$, then the associated nonhomogeneous boundary value problem $L[y] = -f(x)$ with the boundary conditions (IX.2) has a solution if and only if*

$$\int_{x_1}^{x_2} f(\xi) y_0(\xi)\, d\xi = 0.$$

In this case, the solution is given by

$$y = \int_{x_1}^{x_2} \mathbf{G}(x, \xi) f(\xi)\, d\xi$$

where the function $\mathbf{G}(x, \xi)$*—called the generalized Green function—is a solution of*

$$L[\mathbf{G}(x, \xi)] = y_0(x) y_0(\xi), \quad \textit{except at} \quad x = \xi,$$

satisfies the boundary conditions (IX.2), has everywhere a continuous derivative except at $x = \xi$*, satisfies the jump condition*

$$\lim_{\epsilon \to 0} \left. \frac{\partial \mathbf{G}(x, \xi)}{\partial x} \right|_{x=\xi-\epsilon}^{x=\xi+\epsilon} = -\frac{1}{p(\xi)},$$

and is such that

$$\int_{x_1}^{x_2} \mathbf{G}(x, \xi) y_0(x)\, dx = 0.$$

Problem IX.12

12. Given the boundary value problem

$$y'' + y = -f(x),$$
$$y(0) = 0, \quad y(a) = 0.$$

(a) Choose a such that Green's function exists.

(b) Choose a such that the generalized Green function exists and $f(x)$ such that the problem has a solution.

2. Examples

Let us now illustrate the construction of the generalized Green function and the subsequent solution of a nonhomogeneous boundary value problem with some examples:

Let us first consider the relatively simple example

$$y'' = -f(x)$$

with the boundary conditions

$$y(0) = 0, \quad y(1) - y'(1) = 0.$$

Clearly, the associated homogeneous problem has the nontrivial solution ax, which becomes upon normalization

$$y_0 = \sqrt{3}x.$$

Now, according to our theory, we have to solve the equation

$$y'' = \sqrt{3}x\sqrt{3}\xi.$$

The solution is

$$y = \frac{x^3\xi}{2} + a(\xi)x + b(\xi).$$

Hence,

$$G_1(x, \xi) = \frac{x^3\xi}{2} + a_1(\xi)x + b_1(\xi),$$

$$G_2(x, \xi) = \frac{x^3\xi}{2} + a_2(\xi)x + b_2(\xi).$$

The condition $G_1(0, \xi) = 0$ yields $b_1 = 0$ and the condition $G_2(1, \xi) - G_2'(1, \xi) = 0$ yields $b_2 = \xi$. Thus,

$$G_1(x, \xi) = \frac{x^3\xi}{2} + a_1(\xi)x,$$

$$G_2(x, \xi) = \frac{x^3\xi}{2} + a_2(\xi)x + \xi.$$

From the condition $G_1(\xi, \xi) = G_2(\xi, \xi)$ we obtain $a_1(\xi) = a_2(\xi) + 1$ and condition $\int_0^1 \mathbf{G}y_0\, dx = 0$ yields finally $a_2(\xi) = (\xi^3/2) - (9\xi/5)$. Hence

$$\mathbf{G}(x, \xi) = \begin{cases} \dfrac{x^3\xi}{2} + \dfrac{x\xi^3}{2} - \dfrac{9x\xi}{5} + x & \text{for } 0 \le x < \xi, \\[2ex] \dfrac{x^3\xi}{2} + \dfrac{x\xi^3}{2} - \dfrac{9x\xi}{5} + \xi & \text{for } \xi \le x \le 1, \end{cases}$$

and if $\int_0^1 xf(x)\, dx = 0$, then the solution of our problem is given by

$$y = \int_0^\xi \left(\frac{x^3\xi}{2} + \frac{x\xi^3}{2} - \frac{9x\xi}{5} + x\right) f(\xi)\, d\xi$$
$$+ \int_\xi^1 \left(\frac{x^3\xi}{2} + \frac{x\xi^3}{2} - \frac{9x\xi}{5} + \xi\right) f(\xi)\, d\xi.$$

Next, we consider Legendre's equation of zeroth order with a disturbing force:

$$\frac{d}{dx}[(1 - x^2)y'] = -f(x)$$

and require that the solutions satisfy the boundary conditions $|y(-1)| < \infty$ and $|y(1)| < \infty$.

Clearly, the associated homogeneous boundary value problem

$$\frac{d}{dx}[(1 - x^2)y'] = 0$$

has a nontrivial normalized solution which satisfies the regularity conditions at -1 and 1, namely

$$y_0(x) = \frac{1}{\sqrt{2}}.$$

Hence, $y_0(x)y_0(\xi) = \frac{1}{2}$ and we have to solve the equation

$$\frac{d}{dx}[(1 - x^2)y'] = \frac{1}{2},$$

which has the general solution

$$y = -\tfrac{1}{4}\log|1 - x^2| + a(\xi)\log\left|\frac{1 + x}{1 - x}\right| + b(\xi),$$

or, as we can also write,

$$y = (a - \tfrac{1}{4})\log|1 + x| - (a + \tfrac{1}{4})\log|1 - x| + b(\xi).$$

We consider now the functions

$$G_1(x, \xi) = (a_1 - \tfrac{1}{4})\log|1 + x| - (a_1 + \tfrac{1}{4})\log|1 - x| + b_1,$$
$$G_2(x, \xi) = (a_2 - \tfrac{1}{4})\log|1 + x| - (a_2 + \tfrac{1}{4})\log|1 - x| + b_2,$$

and determine a_1 such that $|G_1(-1, \xi)| < \infty$ and a_2 such that $|G_2(1, \xi)| < \infty$. Obviously, we obtain

$$a_1 = \tfrac{1}{4}, \quad a_2 = -\tfrac{1}{4}.$$

Hence

$$G_1(x, \xi) = -\tfrac{1}{2}\log|1 - x| + b_1(\xi),$$
$$G_2(x, \xi) = -\tfrac{1}{2}\log|1 + x| + b_2(\xi).$$

Our next step is to determine $b_2(\xi)$ such that

$$G_1(\xi, \xi) = G_2(\xi, \xi)$$

and we obtain

$$b_2(\xi) = b_1(\xi) + \tfrac{1}{2}\log\left|\frac{1 + \xi}{1 - \xi}\right|.$$

Thus,

$$G_1(x, \xi) = -\tfrac{1}{2}\log|1 - x| + b_1(\xi),$$

$$G_2(x, \xi) = -\tfrac{1}{2}\log|1 + x| + b_1(\xi) + \tfrac{1}{2}\log\left|\frac{1 + \xi}{1 - \xi}\right|.$$

Now, let us determine $b_1(\xi)$ such that (IX.23) is satisfied

$$\int_{-1}^{\xi} (-\tfrac{1}{2}\log|1 - x| + b_1)\,dx + \int_{\xi}^{1}\Big(-\tfrac{1}{2}\log(1 + x) + b_1 + \tfrac{1}{2}\log\left|\frac{1+\xi}{1-\xi}\right|\Big)\,dx = 0:$$

$$\tfrac{1}{2}[(1 - \xi)\log|1 - \xi| - (1 - \xi) - 2\log 2 + 2] + b_1(\xi)(\xi + 1)$$
$$- \tfrac{1}{2}[2\log 2 - 2 - (1 + \xi)\log|1 + \xi| + (1 + \xi)] + b_1(\xi)(1 - \xi)$$
$$+ \tfrac{1}{2}\log\left|\frac{1+\xi}{1-\xi}\right|(1 - \xi) = 0,$$

$$- 2\log 2 + 1 + 2b_1(\xi) + \log|1 + \xi| = 0.$$

Hence

$$b_1(\xi) = \log 2 - \tfrac{1}{2} - \tfrac{1}{2}\log|1 + \xi|$$

and we have finally

$$\mathbf{G}(x, \xi) = \begin{cases} -\tfrac{1}{2}\log|1 - x|\,|1 + \xi| + \log 2 - \tfrac{1}{2} & \text{for } -1 \le x < \xi, \\ -\tfrac{1}{2}\log|1 + x|\,|1 - \xi| + \log 2 - \tfrac{1}{2} & \text{for } \xi \le x \le 1. \end{cases}$$

Now, if $f(x)$ is such that

$$\int_{-1}^{1} f(x)\,dx = 0,$$

then,

$$y = \int_{-1}^{1} \mathbf{G}(x, \xi) f(\xi)\,d\xi$$

is the solution of our problem.

Problems IX.13–18

13. Find the solution of the latter example in subsection 2 of this section for a suitably chosen function $f(x)$.

14. Find Green's function for the boundary value problems:
(a) $y'' = -f(x)$, $y(0) = 0$, $y'(1) = 0$.
(b) $y'' = -f(x)$, $y(-1) = 0$, $y(1) = 0$.

***15.** Find Green's function for the boundary value problem

$$xy'' + y' = -f(x), \quad y(1) = 0, \quad |y(0)| < \infty.$$

***16.** Find Green's function for the boundary value problem

$$\frac{d}{dx}(xy') - \frac{\mu^2}{x}y = -f(x), \qquad y(1) = 0, \quad |y(0)| < \infty.$$

***17.** Find Green's function for the boundary value problem

$$y'' = -f(x),$$
$$y(-1) - y(1) = 0,$$
$$y'(-1) - y'(1) = 0.$$

18. In problems 15 and 16 the boundary condition at one end point is not homogeneous. Explain, why the discussion leading to lemma IX.1 is still valid in these cases, especially the application of lemma V.5.

§3. FURTHER ASPECTS OF GREEN'S FUNCTION

1. Green's Function and the Sturm-Liouville Boundary Value Problem

Let us write equation (IX.1) in the more elaborate form

$$L[y] + \lambda R(x)y = -f(x) \qquad \text{(IX.24)}$$

where $L[y]$ is, as in Chapters V and VII, the self-adjoint differential expression

$$L[y] = \frac{d}{dx}[P(x)y'] + Q(x)y.$$

As mentioned already on several occasions, the left side of (IX.24) is also self-adjoint if $L[y]$ is self-adjoint. We leave the boundary conditions (IX.2) unchanged.

We can now state the following theorem regarding the solvability of (IX.24), (IX.2):

Theorem IX.5. *If $\lambda_1, \lambda_2, \lambda_3, \cdots$ are the eigenvalues of the Sturm-Liouville problem*

$$L[y] + \lambda Ry = 0$$

with the boundary conditions (IX.2) (which is clearly the homogeneous problem associated with (IX.24), (IX.2)), then the nonhomogeneous problem (IX.24), (IX.2) has a unique solution for all values $\lambda \neq \lambda_k$ ($k = 1, 2, 3, \cdots$) and this solution can be represented by means of Green's function (which is in this case also a function of λ):

$$y = \int_{x_1}^{x_2} G(x, \xi, \lambda)f(\xi)\, d\xi, \qquad \lambda \neq \lambda_k.$$

However, if $\lambda = \lambda_k$, then a solution of (IX.24), (IX.2) exists if and only if

$$\int_{x_1}^{x_2} f(x)y_k(x)\, dx = 0$$

whereby y_k is the eigenfunction associated with the eigenvalue λ_k.

The *proof* of this theorem can be established by piecing together theorems V.7, V.8, IX.1, IX.3, and IX.4.

A general theorem regarding the convergence of the Fourier expansion of a given function in terms of eigenfunctions (as developed in Chapter V, §2.5 from a formal viewpoint) can be obtained via Green's function, as we shall point out without going into detail:

We can assume without loss of generality that $\lambda = 0$ is not an eigenvalue of

$$L[y] + \lambda Ry = 0$$

with the boundary conditions (IX.2).[1]

If $f(x)$ is twice differentiable and satisfies the boundary conditions (IX.2), then $y = f(x)$ is a solution of

$$L[y] + \lambda Ry = L[f(x)]$$

for $\lambda = 0$. By means of Green's function that is associated with this particular problem, we can represent $f(x)$ by the following integral equation:

$$f(x) = -\int_{x_1}^{x_2} G(x, \xi, 0)L[f(\xi)]\, d\xi. \tag{IX.25}$$

The problem of expanding $f(x)$ into a uniformly convergent series in terms of eigenfunctions thus appears reduced to the problem of expanding $G(x, \xi, 0)$ into a uniformly convergent series in terms of eigenfunctions. If λ_n are the eigenvalues of the Sturm-Liouville problem and y_n the associated eigenfunctions, then one can show that the following expansion holds and converges uniformly:

$$G(x, \xi, 0) = \sum_{n=1}^{\infty} \frac{y_n(x)y_n(\xi)}{\lambda_n}, \tag{IX.26}$$

whereby

$$\int_{x_1}^{x_2} R(x)y_n^2(x)\, dx = 1$$ [2]

We do not attempt to establish (IX.26) by proof because such undertaking would be slightly beyond the scope of this treatment. However, we wish to point out how one can obtain from (IX.26) a uniformly convergent expansion of $f(x)$ as was claimed in Chapter V, §2.6.

In view of (IX.26), we can write (IX.25) in the form

$$f(x) = -\int_{x_1}^{x_2} \sum_{k=1}^{\infty} \frac{y_k(x)y_k(\xi)}{\lambda_k} L[f(\xi)]\, d\xi.$$

[1] See E. A. Coddington and N. Levinson: *Theory of Ordinary Differential Equations*, McGraw-Hill Book Company, New York, 1955, p. 193.

[2] See R. Courant and D. Hilbert: *Methods of Mathematical Physics*, pp. 134 and 360.

Because of the uniform convergence of (IX.26), we can interchange summation and integration and obtain

$$f(x) = -\sum_{k=1}^{\infty} \frac{y_k(x)}{\lambda_k} \int_{x_1}^{x_2} y_k(\xi) L[f(\xi)]\, d\xi.$$

Both $y_k(x)$ and $f(x)$ satisfy the boundary conditions (IX.2), hence

$$\int_{x_1}^{x_2} y_k(\xi)L[f(\xi)] - f(\xi)L[y_k(\xi)]\, d\xi = 0$$

in view of lemma V.5 and consequently

$$\int_{x_1}^{x_2} y_k(\xi)L[f(\xi)]\, d\xi = \int_{x_1}^{x_2} f(\xi)L[y_k(\xi)]\, d\xi.$$

Hence

$$f(x) = -\sum_{k=1}^{\infty} \frac{y_k(x)}{\lambda_k} \int_{x_1}^{x_2} f(\xi)L[y_k(\xi)]\, d\xi.$$

Now

$$L[y_k] = -\lambda_k R y_k$$

because y_k is an eigenfunction of $L[y] + \lambda R y = 0$ under the boundary condition (IX.2), and we see finally that

$$f(x) = \sum_{k=1}^{\infty} y_k(x) \int_{x_1}^{x_2} R(\xi) f(\xi) y_k(\xi)\, d\xi = \sum_{k=1}^{\infty} c_k y_k(x)$$

with

$$c_k = \int_{x_1}^{x_2} R(\xi) f(\xi) y_k(\xi)\, d\xi,$$

which is identical with the result that was obtained in Chapter V, §2.5.

As an illustration, let us expand the function

$$f(x) = x(1 - x)$$

into a Fourier sine series in the interval $0 \le x \le 1$. If we proceed according to Chapter IV, we have to extend the definition of our function such that the extended function is odd and has the period 2, i.e.,

$$f(x) = \begin{cases} x(1 - x) & \text{for } 0 \le x \le 1, \\ (x - 1)(x - 2) & \text{for } 1 < x \le 2, \end{cases}$$

where $f(x + 2) = f(x)$.

Then, we obtain the following Fourier expansion:

$$f(x) = \sum_{k=1}^{\infty} b_k \sin k\pi x$$

with

$$b_k = 2\int_0^1 \sin k\pi x(x - x^2)\,dx = \begin{cases} \dfrac{8}{\pi^3 k^3} & \text{for } k = 2\nu + 1, \\ 0 & \text{for } k = 2\nu. \end{cases}$$

Consequently

$$f(x) = \frac{8}{\pi^3}\sum_{\nu=0}^{\infty}\frac{\sin(2\nu + 1)\pi x}{(2\nu + 1)^3}. \tag{IX.27}$$

If we proceed now along the lines as developed in this subsection, we have to consider a Sturm-Liouville problem with the boundary conditions

$$y(0) = 0,$$
$$y(1) = 0$$

(these are the conditions satisfied by $f(x)$), which has

$$y_k = \sqrt{2}\sin k\pi x$$

as normalized eigenfunctions. Clearly

$$y'' + \lambda y = 0$$

will serve the purpose and we obtain as eigenvalues $\lambda = k^2\pi^2$ ($k = 1, 2, 3, 4, \cdots$). Two solutions of this equation with $\lambda = 0$ which satisfy either the first or the second boundary condition are

$$y_1 = x, \quad y_2 = 1 - x$$

and Green's function is, according to (IX.17),

$$G(x, \xi, 0) = \begin{cases} x(\xi - 1) & \text{for } 0 \le x < \xi, \\ \xi(x - 1) & \text{for } \xi \le x \le 1, \end{cases}$$

which has the expansion

$$G(x, \xi, 0) = 2\sum_{k=1}^{\infty}\frac{\sin k\pi x \sin k\pi\xi}{k^2\pi^2}$$

according to (IX.26).

Now,

$$L[f(x)] = f''(x) = -2$$

and consequently, in view of (IX.25),

$$f(x) = \frac{4}{\pi^2}\sum_{k=1}^{\infty}\frac{\sin k\pi x}{k^2}\int_0^1 \sin k\pi\xi\,d\xi.$$

Since $\int_0^1 \sin k\pi\xi\,d\xi = 2/k\pi$ for $k = 2\nu + 1$ and 0 for $k = 2\nu$, we arrive at (IX.27).

Problems IX.19–23

***19.** Prove that $|P_n(x)| \leq 1$ for $|x| \leq 1$, where $P_n(x)$ represents a Legendre polynomial.

20. Prove that

$$\sum_{n=1}^{\infty} \frac{P_n(x)P_n(\xi)}{n(n+1)}$$

converges uniformly in $|x| \leq 1$, $|\xi| \leq 1$.

21. Prove that

$$\sum_{k=1}^{\infty} \frac{\sin k\pi x \sin k\pi\xi}{k^2\pi^2}$$

converges uniformly everywhere.

22. Find the expansion of Green's function for the problem

$$\frac{d}{dx}(xy') - \frac{\mu^2}{x} = -f(x), \qquad \mu \geq 1,$$

$$y(0) = 0, \quad y(1) = 0.$$

23. Expand the function $f(x) = (x - 1)(x + 1)$ in $|x| \leq 1$ into a Fourier series via Green's function and check your result by using the method of Chapter IV.

2. Green's Function for Partial Differential Equations

Let us consider the equation of a membrane in equilibrium state under the influence of a time-independent force. Let the external force density (force per unit area) be represented by $f(x, y)$ and assume that this function is continuous. Then, equation (II.34) will reduce to

$$\frac{\partial^2 u}{\partial x^2} + \frac{\partial^2 u}{\partial y^2} = -f(x, y) \tag{IX.28}$$

if we choose the units conveniently ($\alpha^2\rho = \tau = 1$).

We will discuss the simple case where the membrane is stretched over a closed simply connected region R which is encompassed by a rectifiable curve C:

$$u(x, y)\big|_{x,y\in C} = 0. \tag{IX.29}$$

In order to obtain a solution for this nonhomogeneous problem, we can reasonably expect that a procedure analogous to the one followed in the preceding discussion will accomplish our purpose. With this preconceived idea in mind, we will assume that there exists a function $G(x, y; \xi, \eta)$ which represents the displacement u of the membrane at the point (x, y)

due to the influence of a unit force exerted upon the point (ξ, η). Then, we conjecture the solution of our problem to be

$$u(x, y) = \iint_R G(x, y; \xi, \eta) f(\xi, \eta)\, d\xi\, d\eta, \tag{IX.30}$$

where R is the region which is enclosed by C. It is reasonable to expect that $G(x, y; \xi, \eta)$ has to satisfy conditions which are analogous to those we have imposed on $G(x, \xi)$ in the one-dimensional case. There is really no difficulty in generalizing these conditions except for (IX.4) or (IX.5). In order to provide a smooth transition, let us ponder a little over (IX.5) and see if we can put it into a form that will be easily accessible to a generalization. Condition (IX.5) reads

$$\left.\frac{\partial G(x, \xi)}{\partial x}\right|_{x=\xi-\epsilon}^{x=\xi+\epsilon} = -\frac{1}{p(\xi)}.$$

We know also that $G(x, \xi)$ has to satisfy $L[G] = 0$ everywhere except at $x = \xi$ (in the case where the homogeneous problem has the trivial solution only) i.e.,

$$\frac{d}{dx}\left[p\,\frac{\partial G(x, \xi)}{\partial x}\right] + qG(x, \xi) = 0 \quad \text{for} \quad x \neq \xi.$$

Hence, in view of (IX.5),

$$\int_{x_1}^{x_2}\left[\frac{d}{dx}\left(p\,\frac{\partial G(x, \xi)}{\partial x}\right) + qG(x, \xi)\right] dx = \lim_{\epsilon\to 0}\int_{\xi-\epsilon}^{\xi+\epsilon}\left[\frac{d}{dx}\left(p\,\frac{\partial G}{\partial x}\right) + qG\right] dx$$

$$= \lim_{\epsilon\to 0}\left[p\,\left.\frac{\partial G(x, \xi)}{\partial x}\right|_{x=\xi-\epsilon}^{x=\xi+\epsilon} + 2q(\bar{x})G(\bar{x}, \xi)\cdot\epsilon\right] = -1$$

where $\xi - \epsilon \leq \bar{x} \leq \xi + \epsilon$.
Thus, we can replace (IX.4) or (IX.5) by the following condition:

$$\lim_{\epsilon\to 0}\int_{\xi-\epsilon}^{\xi+\epsilon} L[G(x, \xi)]\, dx = -1. \tag{IX.31}$$

If we generalize this condition to two dimensions with $L = \Delta$, we obtain

$$\lim_{\epsilon\to 0}\iint_\rho \Delta G(x, y; \xi, \eta)\, dx\, dy = -1,$$

where ρ represents the region $(x - \xi)^2 + (y - \eta)^2 \leq \epsilon^2$.
Application of Green's identity (AI.C4)

$$\iint_R (v\Delta u - u\Delta v)\, dx\, dy = \int_C \left(v\,\frac{\partial u}{\partial n} - u\,\frac{\partial v}{\partial n}\right) ds$$

with $v = 1$ and $u = G(x, y; \xi, \eta)$, renders

$$\lim_{\epsilon \to 0} \iint_{\rho} \Delta G(x, y; \xi, \eta)\, dx\, dy = \oint_{(x-\xi)^2+(y-\eta)^2=\epsilon^2} \frac{\partial G}{\partial n}\, ds,$$

where n is the direction of the outward normal to the circle $(x - \xi)^2 + (y - \eta)^2 = \epsilon^2$. Clearly, n can be replaced by $r = \sqrt{(x - \xi)^2 + (y - \eta)^2}$ since the boundary is a circle and the radius vector of a circle is perpendicular to its circumference, and we obtain

$$\lim_{\epsilon \to 0} \oint_{\gamma_\epsilon} \frac{\partial G(x, y; \xi, \eta)}{\partial r}\, ds = -1$$

where γ_ϵ denotes the circle with radius ϵ and center in (ξ, η). In order to get our bearings in our search for a function which will satisfy this condition, let us consider the effect of a singularity of the first order at the point (ξ, η). We introduce polar coordinates

$$x = \xi + r\cos\theta,$$
$$y = \eta + r\sin\theta$$

with $ds = \sqrt{r^2 + r'^2}\, d\theta$ and consider

$$\int_{\gamma_\epsilon} \frac{ds}{r}.$$

Since the equation of γ_ϵ is now $r = \epsilon$, we have $r' = 0$ and

$$\int_{\gamma_\epsilon} \frac{ds}{r} = \int_0^{2\pi} \frac{\epsilon\, d\theta}{\epsilon} = 2\pi.$$

This shows, that if we choose for $G(x, y; \xi, \eta)$ a function such that its derivative has a singularity of the type $-(1/2\pi r)$, we will be able to satisfy the above-stated condition. We take therefore

$$G(x, y; \xi, \eta) = -\frac{1}{2\pi}\log r + g(x, y; \xi, \eta)$$

where $g(x, y; \xi, \eta)$ and its first- and second-order derivatives are presumed continuous. Clearly,

$$\lim_{\epsilon \to 0} \int_{\gamma_\epsilon} \frac{\partial}{\partial r}\left(-\frac{1}{2\pi}\log r + g(x, y; \xi, \eta)\right) ds = -1.$$

After these preparatory remarks, we are in a position to prove the following

Theorem IX.6. *If the problem $\Delta u = -f(x, y)$ with the boundary condition $u|_C = 0$ where C is a closed rectifiable curve in the x, y–plane encompassing a simply connected region, has a solution, then the solution can be represented in the form*

$$u(x, y) = \iint_R G(x, y; \xi, \eta) f(\xi, \eta)\, d\xi\, d\eta, \tag{IX.32}$$

where R is the region encompassed by C and $G(x, y; \xi, \eta)$ is a function with the following properties:

$G(x, y; \xi, \eta)$ is of the form

$$G(x, y; \xi, \eta) = -\frac{1}{2\pi} \log r + g(x, y; \xi, \eta)$$

where g, $\partial g/\partial x$, $\partial^2 g/\partial x^2$, $\partial g/\partial y$, $\partial^2 g/\partial y^2$ are continuous in R, $G(x, y; \xi, \eta)|_{x,y\in C} = 0$ and $\Delta g(x, y; \xi, \eta) = 0$ everywhere.

(From the latter condition it follows that also $\Delta G = 0$ everywhere except at the point $x = \xi$, $y = \eta$ because $\log r$ is a potential function, i.e., a solution of the potential equation everywhere except at $x = \xi$, $y = \eta$—see problem II.34.)

Proof. Let us substitute u, as represented in (IX.32) into equation (IX.28):

$$\Delta u = \Delta \iint_R G(x, y; \xi, \eta) f(\xi, \eta)\, d\xi\, d\eta.$$

Because of our assumptions on G we can interchange differentiation and integration and obtain

$$\Delta u = \iint_R \Delta G(x, y; \xi, \eta) f(\xi, \eta)\, d\xi\, d\eta$$

and since $\Delta g = 0$ everywhere,

$$\Delta u = -\frac{1}{2\pi} \iint_R \Delta \log r \,.\, f(\xi, \eta)\, d\xi\, d\eta = -\frac{1}{2\pi} \lim_{\epsilon \to 0} f(\bar{\xi}, \bar{\eta}) \iint_\rho \Delta \log r\, d\xi\, d\eta$$

$$= -\frac{1}{2\pi} \lim_{\epsilon \to 0} f(\bar{\xi}, \bar{\eta}) \int_{\gamma_\epsilon} \frac{ds}{r} = -f(x, y),$$

and this concludes our proof.

With the remark that $G(x, y; \xi, \eta)$ also proves to be symmetric in x, y and ξ, η and as such indicates the same reciprocity which we already encountered in the one-dimensional problem, we will conclude our presentation of Green's function and leave it to the reader to pursue the subject by a study of the appropriate literature, which should be easily accessible—and easily comprehensible, once a thorough understanding of the subject as presented in this chapter is reached.

Problem IX.24

***24.** Find the equilibrium state of the membrane which is stretched over a unit circle with center at 0 and exposed to a time-independent external force $f(x, y) = 1 - r^2$, where $r = \sqrt{x^2 + y^2}$.

RECOMMENDED SUPPLEMENTARY READING

R. Courant and D. Hilbert: *Methods of Mathematical Physics*, Vol. I, Interscience Publishers, New York, 1953.

E. A. Coddington and N. Levinson: *Theory of Ordinary Differential Equations*, McGraw-Hill Book Co., New York, 1955.

Ph. Frank and R. von Mises: *Die Differential und Integralgleichungen der Mechanik und Physik*, Vol. I, F. Vieweg und Sohn, Braunschweig, 1930.

APPENDIX

Key for Reference Symbols Used in the Appendix

The capital letter(s) is (are) the code letter(s) for the author as listed below. The number(s) following the code letter(s) is (are) the page number(s), in the quoted publication, on which the reference can be found.

A = T. Apostol: *Mathematical Analysis*, Addison-Wesley Publishing Company, Reading, Mass., 1957.
CI = R. Courant: *Differential and Integral Calculus*, Vol. I, Interscience Publishers, New York, 1934.
CII = R. Courant: *Differential and Integral Calculus*, Vol. II, Interscience Publishers, New York, 1937.
CL = E. A. Coddington and N. Levinson: *Theory of Ordinary Differential Equations*, McGraw-Hill Book Co., New York, 1955.
F = L. R. Ford: *Differential Equations*, McGraw-Hill Book Co., New York, 1955.
H = W. Hurewicz: *Lectures on Ordinary Differential Equations*, Technology Press, MIT, 1958.
K = W. Kaplan: *Ordinary Differential Equations*, Addison-Wesley Publishing Company, 1958.
L = H. Lass: *Vector and Tensor Analysis*, McGraw-Hill Book Co., New York, 1950.
SR = I. S. Sokolnikoff and R. M. Redheffer: *Mathematics of Physics and Modern Engineering*, McGraw-Hill Book Co., New York, 1958.
T = A. Taylor: *Advanced Calculus*, Ginn and Company, Boston, 1955.
W = C. Wylie: *Advanced Engineering Mathematics*, 2nd ed., McGraw-Hill Book Co., New York, 1960.

The reader's attention is called to the fact that the terms used in this appendix for theorems and symbols are at times different from the ones used in the reference material.

I. VECTOR ANALYSIS

A. Vector Operations

1. Terminology:

Let E_2 denote the two-dimensional Euclidean plane, E_3 the three-dimensional Euclidean space. **i** and **j** are unit vectors in the direction of the coordinate axes in E_2, and **i**, **j**, **k** are unit vectors in the direction of the coordinate axes in E_3.

The components of a vector **A** are always denoted by a_1 and a_2 in E_2 and by a_1, a_2, a_3 in E_3. Thus, the components of **F** will be denoted by f_1, f_2, f_3, etc.

2. Absolute value

$$|\mathbf{A}| = \sqrt{a_1^2 + a_2^2} \quad \text{in } E_2 \qquad \text{and} \quad |\mathbf{A}| = \sqrt{a_1^2 + a_2^2 + a_3^2} \quad \text{in } E_3$$

is called the *absolute value* or *Euclidean length* of a vector.

3. Scalar multiplication:

Let α stand for a *scalar* (number or function). For a vector **F** in E_3,

$$\alpha\mathbf{F} = \alpha f_1\mathbf{i} + \alpha f_2\mathbf{j} + \alpha f_3\mathbf{k}$$

and analogously, for a vector in E_2,

$$\alpha\mathbf{F} = \alpha f_1\mathbf{i} + \alpha f_2\mathbf{j}.$$

4. Dot product (inner product):

(A 111, CII 7, L 10, SR 293, 319, T 283, W 462.)

Let **F**, **G** be vectors and θ the angle between them, then

$$\mathbf{F} \cdot \mathbf{G} = |\mathbf{F}||\mathbf{G}| \cos \theta$$

is called the *dot product* (*inner product*).

In E_2 we have $\mathbf{F} \cdot \mathbf{G} = f_1 g_1 + f_2 g_2$, and in E_3 we have $\mathbf{F} \cdot \mathbf{G} = f_1 g_1 + f_2 g_2 + f_3 g_3$.

5. Vector product (cross product):

(A 308, CII 14, L 20, SR 294, T 287, W 463.)

Let **F**, **G** be vectors in E_2; then, the *cross product* (*vector product*) is defined as

$$\mathbf{F} \times \mathbf{G} = \begin{vmatrix} f_1 & f_2 \\ g_1 & g_2 \end{vmatrix}.$$

If **F**, **G** are vectors in E_3, then

$$\mathbf{F} \times \mathbf{G} = \begin{vmatrix} \mathbf{i} & \mathbf{j} & \mathbf{k} \\ f_1 & f_2 & f_3 \\ g_1 & g_2 & g_3 \end{vmatrix}$$

is a vector which is perpendicular to **F** and to **G** and is such that (**F**, **G**, **F** × **G**) have the same orientation relative to each other as the three coordinate axes (x, y, z).

In both cases

$$|\mathbf{F} \times \mathbf{G}| = |\mathbf{F}||\mathbf{G}| \sin \theta$$

where θ is the angle between **F** and **G**; i.e., the absolute value of the cross product is equal to the area of the parallelogram spanned by **F** and **G**. (Note that in the case of E_2, $\mathbf{F} \times \mathbf{G} = |\mathbf{F} \times \mathbf{G}|$.)

B. Differential Operations

1. Tangent and normal vectors:

(G. B. Thomas, *Calculus*, Addison-Wesley, Reading, Mass., 1953, pp. 349, 355, CII 86.)

Let C be a smooth curve in E_2. If $x = x(s)$, $y = y(s)$ is the parametric representation of C, then $x(s)$, $y(s)$ are continuous and have a continuous derivative with respect to s.

Then

$$\mathbf{t} = \frac{dx}{ds}\mathbf{i} + \frac{dy}{ds}\mathbf{j}$$

is the *tangent vector* to C. If s stands for the arc length on C, then $|\mathbf{t}| = 1$.

The vector

$$\mathbf{n} = \frac{dy}{ds}\mathbf{i} - \frac{dx}{ds}\mathbf{j}$$

is the *normal vector* to C.

2. The Nabla (or Del) operator ∇*:*

(A 321, CII 92, L 40, SR 370, T 308, W 480.)

In E_2, the ∇ *operator* is defined as

$$\nabla = \frac{\partial}{\partial x}\mathbf{i} + \frac{\partial}{\partial y}\mathbf{j}$$

and in E_3 as

$$\nabla = \frac{\partial}{\partial x}\mathbf{i} + \frac{\partial}{\partial y}\mathbf{j} + \frac{\partial}{\partial z}\mathbf{k}.$$

3. The gradient:

(A 111, 319, CII 88, L 36, SR 320, T 301, W 481.)

Let $\Phi = \Phi(x, y, z)$ be a scalar function such that the derivatives with respect to the coordinates exist. Then

$$\text{grad}\,\Phi = \nabla\Phi = \frac{\partial\Phi}{\partial x}\mathbf{i} + \frac{\partial\Phi}{\partial y}\mathbf{j} + \frac{\partial\Phi}{\partial z}\mathbf{k}$$

is called the *gradient of* Φ.

$\Phi(x, y, z) = C$ represents a surface S in E_3. Let C be any curve on S and let $x = x(s)$, $y = y(s)$, $z = z(s)$ be the parametric representation of C. Then $d\Phi/ds = 0$ on S. Since

$$\frac{d\Phi}{ds} = \frac{\partial \Phi}{\partial x} \cdot \frac{dx}{ds} + \frac{\partial \Phi}{\partial y} \cdot \frac{dy}{ds} + \frac{\partial \Phi}{\partial z} \cdot \frac{dz}{ds},$$

and

$$\mathbf{t} = \frac{dx}{ds}\mathbf{i} + \frac{dy}{ds}\mathbf{j} + \frac{dz}{ds}\mathbf{k}$$

is the tangent vector to C in the tangent plane to S, we have

$$\text{grad } \Phi \cdot \mathbf{t} = 0,$$

i.e., grad Φ is perpendicular to S.

If $\mathbf{n}^0$ is a unit vector, i.e., $|\mathbf{n}^0| = 1$, in the direction perpendicular to S, then

$$\mathbf{n}^0 = \frac{\text{grad } \Phi}{|\text{grad } \Phi|}.$$

In E_2,

$$\text{grad } \Phi = \frac{\partial \Phi}{\partial x}\mathbf{i} + \frac{\partial \Phi}{\partial y}\mathbf{j}.$$

$\Phi(x, y) = C$ represents a curve C in E_2 and grad Φ is perpendicular to C.

4. The divergence:

(A 322, CII 91, L 42, SR 384, T 306, 457, W 484.)

Let $\mathbf{F} = \mathbf{F}(x, y, z)$ be a vector function in E_3 with differentiable components. Then

$$\text{div } \mathbf{F} = \nabla \cdot \mathbf{F} = \frac{\partial f_1}{\partial x} + \frac{\partial f_2}{\partial y} + \frac{\partial f_3}{\partial z}$$

is called the *divergence of* $\mathbf{F}$. The divergence in E_2 is defined analogously. If $\mathbf{F}$ is a two- or three-dimensional velocity field in a medium of density 1, then div $\mathbf{F}$ represents the absolute loss of matter from a unit area (in E_2) or a unit volume (in E_3) per unit time.

5. The Laplacian Δ:

(A 325, CII 93, L 45, SR 413, T 309, W 489.)

Let $\Phi = \Phi(x, y, z)$ be a scalar function with continuous first- and second-order derivatives. Then

$$\Delta\Phi = \nabla^2\Phi = \text{div grad } \Phi = \frac{\partial^2 \Phi}{\partial x^2} + \frac{\partial^2 \Phi}{\partial y^2} + \frac{\partial^2 \Phi}{\partial z^2}$$

is called the *Laplacian* or *Delta* of Φ.

6. The Laplacian in cylindrical and spherical coordinates:
(A 352, CII 368, 390, L 55, 56, SR 407, 408, W 442, 450.)
If we introduce cylindrical coordinates in subsection 5,

$$x = r\cos\theta,$$
$$y = r\sin\theta,$$
$$z = z,$$

then

$$\Delta\Phi = \frac{1}{r}\frac{\partial}{\partial r}\left(r\frac{\partial\Phi}{\partial r}\right) + \frac{\partial^2\Phi}{r^2\,\partial\theta^2} + \frac{\partial^2\Phi}{\partial z^2},$$

and for spherical coordinates,

$$x = r\cos\theta\sin\phi,$$
$$y = r\sin\theta\sin\phi,$$
$$z = r\cos\phi,$$

we obtain

$$\Delta\Phi = \frac{1}{r^2}\frac{\partial}{\partial r}\left(r^2\frac{\partial\Phi}{\partial r}\right) + \frac{1}{r^2\sin^2\phi}\frac{\partial^2\Phi}{\partial\theta^2} + \frac{1}{r^2\sin\phi}\cdot\frac{\partial}{\partial\phi}\left(\sin\phi\frac{\partial\Phi}{\partial\phi}\right).$$

7. The curl:
(A 320, CII 91, 404, L 45, SR 396, T 312, W 484.)
Let **F** be a vector in E_3 with differentiable components. Then

$$\text{curl}\,\mathbf{F} = \nabla\times\mathbf{F} = \begin{vmatrix} \mathbf{i} & \mathbf{j} & \mathbf{k} \\ \dfrac{\partial}{\partial x} & \dfrac{\partial}{\partial y} & \dfrac{\partial}{\partial z} \\ f_1 & f_2 & f_3 \end{vmatrix}$$

is called the *curl of* **F**.

There exists a function $\Phi(x, y, z)$ such that $\mathbf{F} = \text{grad}\,\Phi$ if and only if curl $\mathbf{F} = 0$. An analogous theorem holds for the E_2, where the curl of **F** is defined as the scalar

$$\text{curl}\,\mathbf{F} = \begin{vmatrix} \dfrac{\partial}{\partial x} & \dfrac{\partial}{\partial y} \\ f_1 & f_2 \end{vmatrix}.$$

8. Directional derivative:
(A 111, CII 64, L 38, SR 369, T 419, W 481.)
Let $\mathbf{A}^0$ be a unit vector, i.e., $a_1^2 + a_2^2 + a_3^2 = 1$. Then, the derivative of $\Phi(x, y, z)$ in the direction of $\mathbf{A}^0$ is given by

$$\frac{\partial\Phi}{\partial a} = \frac{\partial\Phi}{\partial x}a_1 + \frac{\partial\Phi}{\partial y}a_2 + \frac{\partial\Phi}{\partial z}a_3 = \text{grad}\,\Phi\cdot\mathbf{A}^0.$$

C. Integral Theorems

1. Independence of a line integral:
(A 292, L 105, SR 378, W 514.)

$$\oint_C (f_1\,dx + f_2\,dy + f_3\,dz) = \oint_C \mathbf{F}\cdot\mathbf{t}\,ds = 0$$

if and only if curl $\mathbf{F} = 0$, where C is a closed rectifiable curve in E_3.

2. Green's theorem:
(A 283, CII 360, 364, 366, SR 342, T 420, W 502.)
Let R be a simply connected region in E_2 encompassed by the rectifiable curve C which is oriented in the positive direction. Let C be represented by $x = x(s)$, $y = y(s)$. Let $\mathbf{F}(x, y)$ represent a vector field which may be interpreted as a velocity field. Then, in a medium of density 1, the absolute loss of matter from an area element ΔR per unit time is given by

$$\operatorname{div}\mathbf{F}\cdot\Delta R.$$

On the other hand, the amount of matter passing through an element Δs of the curve C in the direction of the outward normal is given by

$$\mathbf{F}\cdot\mathbf{n}^0\,\Delta s$$

where $\mathbf{n}^0$ is the unit vector in the direction of the outward normal. Hence

$$\iint_R \operatorname{div}\mathbf{F}\,dx\,dy = \oint_C \mathbf{F}\cdot\mathbf{n}^0\,ds,$$

or in components

$$\iint_R \left(\frac{\partial f_1}{\partial x} + \frac{\partial f_2}{\partial y}\right) dx\,dy = \oint_C f_1\,dy - f_2\,dx,$$

or in view of $f_1\,dy - f_2\,dx = \mathbf{F}\times d\mathbf{s}$:

$$\iint_R \operatorname{div}\mathbf{F}\,dx\,dy = \oint_C \mathbf{F}\times d\mathbf{s}.$$

3. The divergence theorem:
(A 339, CII 386–388, L 114, SR 389, T 449, 453, W 505.)
Let D be a three-dimensional, simply connected domain which is enclosed by the rectifiable surface S.
If $\mathbf{F}$ represents a vector field in E_3, then, in analogy to subsection 2,

$$\iiint_D \operatorname{div}\mathbf{F}\,dV = \iint_S \mathbf{F}\cdot\mathbf{n}^0\,dS$$

where $\mathbf{n}^0$ is the unit vector in the direction of the outward normal to S.

4. Green's formula:
(A 303, CII 390, L 118, SR 493, T 459, W 506.)
Application of the divergence theorem in subsection 3 to the vector field

$$\mathbf{F} = \Phi \operatorname{grad} \Psi,$$

where Φ and Ψ are scalar functions with continuous first- and second-order derivatives, yields

$$\iiint_D \operatorname{div}(\Phi \operatorname{grad} \Psi)\, dV = \iint_S \Phi \operatorname{grad} \Psi \mathbf{n}^0\, dS.$$

Since

$$\operatorname{div}(\Phi \operatorname{grad} \Psi) = \operatorname{grad} \Phi \operatorname{grad} \Psi + \Phi \operatorname{div} \operatorname{grad} \Psi$$

and

$$\operatorname{grad} \Psi \,.\, \mathbf{n}^0 = \frac{\partial \Psi}{\partial n},$$

we obtain, if we also write $\Delta\Psi$ for div grad Ψ,

$$\iiint_D (\operatorname{grad} \Phi \operatorname{grad} \Psi + \Phi\, \Delta\Psi)\, dV = \iint_S \Phi \frac{\partial \Psi}{\partial n}\, dS,$$

and for $\Phi = \Psi$, we have

$$\iiint_D (\operatorname{grad}^2 \Phi + \Phi\, \Delta\Phi)\, dV = \iint_S \Phi \frac{\partial \Phi}{\partial n}\, dS.$$

In two dimensions, the analogous formula

$$\iint_R (\operatorname{grad} \Phi \operatorname{grad} \Psi + \Phi\, \Delta\Psi)\, dx\, dy = \oint_C \Phi \frac{\partial \Psi}{\partial n}\, ds$$

holds where C is a rectifiable curve enclosing the region R, which is simply connected.

Interchange of Φ and Ψ and subtraction of the newly obtained formula from the one above yields *Green's identity:*

$$\iint_R (\Phi\, \Delta\Psi - \Psi\, \Delta\Phi)\, dx\, dy = \oint_C \left(\Phi \frac{\partial \Psi}{\partial n} - \Psi \frac{\partial \Phi}{\partial n}\right) ds.$$

II. CONVERGENCE

A. Definitions

1. Uniform convergence of sequences:
(A 393, CI 392, T 591.)
The sequence $\{u_n(x)\}$ is called *uniformly convergent* toward the limit function $u(x)$ ( ) in the interval $a \leq x \leq b$ if and only if for every

arbitrarily small $\epsilon > 0$ there exists an N_ϵ independent of x such that $|u_n(x) - u(x)| < \epsilon$ ($|u_n(x) - u_m(x)| < \epsilon$) for all $n > N_\epsilon$ (for all $n, m > N_\epsilon$): Weierstrass criterion (Cauchy criterion).

2. Uniform convergence of series:
(A 395, CI 391, T 593.)

The series $\sum_{n=1}^{\infty} u_n(x)$ is called *uniformly convergent* in the interval $a \leq x \leq b$ if and only if for every arbitrarily small $\epsilon > 0$ there exists an N_ϵ independent of x such that $|u_n(x) + u_{n+1}(x) + \cdots| = |R_n(x)| < \epsilon$ ($|u_n(x) + \cdots + u_m(x)| < \epsilon$) for all $n > N_\epsilon$ (for all $n, m > N_\epsilon$): Weierstrass criterion (Cauchy criterion).

3. Absolute convergence of series:
(A 359, CI 369, T 556.)

The series $\sum_{n=1}^{\infty} u_n(x)$ is called *absolutely convergent* at $x = x_0$ if and only if $\sum_{n=1}^{\infty} |u_n(x)|$ converges at $x = x_0$. Otherwise, the series is called *conditionally convergent*, if it is convergent at all.

B. Convergence Tests

1. A necessary condition:
(CI 367.)
It is necessary in order that $\sum_{n=1}^{\infty} u_n(x)$ converge at $x = x_0$ that $\lim_{n\to\infty} u_n(x_0) = 0$.

2. Comparison test (Weierstrass test):
(A 396, CI 392, T 596.)

For the series $\sum_{n=1}^{\infty} u_n(x)$ to converge *uniformly* and *absolutely* in $a \leq x \leq b$ it is sufficient that the series can be *dominated* (*majorized*) by a convergent series of positive constants; i.e., there exists a convergent series $\sum_{n=1}^{\infty} a_n$, where $a_n > 0$ and is constant, such that $|u_n(x)| \leq a_n$ for all n.

C. Theorems

1. Geometric series:
(A 361, CI 34, T 539.)

$\sum_{n=0}^{\infty} q^n$ converges for all $|q| < 1$ and diverges for all $|q| \geq 1$; in the first case,

$$\sum_{n=0}^{\infty} q^n = \frac{1}{1 - q}.$$

2. Harmonic series and generalization:
(A 363, CI 382, T 552.)

The series $\sum_{n=1}^{\infty} \frac{1}{n^{\alpha}}$ converges for all $\alpha > 1$ and diverges for all $\alpha \leq 1$.

3. Uniformly convergent sequences and series:
(A 394, CI 393, T 598.)

If all the functions $u_n(x)$ in the sequence $\{u_n(x)\}$ $\left(\text{series } \sum_{n=1}^{\infty} u_n(x)\right)$ are continuous in $a \leq x \leq b$ and if the sequence (series) converges uniformly in the interval $a \leq x \leq b$, then $\lim_{n\to\infty} u_n(x) = u(x)$ $\left(\sum_{n=1}^{\infty} u_n(x) = U(x)\right)$ is continuous in $a \leq x \leq b$.

4. Integration of sequences and series:
(A 399, 401, CI 394, T 599.)

If all the $u_n(x)$ are continuous in $a \leq x \leq b$ and if the sequence $\{u_n(x)\}$ $\left(\text{series } \sum_{n=1}^{\infty} u_n(x)\right)$ converges uniformly in the interval $a \leq x \leq b$, then

$$\lim_{n\to\infty} \int_a^b u_n(x)\,dx = \int_a^b \lim_{n\to\infty} u_n(x)\,dx \left(\sum_{n=1}^{\infty} \int_a^b u_n(x)\,dx = \int_a^b \sum_{n=1}^{\infty} u_n(x)\,dx\right).$$

5. Differentiation of series:
(A 403, CI 396, T 602.)

If all the $u_n'(x)$ are continuous in $a \leq x \leq b$ and if $\sum_{n=1}^{\infty} u_n'(x)$ converges uniformly in $a \leq x \leq b$, then $\frac{d}{dx} \sum_{n=1}^{\infty} u_n(x) = \sum_{n=1}^{\infty} u_n'(x)$ in $a \leq x \leq b$.

6. Riemann's rearrangement law:
(A 367, CI 373, T 561.)

The terms in the series $\sum_{n=1}^{\infty} u_n(x)$ can be arbitrarily rearranged without changing the value of the sum if and only if $\sum_{n=1}^{\infty} u_n(x)$ converges absolutely.

7. Convergence of a product:
(A 381, CI 421.)

A sufficient condition for the product $\prod_{n=1}^{\infty} (1 + a_n)$ to converge is, that $\sum_{n=1}^{\infty} |a_n|$ converges and that $(1 + a_n) \neq 0$ for all n.

D. Power Series

1. Convergence of a power series:
(A 409, CI 399, T 605, 607.)

If $\sum_{n=0}^{\infty} c_n x^n$ converges for $x = r$, then it converges *absolutely* for $|x| < |r|$ and *uniformly* for $|x| \leq a < |r|$.

2. Termwise integration and differentiation of power series:
(A 412, 413, CI 401, 402, T 608, 611.)

If $\sum_{n=0}^{\infty} c_n x^n$ converges for $|x| < r$, then $f(x) = \sum_{n=0}^{\infty} c_n x^n$ can be integrated and differentiated termwise as often as one pleases in $|x| < r$.

3. Product of power series:
(A 413, CI 403, T 619.)

If $f(x) = \sum_{n=0}^{\infty} a_n x^n$ and $g(x) = \sum_{n=0}^{\infty} b_n x^n$, and both series converge in $|x| < r$, then

$$f(x)g(x) = \sum_{n=0}^{\infty} (a_0 b_n + a_1 b_{n-1} + \cdots + a_n b_0) x^n \quad \text{in} \quad |x| < r.$$

III. ORDINARY DIFFERENTIAL EQUATIONS

A. Theorem on Implicit Functions

(A 144, T 256.)
Let

$$\text{(S)} \qquad \begin{aligned} x &= x(u, v), \\ y &= y(u, v), \end{aligned}$$

where x, y and $\partial x/\partial u$, $\partial y/\partial u$, $\partial x/\partial v$, $\partial y/\partial v$ are continuous in a region that has the point $P_0(u_0, v_0)$ in its interior. Let $x_0 = x(u_0, v_0)$ and $y_0 = y(u_0, v_0)$, and let $\bar{P}_0$ denote the point (x_0, y_0).

If the *Jacobian* (*functional determinant*)

$$\frac{\partial(x, y)}{\partial(u, v)} = \begin{vmatrix} \dfrac{\partial x}{\partial u} & \dfrac{\partial y}{\partial u} \\ \dfrac{\partial x}{\partial v} & \dfrac{\partial y}{\partial v} \end{vmatrix} \neq 0$$

at the point P_0, then there exists a neighborhood N of $\bar{P}_0$ such that the system (S) can be solved uniquely for u, v,

$$\begin{aligned} u &= u(x, y), \\ v &= v(x, y), \end{aligned}$$

and u, v, $\partial u/\partial x$, $\partial v/\partial x$, $\partial u/\partial y$, $\partial v/\partial y$ are continuous in N.

B. Systems of Ordinary Differential Equations of the First Order

1. The initial value problem:

Let $P_0(x_0, y_0, z_0)$ be a given point. The *initial value problem* for a system of two differential equations of the first order consists of finding a system of two functions (*solutions*) $y(x)$, $z(x)$ such that

$$\text{(OD)} \qquad \begin{aligned} y' &= f(x, y, z), \\ z' &= g(x, y, z) \end{aligned}$$

and $y(x_0) = y_0$, $z(x_0) = z_0$. (g and f in (OD) are given functions.)

2. The existence of solutions:

(CL 12, 19, F 121, H 28, K 474, 486.)

If f and g are continuous in a rectangle R: $|x - x_0| \le a$, $|y - y_0| \le b$, $|z - z_0| \le b$, which implies that there exists a constant M such that $|f| \le M$, $|g| \le M$ on R and if f and g satisfy a *Lipschitz condition* with a constant k in R, i.e.,

$$|f(x, \bar{y}, \bar{z}) - f(x, y, z)| \le k(|\bar{y} - y| + |\bar{z} - z|),$$
$$|g(x, \bar{y}, \bar{z}) - g(x, y, z)| \le k(|\bar{y} - y| + |\bar{z} - z|),$$

then there exists a system of two solutions $y = y(x)$, $z = z(x)$ of the initial value problem in subsection 1 such that $y(x_0) = y_0$, $z(x_0) = z_0$ in an interval I: $|x - x_0| \le \alpha$ where $\alpha = \min [a, (b/M)]$. The solutions are continuous on I and have continuous derivatives on I.

(For a proof of this theorem without Lipschitz condition, see CL 6 or H 10.)

3. The uniqueness of the solutions:

(CL 10, 19, F 121, H 28, K 482, 486.)

The solutions of the initial value problem (subsection 1) are unique under the conditions stated in subsection 2.

4. System of two linear differential equations of the first order:

Given

$$y' = a_{11}(x)y + a_{12}(x)z,$$
$$z' = a_{21}(x)y + a_{22}(x)z,$$

where the $a_{ik}(x)$ are continuous functions in $\alpha \le x \le \beta$. Then, the solutions $y = y(x)$ and $z = z(x)$, which assume the initial values $y(x_0) = y_0$, $z(x_0) = z_0$ with x_0 in $\alpha \le x \le \beta$, exist and are unique through the entire interval $\alpha \le x \le \beta$. (This theorem follows from subsections 2 and 3 since $b = \infty$ in this case and the Lipschitz condition is clearly satisfied in view of the continuity of the $a_{ik}(x)$.)

5. Dependence of the solutions on a parameter:
(CL 29, H 31, K 508.)
Let f and g in (OD) be functions of a parameter λ:

$$f = f(x, y, z, \lambda), \quad g = g(x, y, z, \lambda).$$

If the functions f and g satisfy the conditions in subsection 2 and are continuous in λ for $|\lambda - \lambda_0| \leq c$, where λ_0 is a given value, then, the solutions $y = y(x, \lambda)$, $z = z(x, \lambda)$ are also continuous in λ in some neighborhood of λ_0.

If, in particular, f and g are linear functions of λ, then the solutions are continuous in λ for all λ.

6. Linear independence of solutions:
(CL 83, F 64.)
The solutions $\phi_1 = (y_1, z_1)$ and $\phi_2 = (y_2, z_2)$ of the system of linear differential equations in subsection 4 are *linearly independent* in the interval $a \leq x \leq b$ if and only if for some x_0 in $a \leq x \leq b$

$$\begin{vmatrix} y_1 & y_2 \\ z_1 & z_2 \end{vmatrix} \neq 0.$$

ANSWERS AND HINTS TO EVEN NUMBERED PROBLEMS

CHAPTER I

I.2. Yes.

I.4. $\frac{1}{2}$.

I.6. $\dfrac{\partial z(t_1, \epsilon)}{\partial \epsilon} = 0, \dfrac{\partial z(t_2, \epsilon)}{\partial \epsilon} = 0.$

I.8. Shortest path joining (0, 0) and (1, 1).

I.10. (a) $y'' + x^2 y = 0$, (b) $\frac{1}{4}\sqrt{x/y} - y'' = 0$,

(c) $\cos(xy') - \sin(xy')(xy' + x^2 y'') = 0$, (d) $\dfrac{x^2}{(1 + y'^2)^{3/2}} = \text{constant}.$

I.12. $\dfrac{d}{dx} f_{y'} = \dfrac{d}{dy} f_{y'} y'$; $f_y\, dy - df_{y'} y' = 0.$
Since $df = f_y\, dy + f_{y'}\, dy'$, we have $f_y\, dy = df - f_{y'}\, dy'$ and we obtain $df - f_{y'}\, dy' - y'\, df_{y'} = 0$ or $df - d(y' f_{y'}) = 0$.

I.14. $y = C\sqrt{1 + y'^2}$.

I.16. $f_y - \dfrac{d}{dx} f_{y'} + \dfrac{d^2}{dx^2} f_{y''} = 0.$

I.18. $2y'''(1 + y'^2) - 6y'y''^2 = a(1 + y'^2)^4.$

I.20. $u_{xx} + u_{yy} + u_{xx}u_y^2 + u_{yy}u_x^2 - 2u_x u_y u_{xy} = 0.$

I.22. $f_z - \displaystyle\sum_{i=1}^{n} \frac{\partial}{\partial x_i} \frac{\partial f}{\partial z_{x_i}} + \sum_{i=1}^{n} \sum_{k=1}^{n} \frac{\partial^2}{\partial x_i \partial x_k} \frac{\partial f}{\partial z_{x_i x_k}} = 0.$

I.24. $y = a\left(\cosh \dfrac{x - \frac{1}{2}}{a} - \cosh \dfrac{1}{2a}\right).$

I.26. $xy'' + y' - \lambda xy = 0.$

I.28. $\dfrac{d}{dx}(py') + (q - \lambda r)y = 0.$

I.30. Use a two-parameter variation of $\sqrt{\frac{3}{2}}\, x$ and determine one parameter in terms of the other one such that $\displaystyle\int_{-1}^{1} y^2\, dx = 1$ for all variations.

I.32. $t = \displaystyle\int_0^1 \frac{-x(x - 1)y' \pm \sqrt{(1 + y'^2) - x^2(x - 1)^2}}{1 - x^2(x - 1)^2}\, dx.$

I.34. Note that if (x_0, y_0) is a point on C and $M(x_0, y_0) > 0$ where M is continuous, then there exists a δ such that

$$M(x, y) > 0 \quad \text{for all } (x - x_0)^2 + (y - y_0)^2 \leq \delta^2.$$

If $x(s)$, $y(s)$ is the parameter representation of C where s represents the arc length, then we have

$$M(x(s), y(s)) > 0 \quad \text{for all } |s - s_0| \leq \delta$$

if $x(s_0) = x_0$, $y(s_0) = y_0$.

CHAPTER II

II.2. $\dfrac{\partial^2 u}{\partial t^2} = \alpha^2 \dfrac{\partial^2 u}{\partial x^2} - g.$

II.4. $\rho \dfrac{\partial^2 u}{\partial t^2} = -\dfrac{\partial}{\partial x} \dfrac{\dfrac{\partial^3 u}{\partial x^3}(1 + u_x^2) - 3u_x\left(\dfrac{\partial^2 u}{\partial x^2}\right)^2}{(1 + u_x^2)^4}.$

II.6. (a) $X'' - \lambda X = 0$, $Y'' + \lambda Y = 0$; (b) $T' - \lambda T = 0$, $X'' - \mu X = 0$. $Y'' + (\mu - \lambda)\ Y = 0$;
(c) $T'' - \lambda T = 0$, $x^2 X'' + xX' - (\lambda x^2 + \mu)X = 0$, $Y'' + \mu Y = 0$;
(d) $x^2 X'' + 2xX' - \lambda X = 0$, $Y'' + \cot y\, Y' + \lambda Y = 0$.

II.8. $a_0 = \dfrac{1}{\pi}\displaystyle\int_0^{2\pi} \phi(x)\, dx$, $a_1 = \dfrac{1}{\pi}\displaystyle\int_0^{2\pi} \phi(x) \cos x\, dx$, $b_1 = \dfrac{1}{\pi}\displaystyle\int_0^{2\pi} \phi(x) \sin x\, dx.$

II.14. $\dfrac{\partial^2 u}{\partial t^2} = \alpha^2 \dfrac{\partial^2 u}{\partial x^2}$, $u(0, t) = u(l, t) = 0$, $u(x, 0) = 0$,

$$\left.\frac{\partial u}{\partial t}\right|_{t=0} = \begin{cases} ax, & 0 \leq x \leq \dfrac{l}{2}, \\ a(l - x), & \dfrac{l}{2} \leq x \leq l. \end{cases}$$

II.16. $\dfrac{\partial^2 u}{\partial t^2} = \alpha^2 \dfrac{\partial^2 u}{\partial x^2} - g$, $u(0, t) = u(l, t) = 0$, $u(x, 0) = h(x - l)x$, $\left.\dfrac{\partial u}{\partial t}\right|_{t=0} = 0.$

II.18. $\tau_{(x)} = \dfrac{\tau}{\sqrt{1 + u_x^2 + u_y^2}}\left(\dfrac{\partial u}{\partial y}\dfrac{du}{ds} + \dfrac{dy}{ds}\right).$

II.20. 0.

II.22. $(1 - \lambda)\, \Delta\Delta u + \dfrac{\rho}{c} u_{tt} = 0$, where $\Delta\Delta = \dfrac{\partial^4}{\partial x^4} + \dfrac{\partial^4}{\partial x^2 \partial y^2} + \dfrac{\partial^4}{\partial y^4}.$

II.24. $c_1 \cos m\theta + c_2 \sin m\theta = c_1 \cos m(\theta + 2\pi) + c_2 \sin m(\theta + 2\pi)$ leads to the linear system of homogeneous equations

$$c_1(\cos 2\pi m - 1) + c_2 \sin 2\pi m = 0,$$
$$-c_1 \sin 2\pi m + c_2(\cos 2\pi m - 1) = 0.$$

det $= 0$ has as a consequence $m =$ integer.

II.26. Note that $u(r, \theta, 0)$ is a solution of $\Delta u = 0$, $u(1, \theta, 0) = \sin \theta$.

II.28. $\dfrac{\partial u}{\partial t} = \dfrac{k}{\rho\sigma}\dfrac{\partial^2 u}{\partial x^2}$.

II.30. $\dfrac{\partial^2 u}{\partial x^2} + \dfrac{\partial^2 u}{\partial y^2} = 0$.

II.36 $\alpha^2\left[\dfrac{1}{r}\dfrac{\partial}{\partial r}\left(r\dfrac{\partial u}{\partial r}\right) + \dfrac{\partial}{\partial\theta}\left(\dfrac{1}{r}\dfrac{\partial u}{\partial\theta}\right)\right] - g = 0,\ u(1, \theta, t) = 1.$

II.38 $\dfrac{\partial u}{\partial t} = \alpha^2\left[\dfrac{1}{r}\dfrac{\partial}{\partial r}\left(r\dfrac{\partial u}{\partial r}\right) + \dfrac{\partial}{\partial\theta}\left(\dfrac{1}{r}\dfrac{\partial u}{\partial\theta}\right) + \dfrac{\partial^2 u}{\partial z^2}\right],\ u(1, \theta, z, t) = \dfrac{t+1}{t+z^2+1},$

$$u(r, \theta, z, 0) = \frac{1}{z^2+1}.$$

II.40. If $\beta = 0$, then $\alpha\zeta = 0$ on S. Hence

$$\iiint\limits_V 2\zeta\,\Delta\zeta\,dV = \frac{-2k}{\rho\sigma}\iiint\limits_V \operatorname{grad}^2\zeta\,dV \le 0.$$

CHAPTER III

III.2. $\dfrac{\dot{x}y}{\sqrt{\dot{x}^2+\dot{y}^2}} = C,\ \sqrt{\dot{x}^2+\dot{y}^2} - \dfrac{d}{dt}\dfrac{y\dot{y}}{\sqrt{\dot{x}^2+\dot{y}^2}} = 0.$

III.4. Differentiate the first equation with respect to x (y) and the second equation with respect to y (x); then subtract (add) one equation from (to) the other one.

III.10. $x = \xi + \eta,\ y = \xi - \eta,\ \dfrac{\partial^2 u}{\partial\xi^2} - \dfrac{\partial^2 u}{\partial\eta^2} = 0.$

CHAPTER IV

IV.2. Determine a, b, c such that with $\Phi(a, b, c)$

$$= \sum_{k=1}^{n}\left(a\cos\frac{2\pi k}{n} + b\sin\frac{2\pi k}{n} + c - y_k\right)^2,\ \frac{\partial\Phi}{\partial a} = 0,\ \frac{\partial\Phi}{\partial b} = 0,\ \frac{\partial\Phi}{\partial c} = 0.$$

Observe that $\displaystyle\sum_{k=1}^{n}\cos\frac{2\pi k}{n} = 0$ for $n > 2$, since

$$\cos\frac{2\pi k}{n} = \cos\left(\frac{2\pi}{n} + (k-1)\frac{2\pi}{n}\right) \quad\text{and}\quad 2\cos\left(\frac{2\pi}{n} + (k-1)\frac{2\pi}{n}\right)\sin\frac{\pi}{n}$$

$$= \sin\left(\frac{2\pi}{n} + (2k-1)\frac{\pi}{n}\right) - \sin\left(\frac{2\pi}{n} + (2k-3)\frac{\pi}{n}\right) \text{ and consequently}$$

$$\sum_{k=1}^{n}\cos\frac{2\pi k}{n} = \frac{1}{2\sin\pi/n}\left[\sin\left(\frac{2\pi}{n} + (2n-3)\frac{\pi}{n}\right) - \sin\left(\frac{2\pi}{n} - \frac{\pi}{n}\right)\right] = 0.$$

Similarly

$$\sum_{k=1}^{n} \sin \frac{2\pi k}{n} = 0, n > 2 \text{ and } \sum_{k=1}^{n} \cos \frac{4\pi k}{n} = \sum_{k=1}^{n} \sin \frac{4\pi k}{n} = 0, n > 4.$$

Now,

$$\cos^2 \frac{2\pi k}{n} = \tfrac{1}{2}\left(1 + \cos \frac{4\pi k}{n}\right), \sin^2 \frac{2\pi k}{n} = \tfrac{1}{2}\left(1 - \cos \frac{4\pi k}{n}\right)$$

and

$$\sin \frac{2\pi k}{n} \cos \frac{2\pi k}{n} = \tfrac{1}{2} \sin \frac{4\pi k}{n} .$$

Hence

$$\sum_{k=1}^{n} \cos^2 \frac{2\pi k}{n} = \frac{n}{2}, \sum_{k=1}^{n} \sin \frac{2\pi k}{n} \cos \frac{2\pi k}{n} = 0, \sum_{k=1}^{n} \sin^2 \frac{2\pi k}{n} = \frac{n}{2} .$$

Thus, one obtains

$$a = \frac{2}{n} \sum_{k=1}^{n} y_k \cos \frac{2\pi k}{n}, b = \frac{2}{n} \sum_{k=1}^{n} y_k \sin \frac{2\pi k}{n}, c = \frac{1}{n} \sum_{k=1}^{n} y_k.$$

IV.6. $|f + g|^2 = (f \cdot f) + (g \cdot g) + 2(f \cdot g) \leq (f \cdot f) + (g \cdot g) + 2|f||g| = (|f| + |g|)^2$.

IV.14. At $x = 0$: 6 per cent; at $x = \frac{\pi}{3}$; 2.7 per cent.

IV.16. $f(x) \cong \frac{4}{\pi}(\sin x - \frac{1}{9} \sin 3x + \frac{1}{25} \sin 5x - \frac{1}{49} \sin 7x + \frac{1}{81} \sin 9x)$.

IV.18. $f(x) \cong 2(\sin x - \frac{1}{2} \sin 2x + \frac{1}{3} \sin 3x - \frac{1}{4} \sin 4x + \frac{1}{5} \sin 5x)$.

IV.28. Let $m = \max(|a|, |b|)$. Since $m < \pi$, the Fourier series of $s(x)$ converges in $-m \leq x \leq m$ uniformly.

IV.30. $a_{2\nu+1} = \dfrac{4}{\pi(2\nu + 1)^2}, b_{2\nu+1} = -\dfrac{4}{\pi(2\nu + 1)}, a_0 = -\pi, a_{2\nu} = b_{2\nu} = 0.$

IV.32. Observe that

$$\int_{-\pi}^{\pi} f(x) \cos kx \, dx = -\frac{1}{k} \int_{-\pi}^{\pi} f'(x) \sin kx \, dx,$$

$$\int_{-\pi}^{\pi} f(x) \sin kx \, dx = \frac{(-1)^{k+1}}{k} [f(\pi - 0) - f(-\pi + 0)]$$

$$+ \frac{1}{k} \int_{-\pi}^{\pi} f'(x) \cos kx \, dx.$$

IV.34. $$\int_{x_k+\delta}^{x_{k+1}-\delta} \left[(F - P) + \left(P - \frac{a_0}{2} - \sum_{k=1}^{n} a_k \cos kx + b_k \sin kx\right)\right]^2 dx$$

$$\leq \int_{x_k+\delta}^{x_{k+1}-\delta} (F - P)^2 \, dx + \int_{x_k+\delta}^{x_{k+1}-\delta} (P - \cdots)^2 \, dx$$

$$+ 2\sqrt{\int_{x_k+\delta}^{x_{k+1}-\delta} (F - P)^2 \, dx \int_{x_k+\delta}^{x_{k+1}-\delta} (P - \cdots)^2 \, dx}.$$

IV.36. Let $g(x)$ be continuous and have a sectionally continuous derivative. Then, the Fourier coefficients of $g(x)$, α_r, β_r satisfy, according to problem IV.32, the relations

$$|\alpha_r r| \leq M, \quad |\beta_r r| \leq M.$$

One obtains by termwise integration a Fourier series of a function $\int_0^x g(x)\, dx$ which is continuous, has a continuous first derivative, and has a sectionally continuous second derivative. The coefficients of this Fourier series $\alpha_r^{(1)}$, $\beta_r^{(1)}$ satisfy the relations

$$|\alpha_r^{(1)} r^2| \leq M_{(1)}, |\beta_r^{(1)} r^2| \leq M_{(2)}, \quad \text{etc.}$$

IV.40. $f(x)$ continuous, $f'(x)$ sectionally continuous.

IV.42. Note that $\int_0^x \int_0^v s(u)\, du\, dv = \dfrac{x^3}{6}$, etc.

IV.44. $t = \dfrac{(2k+1)\pi}{2\alpha}$.

IV.46. $u(x, t) = \sum_{k=0}^{\infty} A_k \cos k\pi x \cdot e^{-k^2\pi^2 t}$, where $\sum_{k=0}^{\infty} A_k \cos k\pi x = f(x)$.

CHAPTER V

V.2. Let

$$\Phi = \begin{pmatrix} \phi_1^{(1)} & \phi_1^{(2)} \\ \phi_2^{(1)} & \phi_2^{(2)} \end{pmatrix}.$$

Then, $\Phi' = A\Phi$ and $\Phi'^* = \Phi^* A^*$. Now, $(\Phi^*\Phi^{*-1})' = \Phi^*(\Phi^{*-1})' + (\Phi^*)'\Phi^{*-1}$. However, $\Phi^{*-1}\Phi^* = U$; thus $(\Phi^{*-1}\Phi^*)' = 0$ and we have

$$\Phi^*(\Phi^{*-1})'\Phi^* = -\Phi^{*\prime} = -\Phi^* A^*.$$

Therefore $(\Phi^{*-1})' = -A^*\Phi^{*-1}$, i.e.,

$$\Phi^{*-1} = \frac{1}{\Delta}\begin{pmatrix} \phi_2^{(2)} & -\phi_1^{(2)} \\ -\phi_2^{(1)} & \phi_1^{(1)} \end{pmatrix}$$

is a solution of $\psi' = -A^*\psi$.

V.4. Note that if $L = a_0 \dfrac{d^n}{dt^n} + a_1 \dfrac{d^{n-1}}{dt^{n-1}} + \cdots + a_n$, then

$$M = (-1)^n \frac{d^n}{dt^n}(a_0 \cdot) + (-1)^{n-1} \frac{d^{n-1}}{dt^{n-1}}(a_1 \cdot) + \cdots + a_n.$$

V.6. $\dfrac{d}{dx}[e^{\int a_1\, dx} y'] + a_2 e^{\int a_1\, dx} y = 0.$

V.8. $\phi = -x$; $\phi = \dfrac{\pi}{2} - k\pi$ if $x = -\dfrac{\pi}{2} - k\pi$.

$\phi = \frac{1}{2}\arcsin(\tanh 2x)$; $\phi = \dfrac{\pi}{2} + k\pi$ if $\tanh 2x = 0$, i.e., $x = 0$.

V.10. $\sin^2 \phi_1 - \sin^2 \phi_2 = (\sin \phi_1 + \sin \phi_2)\left(\cos \frac{\phi_1 + \phi_2}{2} \sin \frac{\phi_1 - \phi_2}{2}\right).$

Use a similar transformation for $\cos^2 \phi_1 - \cos^2 \phi_2$ and make use of $\lim_{h \to 0} \frac{\sin h}{h} = 1$ in both cases.

V.12. $\delta = \bar{x}_1 + k\pi.$

V.14. $M \leq x \leq M + d; \quad \frac{\pi M}{\sqrt{\lambda}} \leq |x_1 - x_2| \leq \frac{\pi(M + d)}{\sqrt{\lambda}}.$

The larger x, the larger the distance between two consecutive zeros.

V.18. With $L = \frac{d}{dx}\left[(1 - x^2)\frac{d}{dx}\cdot\right] + \lambda$ we have $\int_{-1}^{1} (zL[y] - yL[z])\, dx$

$$= (1 - x^2)(y'z - yz')\Big|_{-1}^{1} = 0.$$

V.20. $y = A\left(\cos \frac{kx}{2} - \frac{2}{k} \sin \frac{kx}{2}\right).$

V.22. $\phi(0) = -\frac{\pi}{4} + \nu\pi, \phi(2\pi) = -\frac{\pi}{4} + \mu\pi; \quad \phi(0) = \frac{\pi}{2} + \nu\pi,$

$\phi(1) = \arctan 2 + \mu\pi.$

V.24. $\left\{\cos n\frac{x}{2}, \sin m\frac{x}{2}\right\}.$

CHAPTER VI

VI.2. Since there exist constants $c_1, c_2 \neq (0, 0)$ such that

$$c_1 y_1(x_0) + c_2 y_2(x_0) = y_0,$$
$$c_1 y_1'(x_0) + c_2 y_2'(x_0) = y_0',$$

it follows that $W[y_1, y_2] \neq 0$ at $x = x_0$.

VI.4. $y = x^\alpha (A \cos (\beta \log x) + B \sin (\beta \log x))$ if $\lambda = \alpha \pm i\beta$ are the solutions of $a_0\lambda^2 + (a_1 - a_0)\lambda + a_2 = 0$.

VI.14. $v(x) = (1 - x^2)^{m/2}$.

VI.16. If $|x| > 1$, then $p(x) = 1 - x^2 < 0$, while $q(x) = n(n + 1) > 0$.

VI.20. Let $f = (x^2 - 1)^n$ and $f_k = \frac{d^k}{dx^k}(x^2 - 1)^n$ for $k = 1, 2, 3, \cdots, n$.

f has zeros of multiplicity n at $x = 1$ and $x = -1$; i.e., f_k has zeros of multiplicity $n - k$ at $x = 1$ and $x = -1$.

Since $f(\pm 1) = 0$, there exists a point $x^{(1)}$ in $-1 < x < 1$ (theorem of Rolle) such that $f_1(x^{(1)}) = 0$.

Now, $f_1(\pm 1) = 0$, $f_1(x^{(1)}) = 0$, hence there exist two points $x_1^{(2)}, x_2^{(2)}$ in $-1 < x < x^{(1)}$ and $x^{(1)} < x < 1$ respectively such that $f_2(x_1^{(2)}) = 0$, $f_2(x^{(2)}) = 0$, etc.

Finally, $f_{n-1}(\pm 1) = 0$, $f_{n-1}(x_i^{(n-1)}) = 0$, $i = 1, 2, 3, \cdots, n-1$. Application of the theorem of Rolle for the last time yields n points $-1 < x_1^{(n)} < x_2^{(n)} < \cdots < x_n^{(n)} < 1$ where $f_n(x_i^{(n)}) = 0$, q.e.d.
(It is left to the reader to discuss the possibility of two or more zeros of the derivative between two zeros of the function.)

VI.22. Since $P_n(x)$ satisfies

$$\frac{d}{dx}[(1 - x^2)P_n' + n(n+1)P_n = 0,$$

we obtain after ν differentiations

$$(1 - x^2)P_n^{(\nu+2)} - 2(\nu + 1)xP_n^{(\nu+1)} + (n - \nu)(n + \nu + 1)P_n^{(\nu)} = 0.$$

Now, suppose that x_1 is a double root of $P_n(x)$. Then $(1 - x_1^2)P_n''(x_1) = 0$ (from Legendre's equation) and since $P_n(\pm 1) = \pm 1$, it follows that $P_n''(x_1) = 0$. Hence, from the differentiated equation for $\nu = 1$, $(1 - x_1^2)P_n'''(x_1) = 0$, etc.
Finally, $P_n^{(n)}(x_1) = 0$, i.e., $P_n(x) =$ constant.

VI.28. $u = c_0$.

VI.30. $\xi^2 \dfrac{d^2\eta}{d\xi^2} + (2\beta + 1)\xi \dfrac{d\eta}{d\xi} + [(\beta^2 - n^2\alpha^2) + a^2\alpha^2\xi^{2\alpha}]\eta = 0.$

VI.32. $\dfrac{d\eta}{d\xi} + \eta^2 + c\xi^2 = 0.$

VI.34. $\lambda = 0: c_\nu = -\dfrac{c_{\nu-1}}{2\nu(2\nu - 1)}$, $\lambda = \dfrac{1}{2} : c_\nu = -\dfrac{c_{\nu-1}}{2\nu(2\nu + 1)}$.

VI.36. $\lambda = 0: c_\nu = \dfrac{c_{\nu-1}}{\nu(\nu + 1)}$.

VI.38. Convergent in $|x| > 1$.

VI.40. Note that $\sum_{n=0}^{\infty} \dfrac{(x^2)^n}{n!} = e^{x^2}$.

VI.48.
$$J_{1/2}(x) = \frac{1}{\Gamma(\frac{1}{2})}\sqrt{2/x}\left(x - \frac{x^3}{3!} + \frac{x^5}{5!} - \frac{x^7}{7!} + \cdots\right),$$
$$J_{-1/2}(x) = \frac{1}{\Gamma(\frac{1}{2})}\sqrt{2/x}\left(1 - \frac{x^2}{2!} + \frac{x^4}{4!} - \frac{x^6}{6!} + \cdots\right).$$

VI.50. $y'' + a_1y' + (a_2 - \epsilon^2)y = 0.$

VI.52. $W = e^{2\lambda x}$.

VI.54. Approximate the integrand in $0 \le t \le \epsilon$ by its tangent line at $(\epsilon, -\log \epsilon \cdot e^{-\epsilon})$ and in $A \le t < \infty$ by its tangent line at $(A, -\log A\, e^{-A})$. Choose $\epsilon = \frac{1}{10}$ and $A = 10$.

VI.56.
$$Y_0(x) = 2\left[\left(\log\frac{x}{2} + \gamma\right)J_0(x) + \left(\frac{x}{2}\right)^2 - \frac{(\frac{1}{2} + 1)(x/2)^4}{(2!)^2} + \frac{(\frac{1}{3} + \frac{1}{2} + 1)(x/2)^6}{(3!)^2} - \frac{(\frac{1}{4} + \frac{1}{3} + \frac{1}{2} + 1)(x/2)^8}{(4!)^2} + \cdots\right].$$

VI.60. $x^2J''_\mu + x^2J_\mu + xJ'_\mu - \mu^2J_\mu = \frac{1}{\pi}\int_0^\pi [x^2\cos^2\theta\cos(\mu\theta - x\sin\theta)$

$+ x\sin\theta\sin(\mu\theta - x\sin\theta) - \mu^2\cos(\mu\theta - x\sin\theta)]\,d\theta$

$$= \frac{1}{\pi}\int_0^\pi [x^2\cos^2\theta\cos(\mu\theta - x\sin\theta) + (\mu x\cos\theta - x^2\cos^2\theta)\cos(\mu\theta - x\sin\theta) - \mu^2\cos(\mu\theta - x\sin\theta)]\,d\theta$$

$$\left[\sin(\mu\theta - x\sin\theta) = u,\ x\sin\theta\,d\theta = dv\right]$$

$$= \frac{\mu}{\pi}\int_0^\pi (x\cos\theta - \mu)\cos(\mu\theta - x\sin\theta)\,d\theta = -\frac{\mu}{\pi}\sin(\mu\theta - x\sin\theta)\Big|_0^\pi$$

$= 0.$

VI.66. $\displaystyle\int_0^\infty\int_0^\infty e^{-(x^2+y^2)}\,dx\,dy = \int_{r=0}^\infty\int_{\theta=0}^{\pi/2} re^{-r^2}\,dr\,d\theta = \frac{\pi}{4}.$

CHAPTER VII

VII.4. (a) $a = 0, \lambda = 2$; (b) $a = \frac{1}{3}, \lambda = 6$.

VII.6. $\displaystyle I[y_1] - I[y_2] = \int_0^1 (y'_1 - y'_2)(y'_1 + y'_2)\,dx \le \max_{[0,1]}|y'_1 - y'_2|\int_0^1 (y'_1 + y'_2)\,dx.$

VII.8. $\displaystyle I[u_n] = \pi - \frac{\pi}{n^2} + \frac{2\pi}{n}\left(\sqrt{4 + \frac{1}{n^2}} - \frac{1}{2}\sqrt{1 + \frac{1}{4n^2}}\right).$

VII.10. Let $I_2[y_1, \eta_1] - \lambda_1C[y_1, \eta_1] = a$ and $I_2[\eta_1] - \lambda_1C[\eta_1] = b$.
$Y(\epsilon) = b\epsilon^2 + 2a\epsilon \ge 0$ for all ϵ if and only if $b \ge 0$ and $Y'(0) = 2a = 0$.

VII.16. $c_1 = 1, c_2 = 0, c_3 = 0$.

VII.22. $\lambda_{1,2} = \dfrac{-5 \pm \sqrt{13}}{2}.$

VII.24. $B^{-1} = -\dfrac{1}{3}\begin{pmatrix} 1 & -2 & -2 \\ -2 & -2 & 1 \\ -2 & 1 & 1 \end{pmatrix}.$

VII.28. $\lambda = 0$: $c_1 = 1, c_2 = 0, c_3 = 0$,
$\lambda = 2$: $c_1 = 0, c_2 = 1, c_3 = 0$,
$\lambda = 6$: $c_1 = \frac{1}{3}, c_2 = 0, c_3 = -1$.

VII.30. $c_1^{(s)}C[f_1, y_1] + c_2^{(s)}C[f_2, y_1] + \cdots + c_n^{(s)}C[f_n, y_1] = 0.$
$n - 1$ coefficients can be chosen arbitrarily in $n - 1$ linearly independent ways $1, 0, 0, \cdots, 0$
$0, 1, 0, \cdots, 0$
. . . .
$0, 0, 0, \cdots, 1.$

VII.36. $\Lambda_1 \cong 2.5$, $\Lambda_2 \cong 25.5$.

VII.38. (a) Expand $\sqrt{x}\, J_{1/3}(\frac{2}{3}\sqrt{\lambda}\, x^{3/2})$, find a series solution of $y'' + \lambda x y = 0$, and compare coefficients, or
(b) Substitute

$$\tfrac{2}{3}\sqrt{\lambda}\, x^{3/2} = \xi,$$

$$\frac{2\sqrt{\lambda}}{3\xi}\, y = \eta^3$$

and show that η satisfies

$$\xi\eta'' + \eta' + \left(\xi - \frac{1}{9\xi}\right)\eta = 0.$$

CHAPTER VIII

VIII.6. $y = (x^2 - 1)^{\mu/2} u$.

VIII.10. Solution for $c_0 = 0, \lambda = -1$ is identical with solution for $c_0 \neq 0, \lambda = 0$.

VIII.16. $\lambda = \dfrac{1 + m}{2}$, $c_\nu = \dfrac{1}{4}\dfrac{[2(m-1) - 4l]c_{\nu-1} + c_{\nu-2}}{\nu(\nu + m)}$, $\nu = 1, 2, 3, \cdots,$

c_0 arbitrary, $c_{-1} = 0$.

VIII.18. $c_n = c_0 \dfrac{(m-l)(m-l+1)\cdots(m-l+n-1)}{(m+1)(m+2)\cdots(m+n)n\,!}$. $\quad (\sigma = m)$.

$$\frac{m+k}{m-l+k-1} = 1 + \frac{l+1}{m-l+k-1}\text{; hence,}$$

$$\frac{1}{c_n} = \frac{1}{c_0}\left(1 + \frac{l+1}{m-l}\right)\left(1 + \frac{l+1}{m-l+1}\right)\cdots\left(1 + \frac{l+1}{m-l+n-1}\right) n!.$$

Choose N such that $0 < \dfrac{l+1}{m-l+k-1} < 1$ for all $k > N$.

Let $\left|\left(1 + \dfrac{l+1}{m-l}\right)\cdots\left(1 + \dfrac{l+1}{m-l+N-1}\right)\right| = \dfrac{c_0}{\gamma}$;
then

$$\left|\frac{1}{c_n}\right| < \frac{1}{\gamma}\left(1 + \frac{l+1}{m-l+N}\right)^n n!.$$

Let

$$1 + \frac{l+1}{m-l+N} = \alpha,\text{ where } 1 < \alpha < 2\text{; then}$$

$$\left|\frac{1}{c_n}\right| < \frac{\alpha^n n!}{\gamma}.$$

VIII.20. $R(z) > z^{(\sigma-1)/2} e^{(-z/2)}(\text{Pol}\,(z) + \gamma e^{z/\alpha}) > \dfrac{1}{\sqrt{z}}\gamma e^{z[(1/\alpha)-1/2]}$ for large z, where $\dfrac{1}{\alpha} - \dfrac{1}{2} > 0$ since $1 < \alpha < 2$.

VIII.26. $P_1^1 = \sqrt{1 - x^2}$, $P_2^1 = 3\sqrt{1 - x^2} \cdot x$, $P_3^1 = \sqrt{1 - x^2}\left(\frac{15x^2}{2} - \frac{3}{2}\right)$,

$P_2^2 = 3(1 - x^2)$, $P_3^2 = (1 - x^2)\,15x$,

$P_3^3 = 15(1 - x^2)^{3/2}$.

VIII.28. $u(r, \theta, 0) = -\dfrac{2\pi^2}{3}$.

CHAPTER IX

IX.4. Determine $c_1(x)$, $c_2(x)$ in $\eta = c_1(x)\cos x + c_2(x)\sin x$ such that $\eta'' + \eta = \sin x$, where $c_1'(x)\cos x + c_2'(x)\sin x = 0$ and determine A and B in

$$y = \eta + A\cos x + B\sin x$$

such that the boundary conditions are satisfied.

IX.6. $\displaystyle\int_0^\pi \cos x\,dx = 0$.

IX.8. $G(x, \xi) = (1 - x)\xi$ for $x > \xi$.

IX.10. $G(x, \xi) = -\frac{1}{2}\log[(1 + \xi)(1 - x)] + C$ for $x \leq \xi$.

IX.12. (a) $a = \dfrac{\pi}{2}$, (b) $a = \pi$, $f(x) = \sin 2x$.

IX.14. (a) $G(x, \xi) = \begin{cases} x & \text{for } x \leq \xi, \\ \xi & \text{for } x > \xi. \end{cases}$ (b) $G(x, \xi) = \begin{cases} -\frac{1}{2}[\xi - x + x\xi - 1] & \text{for } x \leq \xi, \\ -\frac{1}{2}[x - \xi + \xi x - 1] & \text{for } x > \xi. \end{cases}$

IX.16. $G(x, \xi) = \begin{cases} \dfrac{1}{n}\left[\left(\dfrac{x}{\xi}\right)^n - (x\xi)^n\right] & \text{for } x \leq \xi, \\ \dfrac{1}{n}\left[\left(\dfrac{\xi}{x}\right)^n - (x\xi)^n\right] & \text{for } x > \xi. \end{cases}$

IX.18. $p(x) = x = 0$ at $x = 0$; hence, Green's lemma is satisfied.

IX.22. $G(x, \xi, 0) = \displaystyle\sum_{n=1}^{\infty} \frac{1}{\lambda_n} J_\mu(\lambda_n x) J_\mu(\lambda_n \xi)$ where $J_\mu(\lambda_n) = 0$.

IX.24. $$G(x, y;\ \xi, \eta) = -\frac{1}{2\pi}\log\sqrt{(x - \xi)^2 + (y - \eta)^2} + \frac{1}{2\pi}\log\frac{\sqrt{\xi^2 + \eta^2}}{\sqrt{\left(x - \dfrac{\xi}{\xi^2 + \eta^2}\right)^2 + \left(y - \dfrac{y}{\xi^2 + \eta^2}\right)^2}}.$$

INDEX

The letter *A* preceding a page number stands for Appendix.

A CATALOG OF SELECTED

DOVER BOOKS

IN SCIENCE AND MATHEMATICS

A CATALOG OF SELECTED

DOVER BOOKS

IN SCIENCE AND MATHEMATICS

QUALITATIVE THEORY OF DIFFERENTIAL EQUATIONS, V.V. Nemytskii and V.V. Stepanov. Classic graduate-level text by two prominent Soviet mathematicians covers classical differential equations as well as topological dynamics and ergodic theory. Bibliographies. 523pp. 5⅜ × 8½. 65954-2 Pa. $10.95

MATRICES AND LINEAR ALGEBRA, Hans Schneider and George Phillip Barker. Basic textbook covers theory of matrices and its applications to systems of linear equations and related topics such as determinants, eigenvalues and differential equations. Numerous exercises. 432pp. 5⅜ × 8½. 66014-1 Pa. $9.95

QUANTUM THEORY, David Bohm. This advanced undergraduate-level text presents the quantum theory in terms of qualitative and imaginative concepts, followed by specific applications worked out in mathematical detail. Preface. Index. 655pp. 5⅜ × 8½. 65969-0 Pa. $13.95

ATOMIC PHYSICS (8th edition), Max Born. Nobel laureate's lucid treatment of kinetic theory of gases, elementary particles, nuclear atom, wave-corpuscles, atomic structure and spectral lines, much more. Over 40 appendices, bibliography. 495pp. 5⅜ × 8½. 65984-4 Pa. $11.95

ELECTRONIC STRUCTURE AND THE PROPERTIES OF SOLIDS: The Physics of the Chemical Bond, Walter A. Harrison. Innovative text offers basic understanding of the electronic structure of covalent and ionic solids, simple metals, transition metals and their compounds. Problems. 1980 edition. 582pp. 6⅛ × 9¼. 66021-4 Pa. $14.95

BOUNDARY VALUE PROBLEMS OF HEAT CONDUCTION, M. Necati Özisik. Systematic, comprehensive treatment of modern mathematical methods of solving problems in heat conduction and diffusion. Numerous examples and problems. Selected references. Appendices. 505pp. 5⅜ × 8½. 65990-9 Pa. $11.95

A SHORT HISTORY OF CHEMISTRY (3rd edition), J.R. Partington. Classic exposition explores origins of chemistry, alchemy, early medical chemistry, nature of atmosphere, theory of valency, laws and structure of atomic theory, much more. 428pp. 5⅜ × 8½. (Available in U.S. only) 65977-1 Pa. $10.95

A HISTORY OF ASTRONOMY, A. Pannekoek. Well-balanced, carefully reasoned study covers such topics as Ptolemaic theory, work of Copernicus, Kepler, Newton, Eddington's work on stars, much more. Illustrated. References. 521pp. 5⅜ × 8½. 65994-1 Pa. $11.95

PRINCIPLES OF METEOROLOGICAL ANALYSIS, Walter J. Saucier. Highly respected, abundantly illustrated classic reviews atmospheric variables, hydrostatics, static stability, various analyses (scalar, cross-section, isobaric, isentropic, more). For intermediate meteorology students. 454pp. 6⅛ × 9¼. 65979-8 Pa. $12.95

RELATIVITY, THERMODYNAMICS AND COSMOLOGY, Richard C. Tolman. Landmark study extends thermodynamics to special, general relativity; also applications of relativistic mechanics, thermodynamics to cosmological models. 501pp. 5⅜ × 8½. 65383-8 Pa. $12.95

APPLIED ANALYSIS, Cornelius Lanczos. Classic work on analysis and design of finite processes for approximating solution of analytical problems. Algebraic equations, matrices, harmonic analysis, quadrature methods, much more. 559pp. 5⅜ × 8½. 65656-X Pa. $12.95

SPECIAL RELATIVITY FOR PHYSICISTS, G. Stephenson and C.W. Kilmister. Concise elegant account for nonspecialists. Lorentz transformation, optical and dynamical applications, more. Bibliography. 108pp. 5⅜ × 8½. 65519-9 Pa. $4.95

INTRODUCTION TO ANALYSIS, Maxwell Rosenlicht. Unusually clear, accessible coverage of set theory, real number system, metric spaces, continuous functions, Riemann integration, multiple integrals, more. Wide range of problems. Undergraduate level. Bibliography. 254pp. 5⅜ × 8½. 65038-3 Pa. $7.95

INTRODUCTION TO QUANTUM MECHANICS With Applications to Chemistry, Linus Pauling & E. Bright Wilson, Jr. Classic undergraduate text by Nobel Prize winner applies quantum mechanics to chemical and physical problems. Numerous tables and figures enhance the text. Chapter bibliographies. Appendices. Index. 468pp. 5⅜ × 8½. 64871-0 Pa. $11.95

ASYMPTOTIC EXPANSIONS OF INTEGRALS, Norman Bleistein & Richard A. Handelsman. Best introduction to important field with applications in a variety of scientific disciplines. New preface. Problems. Diagrams. Tables. Bibliography. Index. 448pp. 5⅜ × 8½. 65082-0 Pa. $11.95

MATHEMATICS APPLIED TO CONTINUUM MECHANICS, Lee A. Segel. Analyzes models of fluid flow and solid deformation. For upper-level math, science and engineering students. 608pp. 5⅜ × 8½. 65369-2 Pa. $13.95

ELEMENTS OF REAL ANALYSIS, David A. Sprecher. Classic text covers fundamental concepts, real number system, point sets, functions of a real variable, Fourier series, much more. Over 500 exercises. 352pp. 5⅜ × 8½. 65385-4 Pa. $9.95

PHYSICAL PRINCIPLES OF THE QUANTUM THEORY, Werner Heisenberg. Nobel Laureate discusses quantum theory, uncertainty, wave mechanics, work of Dirac, Schroedinger, Compton, Wilson, Einstein, etc. 184pp. 5⅜ × 8½. 60113-7 Pa. $4.95

INTRODUCTORY REAL ANALYSIS, A.N. Kolmogorov, S.V. Fomin. Translated by Richard A. Silverman. Self-contained, evenly paced introduction to real and functional analysis. Some 350 problems. 403pp. 5⅜ × 8½. 61226-0 Pa. $9.95

PROBLEMS AND SOLUTIONS IN QUANTUM CHEMISTRY AND PHYSICS, Charles S. Johnson, Jr. and Lee G. Pedersen. Unusually varied problems, detailed solutions in coverage of quantum mechanics, wave mechanics, angular momentum, molecular spectroscopy, scattering theory, more. 280 problems plus 139 supplementary exercises. 430pp. 6½ × 9¼. 65236-X Pa. $11.95

THE ELECTROMAGNETIC FIELD, Albert Shadowitz. Comprehensive undergraduate text covers basics of electric and magnetic fields, builds up to electromagnetic theory. Also related topics, including relativity. Over 900 problems. 768pp. 5⅜ × 8¼. 65660-8 Pa. $17.95

FOURIER SERIES, Georgi P. Tolstov. Translated by Richard A. Silverman. A valuable addition to the literature on the subject, moving clearly from subject to subject and theorem to theorem. 107 problems, answers. 336pp. 5⅜ × 8½. 63317-9 Pa. $7.95

THEORY OF ELECTROMAGNETIC WAVE PROPAGATION, Charles Herach Papas. Graduate-level study discusses the Maxwell field equations, radiation from wire antennas, the Doppler effect and more. xiii + 244pp. 5⅜ × 8½. 65678-0 Pa. $6.95

DISTRIBUTION THEORY AND TRANSFORM ANALYSIS: An Introduction to Generalized Functions, with Applications, A.H. Zemanian. Provides basics of distribution theory, describes generalized Fourier and Laplace transformations. Numerous problems. 384pp. 5⅜ × 8½. 65479-6 Pa. $9.95

THE PHYSICS OF WAVES, William C. Elmore and Mark A. Heald. Unique overview of classical wave theory. Acoustics, optics, electromagnetic radiation, more. Ideal as classroom text or for self-study. Problems. 477pp. 5⅜ × 8½. 64926-1 Pa. $11.95

CALCULUS OF VARIATIONS WITH APPLICATIONS, George M. Ewing. Applications-oriented introduction to variational theory develops insight and promotes understanding of specialized books, research papers. Suitable for advanced undergraduate/graduate students as primary, supplementary text. 352pp. 5⅜ × 8½. 64856-7 Pa. $8.95

A TREATISE ON ELECTRICITY AND MAGNETISM, James Clerk Maxwell. Important foundation work of modern physics. Brings to final form Maxwell's theory of electromagnetism and rigorously derives his general equations of field theory. 1,084pp. 5⅜ × 8½. 60636-8, 60637-6 Pa., Two-vol. set $19.90

AN INTRODUCTION TO THE CALCULUS OF VARIATIONS, Charles Fox. Graduate-level text covers variations of an integral, isoperimetrical problems, least action, special relativity, approximations, more. References. 279pp. 5⅜ × 8½. 65499-0 Pa. $7.95

HYDRODYNAMIC AND HYDROMAGNETIC STABILITY, S. Chandrasekhar. Lucid examination of the Rayleigh-Benard problem; clear coverage of the theory of instabilities causing convection. 704pp. 5⅜ × 8¼. 64071-X Pa. $14.95

CALCULUS OF VARIATIONS, Robert Weinstock. Basic introduction covering isoperimetric problems, theory of elasticity, quantum mechanics, electrostatics, etc. Exercises throughout. 326pp. 5⅜ × 8½. 63069-2 Pa. $7.95

DYNAMICS OF FLUIDS IN POROUS MEDIA, Jacob Bear. For advanced students of ground water hydrology, soil mechanics and physics, drainage and irrigation engineering and more. 335 illustrations. Exercises, with answers. 784pp. 6⅛ × 9¼. 65675-6 Pa. $19.95

ORDINARY DIFFERENTIAL EQUATIONS, Morris Tenenbaum and Harry Pollard. Exhaustive survey of ordinary differential equations for undergraduates in mathematics, engineering, science. Thorough analysis of theorems. Diagrams. Bibliography. Index. 818pp. 5⅜ × 8½. 64940-7 Pa. $16.95

STATISTICAL MECHANICS: Principles and Applications, Terrell L. Hill. Standard text covers fundamentals of statistical mechanics, applications to fluctuation theory, imperfect gases, distribution functions, more. 448pp. 5⅜ × 8½. 65390-0 Pa. $9.95

ORDINARY DIFFERENTIAL EQUATIONS AND STABILITY THEORY: An Introduction, David A. Sánchez. Brief, modern treatment. Linear equation, stability theory for autonomous and nonautonomous systems, etc. 164pp. 5⅜ × 8¼. 63828-6 Pa. $5.95

THIRTY YEARS THAT SHOOK PHYSICS: The Story of Quantum Theory, George Gamow. Lucid, accessible introduction to influential theory of energy and matter. Careful explanations of Dirac's anti-particles, Bohr's model of the atom, much more. 12 plates. Numerous drawings. 240pp. 5⅜ × 8½. 24895-X Pa. $5.95

THEORY OF MATRICES, Sam Perlis. Outstanding text covering rank, nonsingularity and inverses in connection with the development of canonical matrices under the relation of equivalence, and without the intervention of determinants. Includes exercises. 237pp. 5⅜ × 8½. 66810-X Pa. $7.95

GREAT EXPERIMENTS IN PHYSICS: Firsthand Accounts from Galileo to Einstein, edited by Morris H. Shamos. 25 crucial discoveries: Newton's laws of motion, Chadwick's study of the neutron, Hertz on electromagnetic waves, more. Original accounts clearly annotated. 370pp. 5⅜ × 8½. 25346-5 Pa. $9.95

INTRODUCTION TO PARTIAL DIFFERENTIAL EQUATIONS WITH APPLICATIONS, E.C. Zachmanoglou and Dale W. Thoe. Essentials of partial differential equations applied to common problems in engineering and the physical sciences. Problems and answers. 416pp. 5⅜ × 8½. 65251-3 Pa. $10.95

BURNHAM'S CELESTIAL HANDBOOK, Robert Burnham, Jr. Thorough guide to the stars beyond our solar system. Exhaustive treatment. Alphabetical by constellation: Andromeda to Cetus in Vol. 1; Chamaeleon to Orion in Vol. 2; and Pavo to Vulpecula in Vol. 3. Hundreds of illustrations. Index in Vol. 3. 2,000pp. 6⅛ × 9¼. 23567-X, 23568-8, 23673-0 Pa., Three-vol. set $41.85

ASYMPTOTIC EXPANSIONS FOR ORDINARY DIFFERENTIAL EQUATIONS, Wolfgang Wasow. Outstanding text covers asymptotic power series, Jordan's canonical form, turning point problems, singular perturbations, much more. Problems. 384pp. 5⅜ × 8½. 65456-7 Pa. $9.95

AMATEUR ASTRONOMER'S HANDBOOK, J.B. Sidgwick. Timeless, comprehensive coverage of telescopes, mirrors, lenses, mountings, telescope drives, micrometers, spectroscopes, more. 189 illustrations. 576pp. 5⅜ × 8¼. (USO) 24034-7 Pa. $9.95

THE FOUR-COLOR PROBLEM: Assaults and Conquest, Thomas L. Saaty and Paul G. Kainen. Engrossing, comprehensive account of the century-old combinatorial topological problem, its history and solution. Bibliographies. Index. 110 figures. 228pp. 5⅜ × 8½. 65092-8 Pa. $6.95

CATALYSIS IN CHEMISTRY AND ENZYMOLOGY, William P. Jencks. Exceptionally clear coverage of mechanisms for catalysis, forces in aqueous solution, carbonyl- and acyl-group reactions, practical kinetics, more. 864pp. 5⅜ × 8½. 65460-5 Pa. $19.95

PROBABILITY: An Introduction, Samuel Goldberg. Excellent basic text covers set theory, probability theory for finite sample spaces, binomial theorem, much more. 360 problems. Bibliographies. 322pp. 5⅜ × 8½. 65252-1 Pa. $8.95

LIGHTNING, Martin A. Uman. Revised, updated edition of classic work on the physics of lightning. Phenomena, terminology, measurement, photography, spectroscopy, thunder, more. Reviews recent research. Bibliography. Indices. 320pp. 5⅜ × 8¼. 64575-4 Pa. $8.95

PROBABILITY THEORY: A Concise Course, Y.A. Rozanov. Highly readable, self-contained introduction covers combination of events, dependent events, Bernoulli trials, etc. Translation by Richard Silverman. 148pp. 5⅜ × 8¼. 63544-9 Pa. $5.95

THE CEASELESS WIND: An Introduction to the Theory of Atmospheric Motion, John A. Dutton. Acclaimed text integrates disciplines of mathematics and physics for full understanding of dynamics of atmospheric motion. Over 400 problems. Index. 97 illustrations. 640pp. 6 × 9. 65096-0 Pa. $17.95

STATISTICS MANUAL, Edwin L. Crow, et al. Comprehensive, practical collection of classical and modern methods prepared by U.S. Naval Ordnance Test Station. Stress on use. Basics of statistics assumed. 288pp. 5⅜ × 8½. 60599-X Pa. $6.95

DICTIONARY/OUTLINE OF BASIC STATISTICS, John E. Freund and Frank J. Williams. A clear concise dictionary of over 1,000 statistical terms and an outline of statistical formulas covering probability, nonparametric tests, much more. 208pp. 5⅜ × 8½. 66796-0 Pa. $6.95

STATISTICAL METHOD FROM THE VIEWPOINT OF QUALITY CONTROL, Walter A. Shewhart. Important text explains regulation of variables, uses of statistical control to achieve quality control in industry, agriculture, other areas. 192pp. 5⅜ × 8½. 65232-7 Pa. $6.95

THE INTERPRETATION OF GEOLOGICAL PHASE DIAGRAMS, Ernest G. Ehlers. Clear, concise text emphasizes diagrams of systems under fluid or containing pressure; also coverage of complex binary systems, hydrothermal melting, more. 288pp. 6½ × 9¼. 65389-7 Pa. $10.95

STATISTICAL ADJUSTMENT OF DATA, W. Edwards Deming. Introduction to basic concepts of statistics, curve fitting, least squares solution, conditions without parameter, conditions containing parameters. 26 exercises worked out. 271pp. 5⅜ × 8½. 64685-8 Pa. $7.95